Battling the Elements

Published in cooperation with
the Center for American Places,
Harrisonburg, Virginia

Battling the Elements

WEATHER AND TERRAIN
IN THE CONDUCT OF WAR

Harold A. Winters
with Gerald E. Galloway Jr.
William J. Reynolds
and David W. Rhyne

The Johns Hopkins University Press

Baltimore and London

© 1998 The Johns Hopkins University Press
All rights reserved. Published 1998
Printed in the United States of America on acid-free paper
9 8 7 6 5 4 3 2 1

The Johns Hopkins University Press
2715 North Charles Street
Baltimore, Maryland 21218-4363
The Johns Hopkins Press Ltd., London
www.press.jhu.edu

Library of Congress Cataloging-in-Publication Data will be found at the end of this book.

A catalog record for this book is available from the British Library.

ISBN 0-8018-5850-X

Contents

Preface and Acknowledgments

While a visiting professor in the Department of Geography and Computer Science at the U.S. Military Academy during the 1982–83 academic year, I sat in on two outstanding courses, History of Military Art and Military Geography. During that time I also benefited greatly from continuing discussions with Cols. Gerald Galloway Jr. and Gilbert Kirby, and Lt. Col. William Reynolds (all of the Department of Geography and Computer Science) on the relationships between warfare and geography with special attention to environmental factors in combat. These discussions, plus conversations with other faculty, ideas generated from the two USMA courses, a long-term interest in military geography, and access to the departmental and Academy libraries, provided a basis for my review of numerous military operations, the purpose being to identify an array that illustrates varying relationships between battle and five basic elements of physical geography—terrain, weather, climate, soil, and vegetation.

In October 1984 I prepared an outline for a proposed book of fourteen chapters with the major topics listed in the preceding table of contents. Over the next several years I wrote six (the introduction, chapters 1, 2, 6, 8, and the conclusion), Galloway four (5, 7, 11, and 12), Reynolds three (3, 9, and 10), and Rhyne one (4). In 1992 I returned to the U.S. Military Academy as a visiting professor in the Department of Geography and Environmental Engineering. During that academic year I reviewed all of the manuscript and, with financial support from Michigan State University, employed Winfield Swanson of Washington, D.C., to technically edit the twelve central chapters. Revisions in the text continued after my return to Michigan State University in 1993 and during subsequent visiting professorships at the University of South Carolina and Arizona State University. The manuscript was completed in Las Vegas in 1997.

Meanwhile, in 1995 George F. Thompson, president of the Center for American Places, expressed interest in the publication of the manuscript, evaluated a preliminary draft, arranged for its external review, and expedited a contract with Johns Hopkins University Press in 1996. George Thompson's technical expertise, sound advice, and continuing good will were invaluable in completing this project. The book has also benefited from the copyediting by Therese D. Boyd. I thank the staff at the Johns Hopkins University Press for their work on the project, especially Anita Walker Scott, design and production manager, and Julie McCarthy, production editor.

Several organizations and many people have helped in the development of this book. At the U.S. Military Academy these include Todd Bacastow, Palmer Bailey, John Baker, Kent Butts, Frank Caravella, William Doe, Paul Foley, Robert Getz, Jack Grubbs, Stephen Houston, Jeffrey LaMoe, Michael Smith, and John Talbot, all involved at one time or another in the geographical instruction program at West Point. Others there who gave assistance include Leonard Fullenkamp, Robert Doughty, Roy Flint, and Kenneth Hamburger, all then in the Department of History, and Dawn Lake of the Department of English.

Both U.S. Army Engineers Topographic Engineering Center (TEC) and the U.S. Army Center of Military History (CMH) provided early support for this effort, specifically Dave Maune and Walt Boge at TEC and, at CMH, Douglas Kinnard, Jeffrey Clarke, John Ellsburg, Ernst Fisher, Jeff Greenhut, Bruce Hardcastle, Brooks Kleber, Harold Nelson, Francis Robertson, and Glen Robertson. Individuals at the U.S. Army War College (Carlisle), including Fred Beck, Roger Cirillo, Charles von Luttichau, Jay Luvas, John Schlight, Richard Sommers, and David Trask, also offered constructive suggestions and comments.

A number of U.S. geographers aided in the improvement of the manuscript in a variety of ways. These include Anthony Brazel, Sandra Brazel, the late Melvin Marcus, Barbara Trapido, and Russell Vose, all at Arizona State University; Mark Cowell at Indiana State University; Harm de Blij of Chatham, Massachusetts, and Boca Grande, Florida; David DiBiase at the Pennsylvania State University; John Hehr at the University of Arkansas; the late Albert Jackman of Western Michigan University; Charles Johnston at the University of Hawaii; Charles Kovacik of the University of South Carolina; Norman Meek at California State University, San Bernardino; and Richard Rieck of Western Illinois University. Richard Eaton (Brig. Gen., USA, Ret.) of Gainesville, Georgia, gave me some especially valuable insights into many aspects of military geography.

At Michigan State University I benefited from numerous discussions with Michael Rip. Michael Lipsey repeatedly and patiently helped me with computing needs, and Ellen White and Jim Hanlon of the MSU Cartography Center drafted many of the maps. In addition to preparing several additional illustrations, during 1996 and 1997, Catherine Van Vliet Mey of Royal Oak, Michigan, expertly converted all maps and figures into the needed format while also making editorial changes in the drawings. Sincere and lasting appreciation is extended to Marilyn Bria, Department of Geography Office Supervisor at Michigan State University, who for many years helped me with word processing and the production of the manuscript. Finally, I want to thank my wife, Marjorie Winters, for support, patience, and always wise advice.

My coauthors and I wish to extend special acknowledgments to Gilbert W. Kirby (Brig. Gen., USA, Ret.) and John B. Garver (Col., USA, Ret.) who were central to the establishment of a modern geography program at West Point. As head of the Department of Geography and Computer Science and later the Department of Geography and Environmental Engineering, for nearly a generation Gil Kirby skillfully directed and nurtured the overall academic program. Meanwhile, as deputy head during the 1970s, John Garver guided the geography curriculum (including the establishment of a fine course in military geography) with insight and intellect. Their long and tireless efforts on behalf of the Academy and the discipline are, without question, exemplary. Kirby, who was both supportive and instrumental in the early development of this book, once said, when offering his sagacious counsel and remarkable style of encouragement, "If you shovel enough coal in the boiler they may let you blow the whistle." On your behalf, Gil and John, we pull the cord.

Battling the Elements

Introduction

War and Geography

Those who do not know the conditions of mountains and forests, hazardous defiles, marshes and swamps, cannot conduct the march of an army.

SUN TZU, *THE ART OF WAR*

There has always been a potent and omnipresent synergy between the environment, or physical geography, and battle.[1] Over the centuries efforts to dominate an adversary by military force have led to a grim progression of weapons, techniques, and principles, all designed more or less to gain control of space and lines of communication. Furthermore, aggressive human endeavor and the resulting repetition of conflict make it a near certainty that even more formidable and destructive wartime devices will appear in the future. Through all of this, past, present, and future, it is clear that basic elements of geography are always important factors in the conduct and outcome of battle.

Thus, by examining connections between a number of military operations and their basic geographic components, this book shows how seemingly evident factors such as weather, climate, terrain, soil, and vegetation are important, cogent, and sometimes decisive in combat (Map I.1). It also demonstrates that war and the environment are intertwined in a complex, diverse, and often capricious fashion, the result being an ever-varying military cacophony incorporating repetition, counterpoint, juxtaposition, inversion, and reversal, all within the evolving framework of space and time. That is, comparisons of battles affected in a major way by one environmental factor may, in some cases, reveal distinct similarities and parallels, while in other conflicts the setting and outcome may be very different because of changes in circumstance, technology, and judgment. One characteristic, however, remains unchanging; in combat an environmental advantage for one side always means some degree of misfortune for the other, and that situation can easily reverse itself on the next battlefield.

Weather has always been a major factor in war. Two ancient examples are the disastrous effects of cyclones on Kublai Khan's twelfth-century efforts to cross the Korea Strait and invade Japan—storms of such size and importance that they were called "Kamikaze," meaning "the Divine Wind," by the defenders. Six centuries later it was exactly the reverse as two brief, but dissimilar, spells in English Channel weather favored the 1940 British evacuation at Dunkirk and the 1944 Allied landings in Normandy.

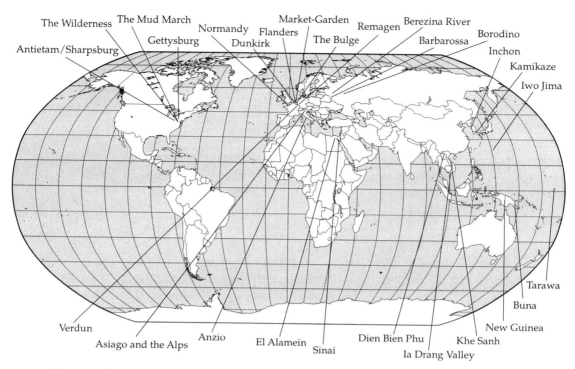

The Wilderness The Mud March Normandy Flanders Market-Garden Remagen Berezina River Borodino
Antietam/Sharpsburg Gettysburg Dunkirk The Bulge Barbarossa Inchon
Kamikaze
Iwo Jima

Tarawa
Buna
Verdun Asiago and the Alps Anzio El Alamein Sinai Dien Bien Phu Ia Drang Valley Khe Sanh New Guinea

MAP I.1. Names and locations for some of the military-environmental relationships considered in this book.

Persistent foul weather has often been used to advantage by an aggressor, such as the repeated foggy days with low clouds and precipitation that contributed greatly to the effectiveness of both the 1944 German attack that evolved into the Battle of the Bulge and the several-month-long Viet Cong siege of Khe Sanh. In contrast, just a few hours of unexpected rain on clay-rich Virginia soil completely stopped a large Union army offensive well before a planned engagement with the enemy, resulting in the infamous 1863 Mud March of the American Civil War.

Climate plays many roles in battle. The relatively temperate but constantly humid conditions in coastal northwest Europe made the World War I battlefield in Flanders a miserable four-year muddy morass, yet in the end it offered decisive advantage to neither side. As a vivid counterpoint, the long and frigid winters so typical for the interior of easternmost Europe contributed greatly to the suffering and eventual demise of climatically unprepared armies commanded by Charles XII, Napoleon Bonaparte, and Adolf Hitler in their attacks on Russia. And in a very different way, the heat, aridity, wind, and dust of North Africa presented a continuous tactical problem for opposing commanders in the Western Desert during 1941–43.

Dense mid-latitude vegetation was commonly a formidable factor for both Confederate and Union troops during the American Civil War, two vivid examples being the battles within the Wilderness of Virginia, first in 1863 and then again in 1864. Decades later its tropical counterpart was at least equally formidable in the World War II struggle for New Guinea, at the 1954 French tragedy at Dien Bien Phu, and for the fierce 1965 battle in the Ia Drang Valley in what was then South Vietnam.

Time and again terrain plays a powerful role in conflict. It was central to Gen. Thomas J. "Stonewall" Jackson's 1862 legendary maneuvers in the Shenandoah Valley. And the nature and northeast trend of the Blue Ridge Mountains provided both concealment and opportunity for long marches related to Gen. Robert E. Lee's 1862 and 1863 invasions of Maryland and Pennsylvania. In contrast, the even more deadly 1916 German and French offensives near Verdun, fought over 10 months for control of the rugged, forested, geologically induced hills that form a natural central bastion for the Western Front, brought little change in the line of battle within a terrain that strongly favored the defense.

Even former glaciers have shaped battlefields in an influential fashion, notably at Massachusetts's Bunker Hill and New York's western Long Island. On a much larger scale, the slightly higher terrain that extends from Warsaw to Minsk to Smolensk, past Borodino, and toward Moscow was formed by deposition of sediments at the edge of a massive Eurasian ice-cap that existed many thousands of years ago. During three successive centuries three armies from three different countries (Sweden, France, and Germany) all followed similar paths on this same glacial formation as they marched hundreds of kilometers eastward into Russia toward their eventual demise. Yet in another, very different glacially formed setting, World War I troops from Austria and Italy fought for three years within the same area in the Alps with positions that changed just a few tens of kilometers over many months. Here, high altitude, turbulent weather, and the rugged glacially carved terrain all came together to create the most formidable of battlefields.

Water is always a factor in warfare. Great island battles such as Tarawa and Iwo Jima, along with flanking amphibious assaults at Inchon in Korea (highly successful) and Anzio in Italy (essentially a failure), were profoundly affected by a host of shore-zone conditions including tides, storms, beach material, and the coastal topography. The difficulty in crossing rivers is vividly shown by Napoleon's disaster at the Berezina during his retreat from Russia in 1812 and the "bridge too far" at Arnhem during Market-Garden in 1944. In sharp contrast, the Allies' good fortune in capturing the lone remaining span over the Rhine River at Remagen in early 1945 quickly shattered Germany's defenses along Western Europe's largest river.

These examples, selected from many in this book, illustrate that the physical setting within which battles are fought is neither passive nor presumable. Instead, environmental components present intricately related influences that, whether individually or collectively, have the potential to shape conflict, sometimes becoming a decisive factor in its outcome. But the role of terrain, weather, climate, soil, or vegetation in one battle is by no means a reliable predictor of its effect on the next. Instead, these factors can combine in a number of ways, some anticipated, others unexpected, and a few unprecedented. The fact that no one knows exactly how the next war will be fought is partly explained by the myriad of possible environmental relationships on the battlefield.

To keep things in proper perspective the reader should know that this book is not intended to be a newly definitive work in military history or a technical analysis of contemporary physical geography. Instead it combines established facts and concepts from both disciplines to show that relationships between the environment and combat are highly variable, often unpredictable, and always formidable. Finally, the overarching message of the following chapters is that, despite the evolving technology in warfare, physical geography has a continuous, powerful, and profound effect on the nature and course of combat.

CHAPTER 1

Storms, Fair Weather, and Chance

Kamikazes, Dunkirk, and Normandy

Strategy, power, and materiel notwithstanding, weather can sometimes be equally important in the fortunes of war. Wind and rain figured in the outcome of the Peloponnesian Wars (431–404 B.C.); thunderstorms and shifting winds in the Battle of Long Island in late August 1776; severe cold, the Finno-USSR war; rain-induced mud, Ambrose E. Burnside's second major Civil War offensive in January 1863; dust storms, the World War II North Africa campaigns; fog, the German Ardennes offensive of 1944; typhoons, in the Pacific during World War II; and monsoons, Vietnam.[1] Even fair weather can be deceptive in misleading the observer as to the long-term climatic reality of the setting. For example, in invading Russia, Charles XII of Sweden, Napoleon Bonaparte, and Adolf Hitler all based their judgment for troop movement on summer and early fall conditions, with little or no planning for the following storms and frigid temperatures that cost them so dearly.[2]

Cyclones and Convergence

Since this planet's beginning, its atmosphere has never been at rest. Pressure differences and the earth's rotation keep the atmosphere in constant motion, the moving air lessening the world's temperature extremes by exchanging moist warmth from the tropics with dry cold from the polar regions. Equally important, in some situations the circulating air rises, the lifting process then forcing expansion to distribute the same heat energy throughout a larger volume, thus lowering temperature.[3] Large-scale uplift and subsequent cooling of air may be induced by equatorial convergence zones, tropical and mid-latitude cyclones, topographic barriers (such as mountains and plateaus), and individual convectional cells produced through excess heating of air at the earth's surface.[4] In each of these situations the uplift and cooling may be great enough to reduce temperature below the air's dewpoint (saturation level), thus forcing condensation of water vapor. Under certain conditions the resulting clouds yield precipitation, converting areas that would otherwise be desert into the various types of humid climates that cover about 75 percent of the earth's land areas.

The largest and most frequent storms in the middle latitudes develop where

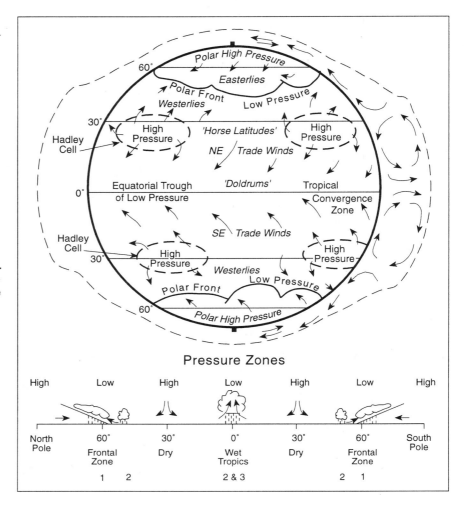

MAP 1.1. General circulation of the atmosphere with the effects of earth-axis inclination and land masses (including orographic barriers) omitted. Air-mass source regions are associated with areas of high pressures and convergence zones are related to low pressure. The pole-to-pole cross-section shows likely precipitation processes at different latitudes. *Note:* 1 = frontal lifting; 2 = areas where convection is common; 3 = intertropical convergence zone.

unlike air masses from different source regions come together. Here a front, or convergence zone, becomes established and, much like a battle line, shifts one way or another according to the power of the atmospheric forces involved.

These large air masses originate in both polar and subtropical areas where high pressure persists (Map 1.1). From these sources air may move both downward and outward in a huge spiraling motion. This type of overall circulation, known as anticyclonic, produces relatively clear weather because the lifting processes needed to form clouds and precipitation are absent. Such conditions are prevalent in polar regions and over extensive areas in the subtropics such as Australia, parts of northern and southern Africa, and in the Middle East.

Huge air masses regularly drift from these areas of high pressure toward adjacent subpolar and equatorial convergence zones where conditions induce upward-moving air currents that, in turn, form a widespread and lasting cloud cover. For example, the predominantly windy and overcast weather of Scotland and the Falkland Islands (Islas Malvinas) is typical of places in the upper mid-latitudes which are influenced much of the time by the boundary, or front, between polar and subtropical air. In contrast, the rain that created major problems in the southwest Pacific during World War II and clouds over extensive areas in Central America so persistent that the sun rarely shines illustrate weather associated with intertropical convergence and related uplift.

The areas of lowest pressure within a zone of converging air tend to have the most unsettled weather. Known as cyclones (not to be confused with tornadoes), these storms are common in both the humid tropics and the middle latitudes, their size and turbulence ranging from minor disturbances with localized clouds and the slight possibility of scattered precipitation to huge storms that may bring extensive flooding and massive destruction.

All middle-latitude wave cyclones tend to move eastward and generally bring widespread overcast skies, gusty and gradually shifting winds, and varying amounts of precipitation.[5] Rain and possibly snow may occur where the warmer, less dense, moisture-laden air is forced aloft along the cold and warm fronts, the intensity generally being greatest in the more complex, occluded section that forms where the two fronts join (Fig. 1.1).[6]

These wave cyclones progress through a cycle, which begins with cyclogenesis, followed by maturity. Eventually the cyclone self-destructs in a phase known as occlusion. In doing so, the storm's components change in location and magnitude. Thus the cyclone is ever-changing and, even with modern technology, aspects of the system remain difficult or impossible to predict with certainty. But one prediction is reliable: the weather will change markedly as these disturbances move over an area.

The tropical depression is the equatorial counterpart to the middle-latitude wave cyclone. Each year some grow large enough to become typhoons or hurricanes.[7] Developing over warm oceans, these storms lift enormous volumes of moist air far upward to produce towering clouds, large amounts of rain, and winds that swirl around their "dead eyes" faster than 120 km/h (75 MPH).

Meanwhile, funnels of fast-moving air in the upper atmosphere form shifting jet streams that guide the path of the storms below.[8] Jet streams in the tropics tend to move east to west, and in the middle latitudes west to east. As a result, typhoons and hurricanes advance westward, eventually curving toward their hemispheric pole, while the middle-latitude wave cyclones migrate eastward along the polar front (see Map 1.1 above).

Yearly variations in earth-sun relationships force the mid-latitude polar fronts in each hemisphere and the equatorial tropical convergence zone to shift gradually north and south. During the winter season of either hemisphere, the mean position of the polar front shifts toward the equator, increasing temperature contrasts between the cold and warm air. This in turn favors increased atmospheric lifting and turbulence, thus enlarging the potential size and intensity of winter wave cyclones. In contrast, typhoons and hurricanes, whose for-

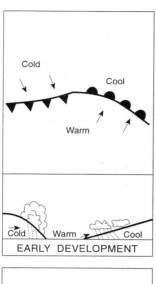
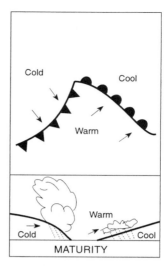
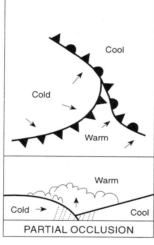
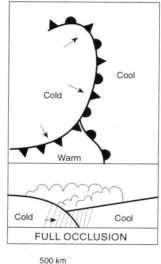
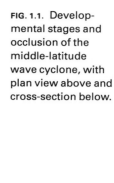

FIG. 1.1. Developmental stages and occlusion of the middle-latitude wave cyclone, with plan view above and cross-section below.

mation depends on maximum heat from the ocean, develop most frequently in late summer and early fall when tropical water temperatures are highest.

MILITARY CONSEQUENCES

Whether in the tropics or middle latitudes, these shifts and changes in stormy weather zones affect widespread areas and produce fairly distinct seasonal differences that may be of huge military importance. Consider, for example, the alternating wet and dry monsoons of southeast Asia, the long summer drought followed by overabundant winter cyclonic rains in Italy, and the alternating heat and cold of European Russia, all vivid examples of weather's importance on major mid-twentieth-century battlegrounds.

Large-scale weather patterns, which are now often well documented, must be considered carefully in any sound military plan. More problematic are localized and less obvious conditions that may significantly influence weather. These include the effects of nearby terrain or water bodies on cloud formation and related precipitation, shore-zone orientation that seasonally presents meteorological-induced advantages or problems relating to wave heights and storm paths, and periodic debilitating winds related to mountains, deserts, or certain coasts.[9] In addition, there are unexpected singular events—such as passing thunderstorms, short-lasting winter thaws, hazardous icing conditions, blizzards, dense fog, and even destructive tornadoes—all of which are especially difficult or even impossible to predict because of their transient nature.

Even so, a systematic assessment of known and potential components of large storms serves the commander well. Especially important cyclonic attributes include the stage of development, rate of movement, path of the storm, possible ramifications for offensive and defensive action, and implications regarding communication, reinforcements, and resupply. For many soldiers an evolving weather pattern has proved to be a valuable ally. But just as often it can create difficult conditions that can be neither predicted nor controlled. Detailed plans for military operations are generally made well in advance but in doing so it is impossible to predict weather conditions and their influences with certainty.

For example, twice within seven years in the thirteenth century storms contributed to the defeat of a Mongol force that had crossed the Korea Strait to invade Japan. In contrast, in the spring of 1940, prolonged cloud cover and the absence of major storms over the English Channel facilitated an amphibious withdrawal at Dunkirk that saved an army from destruction. And four years later a fleeting period of fair weather facilitated the Normandy landing, so essential for the Allies' liberation of Western Europe.

If the weather related to these four operations had been otherwise, their nature would certainly have been somewhat different, possibly affecting history significantly. What would our world be like today if, through the thirteenth-century actions of Kublai Khan, Japan had become more open to the influences of East Asia rather than remaining independently isolationist for another 500 years? How would the war in Europe have differed if more than 300,000 Allied soldiers, including the heart of the British army, had been captured by the Germans at Dunkirk in 1940? And if the Normandy invasion of 1944 had been long-delayed, or failed completely because of bad weather, thereby prolonging the war, is it possible that the first atomic bombs would have been dropped on Germany instead of Japan?

THIRTEENTH-CENTURY JAPAN AND THE KAMIKAZE

Those World War II Japanese kamikaze pilots who died trying to crash their explosive-laden planes into U.S. ships took their name from high winds attributed to divine origin which, about 700 years before, twice destroyed Kublai Khan's shipboard invasion force. Those gales were associated with cyclones

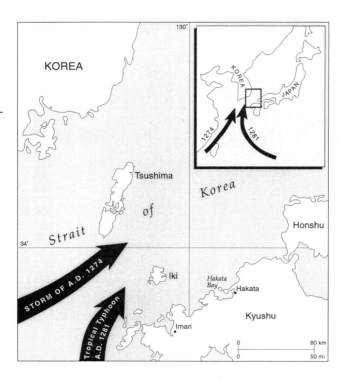

MAP 1.2. Strait of Korea, Kublai Khan's invasion sites, and probable paths for the storms of A.D. 1274 and 1281.

originating hundreds of kilometers away that just happened to follow a course that led to the Korea Strait. It was, for Japan, the fortunate coincidence of timing and turbulence that made these two storms the forever legendary Kamikaze. But where did they come from and how is it that they twice brought destruction to the invading Asian horde?

Kublai Khan (1216–94) first went into battle at the age of 10 with his Mongol grandfather, Genghis. By 1259 Kublai had assumed the throne of a family empire that extended from the borders of Europe, through the vastness of Asia, across northern China, to the east coast of the continent. He did not, however, rule Japan, a small but wealthy kingdom almost within sight across the narrow strait from his holdings in Korea.

Kublai Khan first attempted to conquer the Japanese through diplomacy. In 1268 he sent envoys via Korea to deliver a letter to the Japanese emperor, explaining that subservience to the heaven-decreed emperor of Mongolia would avoid war.[10] The emissaries presented their message at the emperor's court in Kyoto but the Japanese did not respond, even though they were well aware of the Mongol leader's power.

As a result Khan ordered the king of Korea to construct 1000 vessels, assemble 40,000 troops, and provide supplies in preparation for an attack upon Japan.[11] For five years Kublai Khan continued to send envoys across the strait with hopes for subjugation rather than warfare. Ignoring the Mongol leader's demands, the Japanese responded by fortifying positions at anticipated attack points on their southern island of Kyushu (Map 1.2).

By 1274 the 1000 Korean vessels were ready and, having no satisfactory response to his demands, Khan, with possibly as many as 40,000 Chinese, Korean, and Mongol troops, gave orders to invade Kyushu.[12] Some Korean sailors may have been relieved to learn that the force would not sail until late November 1274, "when the typhoon season was well over."[13] Others more experienced likely had second thoughts, knowing that late fall storms are not uncommon in the Strait of Korea and all can be dangerous and life-threatening.

The invasion commenced with the occupation of Tsushima in the strait. A few days later, Iki, an island even nearer to the Kyushu coast, was secured. These served as stepping stones toward the larger islands of Japan. On or about 20 November, Khan's force sailed for Kyushu, landing unopposed in Hakata Bay.

Anticipating the invasion, the Japanese leadership acquired defenders from a number of the island's feudal lords, eventually assembling a force that was probably larger than the Mongol army. As the battle commenced it quickly became apparent that the Japanese ritualistic samurai ethic of formally announcing personal deeds and ancestry when challenging a single combatant was not a suitable tactic against attackers who fought in groups and at will. As a result, on the first day of fighting the invading army forced the Japanese to withdraw into ancient stone fortifications. As evening approached the Japanese remained undefeated and in a favorable defensive position. In contrast, Khan's troops were exposed to the enemy and, possibly fearing a night counterattack, were ordered to withdraw first to the shore and then to the apparent safety of their ships in the harbor.

This decision was disastrous for the invaders. That evening a large storm moved northeastward into the Korea Strait and soon destroyed much of Khan's fleet now crowded with warriors who, ironically, had been ordered to board the ships for protection. From where did this storm come?

The progressively shorter days of autumn result in the gradual lowering of average temperature for air centered over Siberia. This cooling increases the extent and pressure of the air mass and by late fall it influences much of the continent. Also produced are winds that blow from the land and toward the Indian and Pacific Oceans. This monsoonal circulation persists during the winter and produces months of drought for much of Southeast Asia.

Furthermore, expansion of high atmospheric pressure over Siberia forces the southward migration of the polar front that forms the boundary between cold polar and moist subtropical air. During November, the month of Khan's scheduled invasion, Japan is just far enough removed from the Siberian high-pressure center to be affected by the fluctuating boundary between the cooler Asian winds and the warm, moist air over the ocean. Most important, storms are guided north and east along the trend of this front which, in turn, is positioned by the upper-air jet stream.[14]

As a result, two types of atmospheric disturbances tend to approach Japan, both from the south and west. The first is a mid-latitude wave cyclone that develops along the polar front. These may exceed 1000 km (620 mi) in diameter, increase in frequency during late fall and winter, and bring widespread clouds and the likelihood of precipitation. They are often followed by strong winds out of the west and northwest. However, their exact path, rate of movement, and degree of development tend to vary greatly.

The second type of storm is much less frequent but more life-threatening. These cyclones originate in the tropics and, as they move westward, a few grow in magnitude to become typhoons. Most common in late summer and early fall, they may form over the Pacific Ocean as late as December, as was so apparent from the huge damage and disorder imposed on a large U.S. naval task force in 1944. All of these storms move westward eventually to curve poleward in their respective hemispheres. Those that form in the tropics north of the equator over the Pacific tend to turn northward in the western part of that ocean. In doing so, they may affect the Philippines, coastal Southeast Asia, and Japan.

In late November 1274, either a very large mid-latitude cyclone or an even more powerful typhoon moved directly toward Kyushu. Then, in a coincidence of enormous proportions, the storm reached the coast just a few hours after Khan's troops had boarded his ships in Hakata Bay.

That night the full force of the cyclone was directed on Khan's fleet. Rain, high winds, and increasingly rough water made navigation difficult in the harbor and perilous at sea. Many ships struck rocks and sank. Others foundered offshore. Before the storm had passed, as many as 300 vessels and 13,000 men, or about one-third of the Mongol army, were lost at sea.[15] Meanwhile, the storm that destroyed much of Khan's force became reverently known as the Kamikaze, or Divine Wind, throughout Japan.

Khan, who likely considered himself an earthly god, had never before lost a campaign. To avenge the defeat, he formed the "office of chastisement of Japan."[16] But he was also trying to conquer the Sung Dynasty, which controlled a formidable Chinese empire south of the Yangtze River. Lacking resources to fight both a land war to the south and an amphibious operation to the north, Khan again tried his brand of diplomacy with the Japanese. The envoys he sent in the late 1270s were not only ineffective, they never returned. The Japanese took to beheading them, usually where they had disembarked.

Kublai Khan had much better results from his encounters with the Chinese; their resistance collapsed by 1279. Now, five years after his first attempt, he could again move against Japan. The plan was to combine an army of 20,000 soldiers and 1000 junks in Korea with the "South of Yangtze Force" of 100,000 men and 3000 vessels in southern China. By early 1280 the troops in Korea were assembled and ready for battle.

The gathering of the southern forces, however, was delayed first through summer and then into late fall. By that time the prevailing monsoon winds, blowing from land to sea, made northward navigation toward Korea impossible for many of Khan's ships. As expected, the following spring brought diminished winds, but calm conditions then persisted through April as the warmer weather slowly lessened the atmospheric pressure over Siberia (this being a common transition previous to the monsoon reversal). Not until May 1281, when the winds finally blew from the sea to the land, could the bulk of the southern forces move north. Then, as June approached, so did the beginning of the typhoon season.

It was logistically impossible to quarter the large Mongol forces in Korea until the typhoon threat ended late that year.[17] To carry out the invasion Khan had to once again move his troops across the Korea Strait—even if it was approaching the height of the typhoon season in the western Pacific.

Meanwhile the Japanese, aware of a second invasion possibility, adjusted their samurai battle tactics and built a shoreline wall 4.5 m (15 ft) high and 40 km (25 mi) long at Hakata Bay, the expected point of attack. They also established a fleet of small sea craft to attack the enemy offshore.

By June 1281 all of Khan's large southern force had arrived in Korea and preparations for a combined assault were finalized. First the smaller northern fleet sailed for Hakata Bay, the same site as the attempted invasion of 1274. Arriving late in June, they fought fiercely but could not overwhelm the strong Japanese defensive positions. After withdrawing to an island just offshore to regroup, they then landed on another nearby Kyushu beach, but here too they were repulsed, finally retreating to their ships.

During mid-July ships brought the first elements of the larger southern force to offshore islands and by early August some 150,000 men were assembled. These troops were then put ashore at Hakata Bay and Imari Bay, about 50 km (30 mi) to the southwest (see Map 1.2 above). The Japanese, however, having modified the rituals of samurai combat in favor of effective group tactics, were formidable and fierce on the battlefield. As a result, the Mongol generals, perhaps with a plan to attack elsewhere, ordered their troops to withdraw and reembark the waiting ships. Once again their timing could not have been worse.

Each year about 600 minor disturbances form to move westward over tropical oceans. Fewer than 10 percent of these develop into typhoons (Map 1.3). In early August 1281 a small, moisture-laden storm developed 5 to 10 degrees north of the equator over the Pacific Ocean. Originating well east of the Philippines, it moved westward along with the general circulation pattern and gradually received enough heat from the warm ocean to grow into a major tropical cyclone. As the Japanese priests thousands of miles away prayed to their gods for deliverance from the second Mongol invasion, the disturbance reached typhoon size and began, as two out of three do in this area, to curve northward toward the coast of China.

The storm, now with winds most likely in excess of 120 km/h (75 MPH), continued veering toward the right, a course that eventually brought it directly in line with southern Japan. The huge system reached the coast on 15 August 1281, shortly after more than 100,000 Mongol troops had boarded their ships.

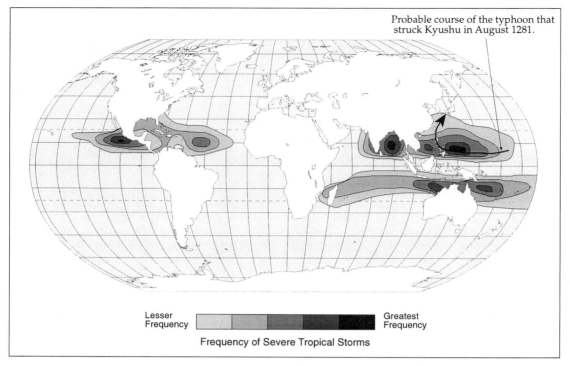

Probable course of the typhoon that struck Kyushu in August 1281.

Lesser Frequency

Greatest Frequency

Frequency of Severe Tropical Storms

MAP 1.3. Source areas and frequencies for especially severe tropical storms known as typhoons (western Pacific Ocean), hurricanes (Atlantic Ocean, and eastern Pacific Ocean), and cyclones (Indian Ocean).

To the Japanese emperor and priests, their prayers for a second Kamikaze had been answered. "A green dragon had raised its head from the waves . . . sulfurous flames filled the firmament" to destroy the enemy.[18]

Few ships could maneuver effectively with the high winds, large waves, and strong tidal surges.[19] As navigation became futile, hundreds of the cumbersome junks carrying thousands of soldiers were "smashed . . . to matchwood."[20] Japanese records indicate that after the winds calmed, the loss of ships was so great at the entrance to Imari Bay that "a person could walk across from one point of land to another on a mass of wreckage."[21] In all, an estimated 4000 ships and 100,000 Mongol lives were lost at sea. And survivors who struggled to shore were quickly killed by the avenging Japanese.

Now we know that, rather than divine winds, the Mongols were victims of two intense cyclones that originated far from the shores of Japan. But were their defeats simply misfortune or did Khan make an unwise gamble with the elements?

Wave cyclones vary greatly in intensity and trajectory; they are also somewhat periodic, being generally separated from one another by hundreds of kilometers in distance and several days' duration. Over the centuries thousands of gales have struck Japan. Furthermore, the largest of these, the typhoon, is more frequent in the western Pacific than in any other ocean. As a result, Kyushu experiences an average of one typhoon every two years, suggesting that in 1281 Khan was simply unfortunate. But the peak season for these storms is from July through September, the exact time of year for the prolonged amphibious operation associated with Khan's second invasion. The longer his troops remained in the Korea Strait without securing safe harbors for his ships and troops, the more he tempted fate.

It is not clear if the November storm of 1274 was a late-season typhoon that originated in the tropics or one of the mid-latitude cyclones associated with the polar front which are increasingly common in the late fall. Either way the results were the same for Khan's army.

There is no way of knowing what the outcome of battle would have been if Mongol forces had fully met the Japanese on Kyushu in the thirteenth century. Khan obviously believed that the invasions of Japan would be successful and likely never considered the notion that his biggest enemy would be storms. Whether in July or November, expecting to avoid completely such disturbances for any length of time was unrealistic.

To the Japanese, the victories were not simply the result of brave warriors who had fought so effectively against an apparently invincible force. Instead it came from God-given divine winds that appeared to answer their prayers. Because of this, the Kamikaze legend became firmly established in Japanese mythology as "a weapon from heaven and from then on was considered a symbol that the land was divinely protected."[22] It was this belief that was to produce deadly heroics more than six centuries later, during World War II, when the Japanese faced its U.S. adversary who approached not from Korea but, ironically, from the same direction as the Kamikaze.

Twentieth-Century Europe

If engaged with a powerful enemy, the only large-scale military operation more difficult than an amphibious landing is a maritime withdrawal. In either case the relationship between weather and water can foster success or hasten defeat. For more than a week in 1940, a persistent late-spring weather pattern over Europe's northwest coast aided the cross-Channel escape of a large Allied force from Dunkirk. Exactly four years later, not far away along the shore in Normandy, quite different short-term atmospheric conditions combined with aspects of the tide and moon to favor what may have been the most important amphibious landing in the history of warfare. The weather associated with these two operations had enormous and lasting implications on the nature and aftermath of World War II.

In the simplest sense, Europe can be described as a large peninsula with many smaller peninsulas that extend westward from Asia. Because of its shape and the adjacent water bodies it has the most maritime climate of all the continents. The major weather factors influencing this region are:

a persistent storm-producing low-pressure cell near Iceland;

the warm Azores high which is a fair-weather, subtropical, anticyclonic system centered on the eastern North Atlantic Ocean;

the westernmost influences of the frigid Siberian high that prevails during winter and early spring;

relatively warm offshore water of the North Atlantic Drift; and

low-lying terrain that does not block circulation eastward.[23]

Dunkirk and the nearby Normandy coast are about 50 degrees north of the equator, making them central to evolving mid-latitude weather patterns of Western Europe (Map 1.4). Location and strong maritime influences favor the year-round passage of east-moving wave cyclones that bring clouds, wind, high humidity, rapid weather changes, and pronounced seasonal differences.

During the spring it is not uncommon for periods of fair weather to linger over the area. These are induced by air masses from the Azores high which drift toward the northeast. When in place over Normandy and Flanders, this air forms a "blocking high" that, along with a split in the upper-air jet stream, guides surface storms along two divergent paths: one to the north leading toward Scandinavia, the other to the south extending to the Mediterranean Sea. On average, this fair-weather pattern prevails about 40 percent of the time in April, decreases to 25 to 30 percent chance in late May, and is even less common during early June. These probabilities become more important when it is recalled that the bulk of the Dunkirk evacuation lasted from 26 May to 4 June 1940, and that the Normandy landing was on 6 June 1944.

The slow deterioration of northwest European weather from April through June results partly from the decreasing pressure of the Siberian high, which, in turn, permits a greater number of cyclones to track through the area, bringing storms and rough seas. Even so, there are occasions in late spring when air from the Azores shifts into the area. The resulting weather will initially be clear, warm, and dry with southwest winds, good visibility, and considerable sunshine.

Curiously, if this same fair-weather anticyclonic system remains for several days, the colder land and water surfaces chill the lowermost part of the atmosphere. As the air closest to the earth's surface cools, and if it falls below the dewpoint temperature, extensive fog will form, even though the atmosphere is stable, or nonstormy. As basal cooling migrates upward, so does condensation to form low clouds that may extend a few thousand meters in altitude. Because most sunlight is then reflected and absorbed by the clouds, the weather along the coast becomes chilly, damp, overcast, generally unpleasant, and unchanging for hours or even days. Meanwhile, the winds are light and variable, storms are absent, visibility is limited, the seas are navigable, and flying conditions are poor.

Thus it can be anticipated, but not definitely predicted, that the coast's most favorable spring weather will most likely occur in April. As the season progresses, the weather can be expected to be more changeable and stormy as middle-latitude cyclones become more frequent.[24] All of this means that very

MAP 1.4. Northwest Europe showing Dunkirk and the Normandy Coast. (Map modified from diagram by A. K. Lobeck in Army Map Service, *Continental Europe—Strategic Maps and Tables,* pt. 1, *Central West Europe.*)

special conditions must have existed to provide more than eight consecutive stormless days dominated by overcast skies and tolerable seas that so favored the evacuation of more than one-third of a million Allied troops from Dunkirk in late spring of 1940. And four years later, persistent stormy conditions gave way to a fleeting episode of weather that was minimally suitable for the airborne and amphibious operations at Normandy. Without favorable weather the evacuation in 1940 would have been much less successful and the Normandy landings in 1944 surely delayed.

DUNKIRK

Some interpret the Allied collapse in Western Europe during the spring of 1940 as a total military disgrace. Others describe the related evacuation at Dunkirk between 26 May and 4 June as a tactical victory bordering on the miraculous. Either way, after a humiliating retreat into Flanders, about 338,000 Allied troops were evacuated while facing a German army that until then had moved upon its enemies along a wide front at a speed never before seen in land warfare.[25]

Certainly no one factor explains the success of the Allied evacuation. All the elements—the final position of these forces in a coastal zone with wide beaches, adjacent protecting dunes, and an inland area containing hundreds of crisscrossing drainage ditches and many canals; some questionable German tactical decisions and their increasing equipment and supply problems; periodic effective Allied resistance along a shrinking and more heavily defended perimeter; and heroic efforts of naval and maritime units, including hundreds of participating English boat-owners—are surely in some way important. Even so, two environmental factors were ascendant for the Allies at Dunkirk.

First, tidal effects, shallow water, limited docking facilities, and severe restrictions of the navigation-ways in the English Channel meant that the troops had to be removed either on relatively small boats or ships that required favorable shore-zone and sailing conditions. Second, because of their exposed positions and very limited air support, Allied forces were highly vulnerable to German aerial attacks, the method Adolf Hitler had chosen to end the fighting in Flanders.

Incredible as it may seem, for much of the time during eight successive days, maritime conditions were adequate for the embarkation of troops and the stability of smaller ships in the English Channel, while mist and low clouds prevailed, hiding much of the evacuation process from German pilots and observers in attacking planes. In terms of meteorology, this prolonged cloudy weather accompanied by moderate seas was unusual but explainable. When considered in the context of evacuating 338,000 from a small beachhead while engaged with a powerful enemy it becomes what Winston Churchill called "a miracle of deliverance."

The German Advance. In late 1939 and early 1940, after the surrender of Poland and during the so-called Phony War, German and Allied armies in westernmost Europe were positioned for battle but there was no combat. Hitler planned to attack in the west in November 1939, but autumn rains and deteriorating field conditions along with advice from his general staff convinced him to wait. In fact, the weather reportedly caused a number of invasion postponements.[26]

Meanwhile, through intelligence, the German high command learned that if conflict did occur, Allied orders would direct troops in the west to move into Belgium, keeping the front away from the industrialized region of northern France, while at the same time holding defensive positions along the Maginot Line in the east. To counter this, the Germans planned—using blitzkrieg tactics—to split the enemy forces by advancing through the Ardennes in Belgium, an underpopulated region of steep hills, dense forest, and narrow valleys. Because the French considered this area to be impenetrable by armored divisions, they had stationed only nonmobilized forces along this part of the frontier.

The Germans believed that after breaking through the Ardennes, their tanks could defeat these sedentary forces and move quickly westward into a battleground of rolling hills, fewer forests, and well-connected roads, an environment well suited for armor. They then envisioned a quick race to the channel coast, causing disruption, separation, and defeat of the French and British armies.

On 10 May 1940, Hitler's attack began. Advancing along a front stretching from the North Sea to the Swiss border, German armies invaded four countries simultaneously—the Netherlands, Belgium, Luxembourg, and France. Eighty combat divisions moved into the Low Countries.

In the Netherlands, with its low and flat terrain marked by many canals, German airborne troops landed behind the Dutch lines, keeping important bridges intact and permitting the army to penetrate quickly. Netherlands' forces retreated toward the northwest, separating themselves from the Allies and giving the Germans access to Rotterdam. On 13 May the Dutch queen and the government sailed for England. The country was then in the hands of Gen. Henri Gerard Winkelman, who, unable to face the prospect of the continued bombing of Rotterdam, surrendered the country the next day.

The success of Hitler's offensive was even more pronounced in the south. Led by such notable generals as Heinz Guderian and Erwin Rommel, this army changed the rules of war. Bursting across southern Belgium and northern Luxembourg, the panzer divisions faced only ineffectual opposition from the sluggish French 9th Army. The supposedly impenetrable Ardennes was quickly crossed—the tanks and armor using the narrow, winding roads, and the troops marching on the adjacent shoulders.

By the eve of the thirteenth, a gap had formed between the slow-moving French 2nd and 9th Armies. The Germans poured through in deep, penetrating tactics with great success. By the morning of the fifteenth, Rommel was 24 km (15 mi) beyond the Meuse River, already behind the French 9th Army and attacking the right flank of their 1st Army, causing it to withdraw toward the northwest.

As General Guderian's forces cleared the Ardennes they entered country more suitable for tanks, thus increasing the rate of advance. By 17 May his armor was nearly halfway from Sedan to the channel coast. A small French counterattack led by Charles de Gaulle was staged that day, but the Luftwaffe kept the French ground forces from gaining much contact with the panzer.

From here the advancing Germans veered northwest (Map 1.5). Already they had achieved two of their major objectives: severing French connections between Paris headquarters and its northern armies; and situating themselves at the right flank and rear of the enemy, a position favorable for complete envelopment. Capture of three channel ports—Boulogne, Calais, and Dunkirk—would assure victory.

The English, however, counterattacked on 21 May at Arras and blunted the first German flanking offensive. The next day the German advance was renewed, reaching the outskirts of Boulogne and Calais. But it would take three and four more days, respectively, to secure these cities.

MAP 1.5. The German advance toward Dunkirk and English Channel evacuation routes.

The final German objective, Dunkirk, continued to elude them. German tanks were only 15 km (10 mi) outside Dunkirk on the twenty-third, and British forces ordered to protect the city had not yet established strong defensive positions. Curiously, with Dunkirk practically in sight and apparently vulnerable, the German vanguard received an order to halt. For three days the armored units waited outside Dunkirk.

Then on 26 and 27 May, a large middle-latitude wave cyclone moved over the coastal area, bringing rains that saturated Flanders. This condition greatly limited panzer mobility and may partially explain German orders to withdraw many of the tanks. On 29 May most of the armor was sent south to join the battle against the French armies defending Paris. As a result, the German army lost its chance for the panzer forces to occupy Dunkirk and eliminate the possibility of a large-scale Allied evacuation.

The halt of the German attack outside Dunkirk has long been of great military interest and debate. Possibly the order resulted from equipment problems or the great extension of the German front; but the terrain of Flanders may also have been an important factor. Flat, swampy, and laced with a complex network of canals and ditches, it is poor country for tanks. Being restricted to rural roads with limited cover, it must have been clear to Guderian that any further advance by his tanks would be difficult. Whatever the reason, Adolf Hitler decided that the German air force would bomb the Allies into submission. In light of the weather pattern that was to come, this decision proved to be the complete reverse of Kublai Khan's amphibious experiences in 1274 and 1281.

The Coast and the Weather. Flanders has perhaps the longest and smoothest stretch of wide sand shore in northwest Europe. The low coastal plain and gently sloping beaches merge with an offshore area of shallow water with numerous sandbars and mud flats. Normal tidal fluctuations can shift the shoreline more than a kilometer (0.6 mi) in some places. In 1940 only the 63-km (39-mi) offshore ship channel, the "Dunkirk Road," was navigable by larger vessels at all times, and with the fall of Calais it became useless because it was easily covered by German shore batteries. That left only two other offshore passes available, the Ruytingen and the Zuydecoote, but both were impassable in heavy seas no matter what the tide (see Map 1.5 above). At high tide in calm weather these two channels did have adequate depth for medium-sized ships. These routes to England were, however, indirect and much longer than the Dunkirk Road.[27] Finally, when subject to high winds and severe storms, the mid-channel is treacherous and rough surf pounds the shoreline. All of these factors made the weather crucial to any evacuation attempt.

By 26 May, German bombing had made the inner harbor at Dunkirk largely useless for embarkation but they could not destroy two long channel-penetrating jetties, or moles. Not being piers, these were actually barriers constructed well into the sea to aid docking navigation and to protect the harbor from storm waves. The shorter west mole was made of stone, as was the shoreward section of the longer east mole, but farther out the east mole consisted of openwork concrete pilings through which the sea flowed. Each mole had a wooden walkway wide enough for four persons and, at regular intervals, stanchions for shiphausers.

The east mole, extending 1600 m (4250 ft) from shore, was especially important in the evacuation. Troops embarking from this jetty could skirt the town and wait on the beach for ships that tied up at the breakwater.[28] At the same time, smaller craft evacuated other soldiers from the sandy beaches. This second task would have been impossible except for the small vessels pool of wooden-hulled boats, fortunately and coincidentally organized only a few weeks earlier to assist in the search for the new German magnetic mines in the English Channel.[29]

On 26 May the evacuation was hampered by the same heavy rain that may have discouraged a final panzer attack. Furthermore, another large wave cyclone was over the mid-Atlantic and appeared to be heading for the English Channel. Such a storm, with high winds and waves, would certainly have slowed the evacuation process further. But by Tuesday, 28 May, the Atlantic storm had shifted its course to move northward along the northwest coast of Ireland toward Scotland. This was in response to the direction of upper-air flow within the jet stream which also reflected the re-establishment of a blocking high over Western Europe.

At dawn on 28 May, the skies at Dunkirk were overcast. A low ceiling of only 100 m (300 ft) limited German air operations. The temperature was mild, indicating that the air was from the Azores high. Even so, the surf remained rough from the extended effects of the storm that was tracking well to the northwest of the coast. As a result, beach evacuation was difficult but it was possible to board people from the east mole. On that day 17,804 troops were evacuated.

Early on 29 May the low clouds remained and there were periods of scattered rain, further limiting German air operations. A breeze produced a low surf and carried streams of smoke from the burning port over the beaches, increasing concealment. As the day progressed the cloud cover gradually thinned and at about 1400 hours conditions were suitable for an attack by German dive-bombers. By 1600 hours the wind had shifted to the north, blowing mist and smoke away from the embarkation area. For the first time since the beginning of the evacuation, German pilots had a clear view of the ships lined up on the east mole. As a result the Luftwaffe, which had previously been concentrating on the harbor, attacked the crowded breakwaters. The result was the loss of many lives and a number of ships on the afternoon of 29 May.

By the following morning, good bombing conditions had disappeared into a veil of clouds, mist, and smoke. There was more scattered light rain, the channel was subdued, and the evacuation continued without threat from German planes again grounded by the weather. In the afternoon the sky became clearer and the German air raids resumed, again concentrating on the crowded east mole. Meanwhile, because of the low surf, many smaller boats evacuated men directly from the beaches. Even with the aerial attacks, on this day more than 29,000 people were removed from the strand while another 24,000 left from the harbor and moles. This was the only day when more men were evacuated from the beaches than from the breakwaters.

At dawn on 31 May a southwest breeze blew over the channel, indicating a renewed influx of air from the Azores high. By noon, as the anticyclone's effect increased, the sky cleared, and gusty winds created a high surf that severely limited embarkation from the beaches. Meanwhile, the Luftwaffe made three major attacks on Allied ships. In spite of the bombing, on that day the process of evacuation was so effective that 68,014 English soldiers were transported home, the highest one-day total for the evacuation. Most of these troops left from the east mole.

By 1 June practically all of the British Expeditionary Force had been evacuated, but the weather was beginning to change to the Germans' favor. A very large anticyclone drifted northeastward toward England and coastal Europe. Its arrival brought a bright, clear morning on 1 June, weather that continued for the next few days. Daytime evacuation became especially dangerous with most removal operations conducted at night. By then, however, the bulk of the troops originally trapped in the Dunkirk pocket had been evacuated. When the last boat left on the evening of 4 June, more than 338,000 Allied troops had been removed, while 40,000 French soldiers remained stranded at Dunkirk.

One can correctly conclude that regardless of atmospheric conditions an evacuation would have occurred at Dunkirk. But what other types of weather might have existed in late spring in Flanders and how might they have affected the withdrawal? Two weather patterns are statistically most likely for this coast during late May. The first is a progression of wave cyclones, each bringing stormy weather and followed by strong northwest winds prolonging rough seas, making embarkation from the moles much more difficult and departure directly from the beach nearly impossible. The second is a more assertive dominance of air masses from the Azores high, a process that would bring fair weather at a time of year when days are the longest. This type of weather, with its clear skies and long daylight hours, would have been ideal for repeated attacks by the Luftwaffe.

By a twist of fate, or better yet a meander of the jet stream, neither of these two most-likely weather patterns dominated Flanders in late May 1940. Instead the one set of conditions that most favored the Allied evacuation persisted for days, the result being that a maximum number of the British army's most experienced personnel could regroup in England and prepare to meet the Germans again the next year in North Africa. Ironically, exactly four years after the last of these soldiers left Dunkirk to disembark in England, another Allied force was climbing aboard ships, boats, and aircraft to land in Normandy under a very different weather pattern.

NORMANDY

Early in 1942 the U.S. War Department began planning for an attack across the English Channel. Concerns for the military survival of the Soviet Union urged a landing as soon as possible while logistics, training needs, and a decision to conduct a major campaign in the Mediterranean area delayed the operation through 1943. It was then decided that a direct amphibious attack on Western Europe would take place in the spring of 1944. Early plans called for landings on France's northernmost coast but the objective was later changed to Normandy.[30]

Tactically it was important that the landing take place at dawn during a low tide that would reveal beach obstacles and avoid prolonged grounding of landing craft so necessary for transporting reinforcements and supplies. About 5 km (3 mi) of visibility was also needed for effective supporting naval gunfire. Clear skies were desirable for accurate bombing and a more effective air cover. A full moon was needed to enhance the planned large-scale nighttime airborne operations. Nonstormy seas (to minimize seasickness, disorganization, and boat accidents) and light winds (to clear fog and smoke) would also be significant assets to the operation. Finally, favorable landing conditions should last at least 36 hours, and preferably several days, to provide enough time to land forces and supplies necessary to secure the beachhead. Predictions for tides and moon phases showed that only a few days of each month were suitable for the landing. Superimposed on these restrictions were day-by-day changes in the weather.

The possibility of making an accurate and reliable long-term weather fore-

TABLE 1.1. Precipitation, cloud, and wind data for Lille, France

Month	Number of Days with Precipitation[a]	Number of Days with Thunderstorms	Number of Days with Fog[b]	Number of Days with Gales[b]	Mean Evaporation (mm)	Mean Cloudiness[b] (oktas)	Mean Sunshine[c] (h)	Most Frequent Wind Direction[b]	Mean Wind Speed (m/sec)[b]
Jan.	18	0.3	8.4	9	—	7.2	59	W	5.5
Feb.	14	0.0	9.2	6	—	6.8	74	SSW	5.0
March	13	0.4	6.8	5	—	6.2	130	NE	4.8
April	14	1.6	3.7	5	67	5.6	176	NE	5.0
May	13	3.4	3.3	4	86	6.1	198	W	4.4
June	12	3.7	4.1	2	79	6.1	205	W	4.0
July	13	4.2	3.5	3	85	6.3	201	W	4.1
Aug.	13	5.6	5.0	3	79	6.2	179	W	3.9
Sept.	14	2.8	6.8	4	70	5.8	150	W	4.0
Oct.	14	0.6	10.5	4	42	6.4	112	W	3.9
Nov.	16	0.1	8.8	4	—	7.4	51	S	4.5
Dec.	17	0.1	10.6	6	—	7.7	39	SSW	4.9
Annual	171	22.8	80.7	55	—	6.5	1574	W	4.5

a. >0.1 mm
b. 1951–1960
c. 1946–1960

TABLE 1.2. Precipitation, cloud, and wind data for Brest, France

Month	Number of Days with Precipitation[a]	Number of Days with Thunderstorms	Number of Days with Fog[b]	Number of Days with Gales[b]	Mean Cloudiness[b] (oktas)	Mean Sunshine[c] (h)	Most Frequent Wind Direction[b]	Mean Wind Speed (m/sec)[b]
Jan.	22	0.8	4.5	10	7.2	66	SW	5.5
Feb.	16	0.9	3.9	10	6.9	85	W	5.4
March	15	1.3	6.5	8	7.0	142	NE	5.7
April	15	0.2	6.5	7	5.9	189	NE	5.4
May	14	1.1	7.7	5	6.2	220	NE	5.2
June	13	1.9	9.0	2	6.5	209	NE	4.5
July	14	1.2	11.0	3	6.5	210	W	4.8
Aug.	15	0.9	9.6	3	6.4	207	SW	4.5
Sept.	16	2.2	7.5	4	6.3	156	SW	4.5
Oct.	19	0.9	7.4	5	6.6	120	SW	4.2
Nov.	20	1.3	3.5	8	7.3	69	W	4.7
Dec.	22	2.0	4.6	11	7.3	56	W	5.2
Annual	201	14.7	81.7	76	6.7	1729	SW	5.0

a. >0.1 mm
b. 1951–1960
c. 1946–1960

TABLE 1.3. Precipitation, cloud, and wind data for Ostend, Belgium

Month	Number of Days with					Most Frequent Wind Direction	Mean Wind Speed (m/sec)[b]
	Precipi- tation[a]	Thunder- storms	Fog	Cloudy Sky[b]	Clear Sky[b]		
Jan.	14	0.0	3	13	3	S	7.1
Feb.	13	0.0	7	14	2	NE	6.7
March	10	0.1	5	10	4	WSW	6.4
April	10	0.3	3	11	2	NNE	6.6
May	9	0.7	1	9	2	NNE	6.7
June	9	0.4	1	8	5	NE	6.3
July	12	1.6	1	10	3	WSW	6.2
Aug.	13	1.0	1	11	2	WSW	6.7
Sept.	10	1.0	2	10	3	NE	6.2
Oct.	14	0.3	4	13	3	S	6.0
Nov.	14	0.1	4	16	2	S	6.2
Dec.	15	0.1	5	15	3	S	6.5
Annual	143	5.6	37	140	34		6.5

a. >0.1 mm
b. >75%
c. <25%

cast for the Normandy coast was nil. Even so, the averaging of meteorological data covering many years does indicate that the best chance for fair spring weather occurs in April or early May when blocking highs are most frequent. Through late May and June the average number of storm days tends to increase with the maximum reached during the first half of July (Tables 1.1–1.4).[31] Thus, for sound meteorological reasons, D-day was originally planned for early spring. However, a decision to increase the size of the invasion force and the need to conduct additional air operations in preparation for the landing forced postponement until June, even though probability indicated that the fairest weather of the spring was giving way to more frequent middle-latitude cyclones.

Then, on the basis of readiness, tides, and moon phase, the invasion was planned for either 4, 5, or 6 June, with the fifth preferable. If for some reason the invasion could not take place on any one of the three days the next alternative date would be 19 June (which turned out to be about the time that Normandy experienced its largest spring storm of 1944). As the early June D-day approached it became increasingly apparent that the time of the invasion depended more upon weather than any other factor.

Months before D-day James M. Stagg of the Royal Air Force was appointed chief meteorological officer of the Supreme Headquarters Allied Expeditionary Forces (SHAEF).[32] As a group-captain, he was responsible for advising Gen. Dwight D. Eisenhower on the weather and did so regularly through the spring. The hope was to provide an accurate five-day forecast because that was

TABLE 1.4. Precipitation, cloud, and wind data for Le Havre, France

Month	Number of Days with			Mean Cloudiness[b] (oktas)	Most Frequent Wind Direction[b]
	Precipitation[a]	Thunderstorms	Fog[b]		
Jan.	18	0.2	5.6	7.2	SW
Feb.	14	0.1	5.7	7.1	SW
March	12	0.5	5.5	6.4	E
April	12	0.4	4.1	5.9	NE
May	13	2.3	2.6	5.8	NE
June	11	2.0	3.9	6.2	W
July	11	1.6	4.0	6.1	W
Aug.	12	2.2	2.4	6.4	W
Sept.	14	1.6	3.4	6.3	W
Oct.	15	0.8	3.8	6.9	SW
Nov.	17	0.4	4.4	7.5	SW
Dec.	17	0.1	6.2	7.6	SW
Annual	166	12.2	51.6	6.6	SW

a. >0.1 mm
b. 1951–1960

the duration needed to carry out the embarkation, make the cross-Channel movement, and firmly establish the beachhead.

On 1 June the weather over the North Atlantic, which was key to the future for the channel, was affected most by an extension of the Azores high southwest of Spain and several well-developed mid-latitude wave cyclones to the north (Map 1.6a). A northeasterly advance of the high-pressure cell would likely bring fair weather to Normandy, but a retreat would open the path for eastward-moving cyclonic storms.

By the afternoon of 3 June 1944, the ridge of high pressure receded westward, and two closely spaced wave cyclones intensified as they approached the British Isles (Map 1.6b). The first of these progressed slowly northeastward and would not affect the invasion beaches directly. The second appeared to be moving along a more southerly path and, if so, would likely create conditions unfavorable for an attempted invasion. It is this pattern that encouraged the Germans to lower their coastal alertness to the point that their patrol boats were ordered to remain in port.

With no immediate signs of improving weather, General Eisenhower postponed the operation very early on 4 June, the date for the planned embarkation of troops if a landing was to be made early on the fifth. As the day progressed it was recognized that the second following storm was turning northeast toward Scotland. Later it became clear that the two storms were gradually merging and intensifying north of the British Isles. These developments favored the rapid westerly movement of a trailing cold front that would sweep far to the south, crossing Ireland, England, and the channel. More

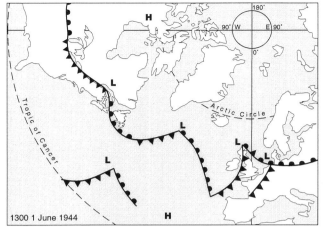

a.

1300 1 June 1944

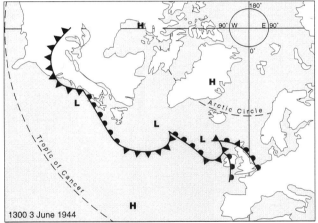

b.

1300 3 June 1944

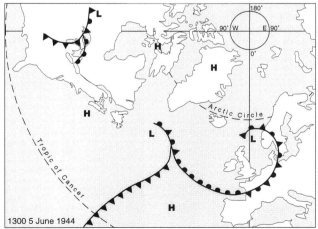

c.

1300 5 June 1944

MAP 1.6. Weather maps. *Note:* L = low pressure; H = high pressure; barbed lines = cold front; half-circle lines = warm front. (Modified from James Martin Stagg, *Forecast for Overlord* [New York: Norton, 1971]; and Karl R. Johannessen, "Hindcasting the Weather for the Normandy Invasion," in *Some Meteorological Aspects of the D-day Invasion of Europe, 6 June 1944,* ed. Roger H. Shaw and William Innes, American Meteorological Society Symposium Proceedings, 19 May 1984, Fort Ord, California.)

important, it was anticipated that a following small ridge of high pressure would bring a short phase of improved weather that might last as long as 48 hours.

At 2130 on 4 June Group-Captain Stagg presented this information to General Eisenhower. Fifteen minutes later, about 30 hours before the first wave of troops would land on the beaches, Eisenhower ordered the landings to take place on 6 June. This decision involved tremendous responsibility and, no matter how the operation turned out, it would affect history in a most profound way. With near stalemate in the Italian campaign and rapid advances by the Red Army on the Eastern Front, it was politically and militarily essential for Allied armies to occupy Western Europe as soon as possible. However, if the landing failed the loss of troops would be staggering, the defeat devastating, and it would be months before it could be attempted again. Furthermore, with an immediate threat from the west greatly diminished, Germany could reinforce its Eastern Front with the Soviet Union which would most likely prolong the war in Europe.

Even a cancellation of an early June invasion in favor of a later date would surely have major ramifications. It is reasonable to conclude that all subsequent operations, including the securing of needed ports, liberation of cities, opening of prison camps, and the crossing of the Rhine would all have been postponed by weeks and possibly months.

Some may have the notion that the nature and outcome of World War II on the Western Front was preordained and certain. In June of 1944, however, success was in no way assured and delay was a definite possibility—if either had happened, the history of that war would surely be far different. For a brief period on 4 June 1944, this enormous effort depended in large part on a short-term weather prediction by a meteorologist and a simple order by a commander.

In the end this decision proved both fortunate and correct. As predicted, the turbulent cold front swept rapidly across the channel and coast on 5 June (Map 1.6c). By the morning of 6 June the weather in Normandy improved markedly, although, to the discomfort of thousands of troops on 5000 ships, boats, and landing craft, the channel remained rough from localized gusty winds and swells generated by distant storms. Clouds had hampered some aspects of the previous nighttime airborne operation, but by dawn these cleared enough to favor Allied air support and naval gunfire. Although the wind picked up during the day, slowing resupply and reinforcements, by evening about 50,000 men had established several beachheads and the Allies had a toehold in Western Europe.

The weather remained at least marginally adequate for the next two days, although wind-driven waves, especially along the eastern landing zones, continued to cause periodic problems. During this 48-hour period the initial landing force was greatly reinforced and supplies essential to hold the position were put ashore. Although unsettled and stormy weather soon returned, by

10 June the beachhead and a fortified perimeter were fairly well established. How fortunate had the Allies been in terms of the landing? A review of mid-June weather reveals at least part of the answer.

If the landing had not taken place on 5, 6, or 7 June, the next earliest suitable date would have been 17 June, but the nineteenth would have been preferable. Both nights, however, would be moonless. Records show that conditions on the seventeenth were adequate for a landing at Normandy but ensuing events show how hazardous the weather can be. During 18 and 19 June, two important things happened. First, the northeast edge of a high-pressure ridge from the Azores drifted over the northern British Isles. Second, a low-pressure center tracking eastward over the Mediterranean Sea greatly intensified. The steep pressure gradient between these two extremes was greatest along the English Channel. This meteorological arrangement produced unusually strong winds out of the north that directly approached the unprotected east-west-trending Normandy coast. The result was a highly destructive four-day storm that interrupted and curtailed the landing of supplies all along the beachhead and did huge damage to the artificially formed Mulberry Harbor so essential to the American sector. Thus a landing on the seventeenth would have been followed immediately by a prolonged storm that would have all but eliminated the

FIG. 1.2. U.S. troops leaving their landing craft on Omaha Beach during the Normandy landing, 6 June 1944. Note considerable distance to the shoreline bluff and that the low tide reveals debris and obstacles on the beach. (National Archives.)

chance for resupply and reinforcements. Furthermore, if a landing had not been made on 17 June it is not likely that it would be attempted on 18, 19, 20, or 21 June because of the then-predictable high-wind period. If this was the case the next suitable landing conditions would not occur until July.

Pondering the detailed effects of such a powerful storm on a mid-June landing may be intriguing, but its actual impact on military operations must remain speculative. Clearly, however, the weather-based decision to land at Normandy on 6 June was most fortunate for the Allies. Its timing also fooled the German forces. Surely German commander Field Marshal Erwin Rommel was well aware that the most favorable landing period during the spring was over and that the next best time for invasion would be in the latter half of July when the potential magnitude and frequency of cyclonic storms could be expected to diminish. Furthermore, informed that the early June short-term forecast was for poor weather, he had left France for a brief visit with his family and an appointment with Hitler. In fact, all German sea-patrol craft were ordered to remain in port on 6 June because of anticipated storms. Obviously neither the absent Rommel nor the coastal commander had the meteorological information that Group-Captain Stagg presented to General Eisenhower.

FIG. 1.3. U.S. troops laboring in the rough surf along the Normandy coast, 6 June 1944. (National Archives.)

The German impression that a landing would not occur during early June and Eisenhower's quick and opportunistic decision to invade based on a short-term weather prediction permitted an enormous, exposed amphibious force to move undetected for many kilometers and land on a foreign shore, achieving the element of surprise against a suspecting, yet unalert, enemy. If the early June upper-air-flow pattern of the North Atlantic had been different—either supporting persistent cyclonic storms, which would have forced an Allied postponement of the landing, or perpetuating fair weather from the Azores high, which would have increased German awareness—the story of Normandy would certainly be different in some ways. As Gen. Omar N. Bradley stated, "In this capricious turn of the weather, we had found a Trojan horse."[33]

FIG. 1.4. Because of the smooth shoreline and absence of large harbors, the Allies sank numerous ships to form a breakwater to facilitate resupply. Note the beach in the upper right and the shallow offshore bottom that has been marked by the passage of larger ships. (National Archives.)

Conclusion

Twice within four years, the Allies had been favored by late spring weather that cost the Germans dearly. At Dunkirk the English evacuation was carried out in front of a dominant but underestimating enemy force on the offensive. In contrast, what was amazing at Normandy was that a fleeting period of fair weather coincided with desirable tidal and lunar conditions to favor an amphibious invasion against the same formidable enemy who had by then shifted to an observant defensive position. The first operation salvaged both hope and substance from defeat. The second led to victory in the west.

In both cases the associated weather patterns were most fortunate for the Allies and could very easily have been quite different. For example, prolonged fair weather would have been ideal for the German Luftwaffe over Dunkirk and would surely have increased German alertness and observations at Normandy. More important, two modern-day North Atlantic storms, vaguely similar to the ancient Kamikazes of 1274 and 1281 in Japan, could have converted the Allies' good fortune into appalling and dreadful events. The two thirteenth-century disasters for Khan and two World War II successes for the Allies are forceful illustrations that, regardless of the century, weather has the potential to be of paramount importance in military operations.

CHAPTER 2

Too Much and Too Wet

The Civil War Mud March and Flanders' Fields

My God, did we really send men to fight in that?

LT. GEN. SIR LAUNCELOT KIGGELL AFTER SEEING THE MUD
AND TERRAIN AT PASSCHENDAELE, FLANDERS, 1917

The mixture of soil and water has been slowing and stopping armies since armies began. A sudden rain or thaw can, within minutes, transform a normally dry and firm battlefield into a quagmire. Elsewhere, mud may be a problem that persists throughout the year. Either way, mud has been a formidable element in every prolonged military campaign, and sometimes it is the decisive factor in a military operation.

Mud is simply a mixture of water and small fragments of decomposed rock. However, the resulting material may vary greatly in composition, thickness, moisture content, distribution, and duration—characteristics that combine in many different ways to produce an infinite variety of situations. Although water in mud originates primarily from precipitation, two other factors must also be considered: moisture-holding capabilities within the soil and its underlying parent material; and prolongation of moisture near and at the surface because of low evaporation rates favored by cool temperatures, high humidity, cloudy conditions, and light winds.

Potential moisture retention within soil and subsurface rock depends upon the relative volume of internal openings (porosity) and their degree of interconnection (permeability), properties that are governed by the shape, arrangement, and packing of constituent particles.[1] That is, rain on porous and permeable material, such as sand, will easily percolate into the subsurface, making the soil moist while it remains stable. In contrast, precipitation on less permeable soils prohibits abundant infiltration, favoring complete saturation, runoff, and standing water. Furthermore, as fine-textured soil becomes increasingly saturated with water, the soil becomes less and less stable, a process that can be markedly accelerated by the type of disturbances commonplace and often widespread in military operations.[2]

Most soil consists of sand, silt, and clay originally derived from the weathering of bedrock. These particles, called clasts, are produced by either chemical decomposition or mechanical disintegration.[3] Either way, the resulting clasts are commonly differentiated on the basis of size. Progressively coarser fragments include sand, granules, pebbles, cobbles, and the largest, boulders.

These larger clasts are most often subangular to rounded in shape, making it impossible for all fragments to fit closely together everywhere. The result is a formation with moderate to high porosity and good permeability. Thus even if moisture is present, land associated with these materials is generally relatively well drained if it is above the highest level of saturation by groundwater.

The smallest products of weathering are silt- and clay-sized particles, most produced by chemical decomposition. These fragments are characteristically microscopic, thin platelets with large surface areas in comparison to their mass or volume. These properties permit more surface-adhering moisture and close-packing, which reduces permeability. The addition of water to soils containing an abundance of these especially small clasts produces mud.

Because the formation of mud requires water, knowledge of the causes and variability of precipitation is useful in assessing its potential effect on military operations. For example, understanding conditions in arid regions that lead to the uncommon precipitation offers a tactical advantage to the well-informed commander.[4] Elsewhere operations can be affected seasonally, such as the persistent winter rains in the Mediterranean area which are induced by a southward-shifting polar front and the related tracks of stormy wave cyclones.[5] The result is the distinct winter wet season with accompanying mud and swollen rivers that twice (1943–44 and 1944–45) frustrated Allied offensives in Italy during World War II. Finally, some areas are continually moist, such as the plain of northwest Europe where mud has hampered marching and fighting forces from well before Agincourt (1415) through the twentieth century.[6]

In this chapter, two planned battles—one that was never fought and another that, to the combatants, seemed never to end—illustrate the challenge of mud.

Military Consequences

THE U.S. CIVIL WAR: BURNSIDE'S MUD MARCH OF JANUARY 1863

By the end of 1862, disillusion permeated most Union troops in the eastern United States. Confederate forces there had, among other things, thwarted Gen. George McClellan's Virginia peninsula campaign; twice within 14 months sent the North into retreat on the same battlefield near Washington, D.C. (Bull Run or Manassas); and in September 1862 invaded Maryland, a Northern state, to fight near Sharpsburg (also known as the Battle of Antietam). Late in the year Pres. Abraham Lincoln replaced McClellan with Gen. Ambrose Burnside, who then advanced into Virginia only to be thoroughly defeated in a mid-December offensive at Fredericksburg.

During late December 1862 and early January 1863, Gen. Robert E. Lee's Army of Northern Virginia strengthened defensive positions along the south bank of the Rappahannock River at and upstream from Fredericksburg.[7] Meanwhile, on the opposite side, Burnside's forces were in bivouac at Falmouth, Virginia (Map 2.1). Here Burnside developed, with reluctant approval

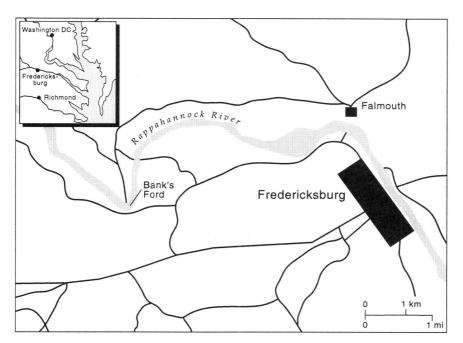

MAP 2.1. Falmouth, Fredericksburg, and the Rappahannock River.

from his superiors and the derision of some associates, plans for his second offensive and scheduled it for 20 January 1863. With the river protecting their left flank, Union troops would march several kilometers upstream (west) on their side (north) of the river, pass the main Confederate force, turn left to cross the stream at two fords, and turn left again to advance eastward toward the enemy's left flank and rear. Burnside anticipated that his maneuver would give him two major advantages. First, it would force Lee's army into a major turning movement as the battle began. Second, once the Southerners were realigned, their backs would be to the Rappahannock River.

As the day for battle approached, Burnside was concerned primarily with command and execution, not knowing that two other unforeseen factors were about to come together and destroy his offensive. The first was the deeply weathered clay-rich soil which, having formed over millions of years, blanketed the countryside. The second was the day-by-day development and eastward movement of a mid-January wave cyclone then far to the west.

The land around Fredericksburg and Falmouth consists of rolling hills dissected by southeast-flowing streams, the Rappahannock being locally the largest. Both towns lie along the approximate boundary between the lower Coastal Plain to the southeast and the rolling Piedmont to the northwest (see Map 6.2). Here, and to the east, the relatively weak sandstones and shales of the Coastal Plain are poorly consolidated and easily weathered. Westward, the upper part of somewhat more resistant and geologically older crystalline formations of the Piedmont are also deeply decomposed from long-term chemical decomposition favored by the abundant moisture and relatively warm climate.[8]

At most places in both tracts, a thick mantle of fine-grained, unconsolidated, reddish-yellow residuum rich in iron and aluminum oxides covers the bedrock. These soils are, for the most part, ultisols, meaning that they are very old and have experienced prolonged weathering, much of it in the form of chemical decomposition that produces innumerable microscopic plate-like fragments.[9] Within the uppermost part of ultisols, many of the smallest particles have moved downward to form a shallow argillic (clay-rich) horizon that, by clogging, inhibits further percolation of water. Thus abundant rain, restricted from downward movement, quickly saturates the upper soil. Any excess moisture either remains as surface pools or drains away into the numerous creeks and streams.

Because the bulk of the military movements in the Eastern Theater were northeast-southwest on the Piedmont, neither the cross-cutting rivers nor the clay-rich soils could be avoided. When the land was dry and rivers low these two terrain factors did not present unusual or unmanageable problems. But both conditions could quickly change if precipitation was abundant, even to the point where they might become decisive elements in a military operation.

Early January of 1863 brought generally fair weather and little precipitation to Virginia, conditions that were ideal for Burnside's plan. With the land dry and firm, mud had not been a factor in preparing his second offensive. In fact, once developed, the only major changes in battle plans (resulting from an intelligence report about Confederate positions south of the river) were to cross the river only at the one closest ford and to delay the attack by 24 hours. Burnside's first decision may have been wise because of logistical efficiency and the developing Confederate defensive posture. The second, however, would soon prove disastrous because of the storm developing far to the west.

The most important factor controlling Virginia's winter weather is the position of the high-altitude, mid-latitude jet stream and its effect on surface storms. Although little was known of meteorology in 1863, daily newspaper reports of temperature, cloud cover, and precipitation provide a basis for "hindcasting" the weather using modern atmospheric principles. The result is a reasonable estimate for evolving positions of the jet stream, air masses, surface fronts, and storm centers.[10] On this basis it is apparent that during early January 1863 the general configuration of the jet stream did not permit major storms to cross Virginia. Such conditions are well illustrated on Saturday, 17 January, when the path of the upper-air jet stream must have extended southward over the western United States, then eastward over the Gulf Coast, and finally northeastward over the Atlantic Ocean (Map 2.2). Within this great arc was an anticyclone consisting of cold Canadian air with high barometric pressure. As this large mass of air moved southward over the Midwest it brought clear skies and frigid temperatures to the Great Lakes area. The high was −10°C (13°F) in Chicago, −11°C (12°F) in Cleveland, and −18°C (below 0°F) in Toronto. By day's end, the front forming the eastern edge of this cold air reached the East Coast, dropping New York City's temperature from about 16°C (60°F) on 16 January to below 0°C (below 32°F) the next day.

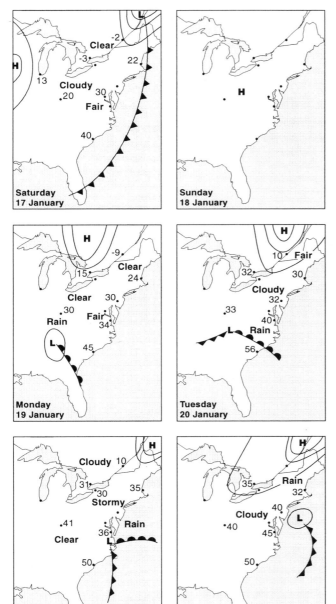

MAP 2.2. Weather patterns over the eastern United States, January 1863. Temperatures are in degrees Fahrenheit as shown on the original newspapers.

By 18 January all of the eastern United States was clear, cold, and ideal for military movements in Virginia. However, the position of the jet stream's arc and its related anticyclone was slowly shifting eastward. Meanwhile, far to the southwest, in an area not influenced by this sprawling high-pressure system, a mid-latitude cyclone, or low-pressure center, grew in size thanks to a strong north-south temperature gradient, abundant moisture carried inland by winds from the Gulf of Mexico, and frontal development.[11] This combination favored converging and rising air currents, which in turn produced clouds and increased the chances for precipitation. Slowly, this system intensified as the gradually shifting jet stream guided the storm eastward across the Confederacy and directly toward Virginia.

At dawn on 20 January, the original day for attack, sunlight shone through a whitish sky. As the day progressed, cloudiness gradually increased, resulting in a complete overcast by afternoon. More ominously, the temperature began to drop as the wind increased and shifted to arrive from the east and northeast. All of these factors indicated the imminent approach of the wave cyclone.[12] Unaware of the nature and size of the oncoming storm, some troops moved toward their assigned positions in preparation for the next day's maneuver and battle. Others were ordered to prepare for the placement of five pontoon bridges across the Rappahannock River at Banks Ford at dawn on the following day. Although the sky was overcast, everything seemed in reasonably good order as twilight passed to darkness on 20 January.

About 2100 hours, however, the first drops of precipitation fell. A light drizzle graduated into a steady rain, driven by a strong east wind, all indicating the full arrival of the wave cyclone. If this storm had been smaller, or its rate of movement faster, or its path more to the north or south, the precipitation might have been light, fleeting, or avoided altogether, making the cyclone of little consequence to Union troops. But unfortunately for Burnside and his plan, this storm was very large, well developed, slow moving, and now tracking directly over southeastern Virginia.

The rain intensified through the night and by dawn on the twenty-first the countryside and the army were drenched. As is common in military operations, the mission was paramount to all other factors. On that basis, and regardless of the weather, Burnside's army of 75,000 began their march across the Piedmont. By this time the soil was saturated. Because of the churning action from moving men, horses, and equipment, all roads and trails quickly became deep muddy tracks. Even cross-country marching by infantry was difficult. As the maneuver toward Banks Ford continued, so did the rain. Conditions steadily worsened as the mud became deeper, more plastic, and adherent. Troop movement slowed. All element of surprise was lost as observing Southerners across the river ridiculed the foundering Union forces attempting to reach Banks Ford. Of the five planned pontoon bridges, only one section of one bridge was ever put in place; the rest were never to reach the river, being stuck in the upland mud of the Piedmont.[13] By evening of 21 January, most of the more than 75,000 soldiers were bogged down and their equipment immobilized. The planned attack had been stopped in its initial phase without even crossing the river, much less facing the enemy.

The rain continued on the twenty-second while temperatures fell as the wind shifted to arrive from the north, both indications that the storm's center had passed just south of Fredericksburg. By then, Burnside accepted the futility of the attack and ordered his troops to remain in place. That evening the steady rain gave way to scattered showers and then ceased altogether. The following morning the army was recalled to its original camp at Falmouth. Ironically, the return march brought gradually thinning clouds followed by sunshine, both signals that the giant wave cyclone had departed North America and continued its swirl northeastward over the Atlantic Ocean.

Weather hindcasting shows that an execution of Burnside's original plan for battle on 20 January would have avoided, at least for a day, the rain and mud that stopped his army on 21 January. This being so, it may also be expected that if the 24-hour delay had not been ordered his troops would have crossed the Rappahannock and moved toward a great military confrontation. Clearly, the potential impact of such an event defies conclusions, but it does invite speculation. Could the North have trapped Lee's army in Fredericksburg with no escape route because of the river at their rear? Or would the Southern general again outmaneuver the Union army to defeat Burnside twice within two months? It is more likely that such a battle would evolve in some other way, reflecting the uncertainty and confusion of war. However, one prediction can be made with certainty. If the battle had been joined on 20 January, any fighting on the next two days would have been done in the rain and mud of the Piedmont.

Speculations aside, Burnside's 24-hour delay coincided precisely with the arrival of one of the winter's worst storms. It also added ominously to a series of unsuccessful Union efforts to defeat the South, this one without even confronting the Confederates. The result was the further demoralization of Northern forces in the east, possibly to their lowest level thus far in the war. At the command level it also illustrated a continuing problem in leadership that increasingly perplexed Lincoln. And finally it forever established Ambrose Burnside's infamous "Mud March" in military history.

WORLD WAR I: FLANDERS

The wide, crescent-shaped North European Plain extends from western France through Belgium and the Netherlands, then eastward into and beyond northern Germany. It was along this lowland that the right flank of invading German forces advanced, first west and then south, in the great, sweeping Schlieffen maneuver that opened World War I in August 1914. The attack quickly resulted in the establishment of the war-long Western Front, extending unbroken from the English Channel to neutral Switzerland.

Flanders, the front's northernmost sector, was of great tactical importance because it forms the central section of the lowland, or military corridor, that extended from Paris to Berlin. It was also vital for controlling access to ports on the Strait of Dover, by far the narrowest stretch of water between England and the Continent. Here, where the fighting went on uninterrupted for more than four years, was also environmentally the worst battlefield on the Western Front. What made it more foul than the rest? The answer lies in the combined

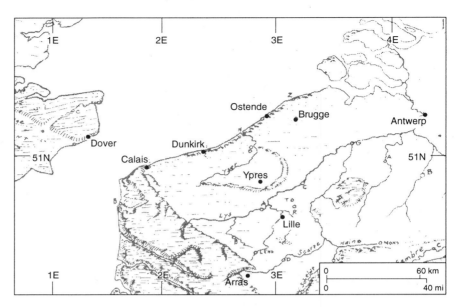

MAP 2.3. Flanders and nearby areas. (Map modified from diagram by A. K. Lobeck in Army Map Service, *Continental Europe—Strategic Maps and Tables,* pt. 1, *Central West Europe.*)

realities of long-term combat and killing with the effects of weather, climate, geology, terrain, soil, and drainage. Altogether they produced a perpetual wartime morass of men, marsh, mud, and pollution.

At 51 degrees North latitude, eastward-moving mid-latitude cyclones affect Flanders year-round (Map 2.3). Air masses moving into this area from the west, having traversed thousands of kilometers of the Atlantic Ocean, tend to have high humidity.[14] This combination of numerous storms and abundant moisture brings year-round frontal precipitation that is especially abundant during cooler and cloudier months when evaporation rates are low. The result is a great surplus of water, which aside from vaporization must seep into the ground, drain off the land, or remain at the surface to form lakes and swamps.

Not long ago in geologic time this area formed the nearly flat bottom of a shallow sea where calcareous deposits accumulated as beds of chalk. These were then buried by a thin layer of sand followed by a thick sequence of clay. The clay was, in turn, covered by another sand deposit (Map 2.4). Although compressed by the weight of the overlying material, none of these sedimentary formations have been well cemented, making them relatively weak rocks.[15]

Within the last several million years tectonic forces uplifted and broadly arched the area immediately to the south (known militarily as the Somme) well above sea level. Being adjacent, this movement also elevated Flanders, but here the land was pushed barely above the level of the North Sea and English Channel. Furthermore, the exposed sediments underlying Flanders were never well consolidated, making them easily weathered and eroded. The result today is a low, flat land, scarcely above sea level except for a few places where the last remnants of the slightly stronger sandy formations are yet to be worn away. These remnants form several low but conspicuous uplands; its central position and commanding view made one, Mont Kemmel, especially important militarily.

The soils of Flanders are most strongly influenced by parent material derived from weathering of underlying bedrock. Since being uplifted above the sea, erosion here has beveled the surface to a near-horizontal plane that cuts obliquely across slightly inclined rock strata of varying composition. This results in two extensive areas of clay-rich soil that are relatively impervious to water seepage, the Clay Belt and the Maritime Belt. These tracts are flanked by somewhat more permeable sandy soils (Northern Sand Belt and Southern Sand Belt), but they too are rather poorly drained because of the flat topography and a groundwater table only a meter or so below the surface. Well-drained soils in Flanders are limited to a few slightly higher erosional remnants in the Clay Belt, a narrow zone of coastal sand dunes, and tracts within the Transition Belt.

Three major rivers in Flanders trend along the dip, or inclination, of the underlying strata, thus they drain northeast, parallel to the coast, rather than directly to the sea. This arrangement results in very low river gradients, making the nearby land especially susceptible to repetitive flooding. The Yser, for example, has been known to go over its banks more than 10 times in a year, and the larger Lys River may, if uncontrolled, inundate vast areas of Flanders.

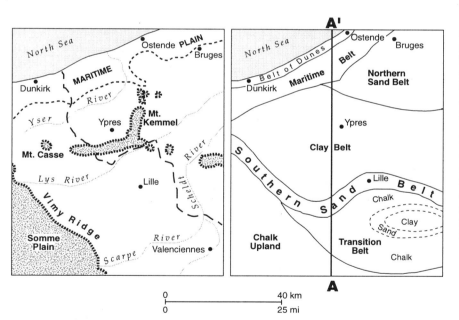

MAP 2.4. Terrain and geologic characteristics of Flanders. (Modified from illustration in Douglas Johnson, *Battlefields of the World War: Western and Southern Fronts: A Study in Military Geography*, American Geographical Society Research Series 3 [New York: Oxford University Press, 1921].)

Flanders has an elaborate system of ditches, canals, dikes, dams, and locks constructed to drain the land for agriculture and to lessen the effects of flooding. In the lowest coastal areas the highest marine tides can actually reverse inland drainage that normally goes to the English Channel. Cleverly, drainage of the land can be improved by opening locks to let the water drain away during low tide and closing them at high tide to prevent the return of sea water. But whoever controls the locks can also reverse this process and intentionally flood extensive tracts, a military tactic first used in October 1914.

Thus in Flanders, abundant year-round precipitation falls on a low-lying, near-horizontal surface to produce poorly drained soils underlain by a high groundwater table. Even in peacetime this combination of factors makes excess moisture a continual problem. But during World War I, the mud and water here were always present, practically unmanageable, and potentially lethal.

Initially Flanders was peripheral to Germany's World War I battle plans. By design, their westward invasion of August 1914 turned southward toward France before reaching the coastal lowlands. With only a relatively small and retreating Belgian army in opposition on their right, the Germans bypassed Flanders in favor of a direct advance toward Paris. By late August, both French and early arriving elements of the British Expeditionary Force briefly engaged the right flank of the German army without success, while Belgian forces that had withdrawn to Antwerp continued to resist (these would soon join English and French forces in Flanders).

Meanwhile the massive German offensive in France first slowed and then stopped completely just short of Paris, in the September 1914 Battle of the Marne. As this battle ended, the German command turned its attention to the only remaining place on the Western Front where a flanking maneuver could still be attempted—in and immediately south of Flanders. A series of Allied-German encounters along this flank quickly extended the Western Front northward from France all the way to the English Channel. Known as the "Race to the Sea," this fighting culminated in Flanders with the mid-October Battle of Yser and the even larger two-phased First Battle of Ypres. As a result, notions of flanking maneuvers gave way to trench warfare and frontal attack. It was also at this time that the use of the machine gun and indirect artillery fire became highly perfected.

By the end of 1914, the Western Front was fully defined and essentially stabilized on Allied soil. Meanwhile the scale of the war became enormous. The extent of the front required elaborate transportation and communication networks, huge supply centers and diverse support services, and numerous headquarters and administrative offices. For the fighting soldier, however, protective and defensive structures at the battle line became increasingly important and elaborate. These generally involved complex trench systems, underground command posts and billets, and tactically positioned observation points. But the land and water rendered what was commonplace elsewhere either difficult or impossible in Flanders.

In addition to the bogs, swamps, tide-affected lowlands, ditches, creeks, and rivers, the high groundwater table (which in much of Flanders is only a meter or two below the surface) precluded standard trenching and underground command posts because even shallow excavations quickly filled with water. Here, digging was commonly done not to form trenches, but to provide material for construction of above-ground breastworks and parapets. Previously constructed embankments for railroads and highways, along with diggings from dikes and ditches, were of immediate tactical importance in planning or conducting military operations.

Movement of heavy equipment, munitions, supplies, and troops was restricted largely to road and rail. Positioning and supply of heavy guns was governed by the road network. Artillery effectiveness was decreased because innumerable shells plunged deeply into the soft earth, each detonation creating muddy craters, many then partly filling with water. The nature of the terrain essentially eliminated cavalry operations. On this battleground the infantry became central and the artillery supreme.

Front-line troops were often knee-deep in mud, even in long-established defensive positions. Wooden catwalks, platforms, and footbridges were everywhere. Any attack generally meant crossing wet and open ground pockmarked with thousands of craters in which a soldier could become entrapped in the muck. Mud slumped and oozed into pits and shallow trenches, and then combined with groundwater to form a dense slurry. Worse yet, diggings often uncovered the decaying remains of humans and animals killed in the conflict. This, along with the enormous refuse and waste of the huge armies, polluted the surface water, creating a continuous health hazard. At many places the groundwater became highly contaminated. Although water seemed everywhere, little of it was suitable for human consumption. Deep wells had to be drilled and extensive pipelines constructed to move potable water toward the front. And where the pipes ended, runners had to carry it forward to the battle lines, outposts, and advancing soldiers. Ironically, all of this simply added more moisture to a place that was already saturated.

Although relatively flat, clumps of trees and patchy forests often limited long-distance observation. As a result, the area's few hills, even if only 5–10 m (15–30 ft) high, became some of the most cherished military positions and common objectives, making them also some of the deadliest battlegrounds. The largest of these, Mont Kemmel and a north-extending spur called Messines Ridge, formed a natural bastion near Ypres that was the central and key position in Flanders. This higher ground permitted widespread observation of the adjacent lowlands. Here deep trenching and underground excavations were feasible, and water and mud were a bit less of a problem. But these same sites were regularly attacked because whoever held the high ground had a major advantage for both offensive and defensive purposes. Better drainage here also offered the opportunity for tunneling for the purpose of detonating huge charges beneath enemy lines. The result was some of the largest infantry-coordinated mine explosions in the history of warfare.

Fighting in these awful conditions went on continuously for more than four years. Tens of thousands were wounded, became sick, or died here. More than 50,000 British and Commonwealth soldiers who fought in the Ypres sector alone were never accounted for. Apparently many were killed or wounded, only to be swallowed by the mud under the weight of their equipment. Even with all the personal suffering and tragedy of the prolonged battle, Douglas Johnson, who wrote the most comprehensive geographical description of the Western Front (1921), concluded, "The awful mud of the Yser [meaning Flanders] is the strongest memory which one carries away from that part of the battlefield [Western Front]."

Conclusion

There is no question that mud can be an important element on the battlefield. The problem is to assess its impact correctly and to plan ways to counter its effect. For Burnside the unexpected and quick presence of mud in mid-January 1863 was both devastating and decisive. His brief delay converted a planned battle into a humiliating defeat even before his troops met the enemy.

The problem with mud was much worse in Flanders, its continual presence over several years bringing enormous hardship to hundreds of thousands of soldiers. Yet it did little in determining the final outcome of the battle. Here, unlike the Union Army's one-day Mud March in Virginia, mud played no favorites. For four years in Flanders it was the constant enemy for all troops, no matter what their duty or which side they were on.

CHAPTER 3

Clouds and Fog
The Bulge and Khe Sanh

Almighty and most merciful father, we humbly beseech Thee, of Thy great
goodness, to restrain these immoderate rains with which we have to
contend. Grant us fair weather for Battle . . .

"PATTON'S PRAYER," BY CHAPLAIN JAMES O'NEILL, DECEMBER 1944, BATTLE OF THE BULGE

It has been said, "The weather is always neutral." But Dwight D. Eisenhower
knew that "nothing could be more untrue. Bad weather is obviously the enemy
of the side that seeks to launch projects requiring good weather, or of the side
possessing great assets such as strong air forces, which depend upon good
weather for effective operations."[1] Periodic inclement weather was a major
problem for all U.S. combat forces in World War II, the Korean conflict, and
Vietnam. Again and again, low clouds and fog, along with their related storms,
affected troop movement, air efforts, and naval operations.

Clouds and fog are simply visible collections of very small water droplets or
minute ice crystals that tend to remain buoyant in the atmosphere. Most
clouds form when air rises and, by expansion, cools below its dewpoint temperature. As mentioned in chapter 1, large-scale ascent of air happens in four
ways: through convection-induced updrafts, by orographic lifting that takes
place wherever air is forced over a high physical barrier such as a mountain
range, in zones of tropical convergence, and along weather fronts in the middle
latitudes. But how do these processes relate to cloud formation?

Convectional activity occurs wherever the earth reradiates solar energy to
the degree that it locally warms a parcel of air immediately above the ground to
a temperature higher than that around it. Like an air bubble in a fish tank, this
warmer and less dense air is forced to rise. The upward movement may progress through thousands of meters and induce great atmospheric cooling. Commonly it produces towering cumulus clouds that sometimes evolve into short-
lived thunderstorms yielding intense rain, high winds, lightning, and violent
weather.

In contrast, orographic clouds associated with air continually moving over
a high physical barrier can persist as long as that circulation prevails. The
result may be days, weeks, or months of continuous clouds and frequent
precipitation.

Some of the largest equatorial clouds are associated with storms in the Inter
Tropical Convergence Zone (ITCZ), and every year a few of these storms grow

to become typhoons or hurricanes. Once established, these tropical depressions and their related clouds tend to follow known patterns of development and movement, making their courses somewhat predictable with adequate data.

For the middle latitudes it is the wave cyclone, with its warm, cold, and occluded fronts, which brings clouds and poor weather. Cold fronts are generally accompanied by a linear zone of upward-billowing clouds, strong winds, heavy rains, the possibility of hail, and winter snowfall. Generally, the leading edge of a cold front passes in three to six hours, but with it comes a distinct change in the weather. Temperatures drop abruptly and winds shift markedly to arrive eventually from a more polar direction. Warm-front weather is generally less turbulent but the cloud cover tends to be more extensive. Passage of these fronts may also bring rapid changes in the weather, including above-average temperatures and gusty winds from the subtropics.

The occluding wave cyclone has the greatest potential impact on military operations in the middle latitudes. In this situation the cold and warm fronts converge to force huge volumes of a warm, moist air mass to swirl gradually upward over denser and colder air. The extent and magnitude of the lifting may be enormous, producing a widespread zone of condensed water droplets often mixed with sublimated ice crystals. Together these may produce a thick and extensive blanket of low clouds, called stratus. Such conditions commonly result in a reduced ceiling that may extend over thousands of square kilometers for days. The related precipitation ranges from persistent rain or snow to a light or moderate drizzle of relatively long duration.

Fog, which is simply a cloud at ground level, forms from four basic situations: places with excess ground radiation, horizontal movement (advection) of moist air over cold land or water, up-slope air migration, and surface evaporation followed by low-level condensation. In literature, as on the battlefield, fog may portray mystery or something sinister. For instance, it was:

> over the great Grimpen Mire that there hung a dense, white fog . . . [that] looked like a great shimmering ice field with the heads of the distant tors borne as rocks upon its surface. . . . My mind was paralyzed by the dreadful shape which had sprung out upon us from the shadows of the fog. A hound it was, an enormous coal black hound, but not such a hound as mortal eyes have ever seen. . . . Never in the delirious dream of a disordered brain could anything more savage, more appalling, more hellish be conceived than that dark form and savage face which broke upon us out of the wall of fog.[2]

Individual storms have also been responsible for disrupting many campaigns and battles. Xerxes' Persian fleet was decimated by stormy seas off the Magnesian Peninsula (c. 482 B.C.), foiling an invasion of Greece. As described in chapter 1, twice within seven years Kublai Khan's forces were destroyed by the weather-born Kamikaze as he attempted to conquer Japan in the thirteenth

Out of the fog, Hannibal's cavalry swept down on the Roman forces of Flaminius at Lake Trasimene in 217 B.C. Hannibal lured the Romans along a defile, appropriately called malpasso, at the base of a steep-sided valley. The Roman column advanced southward without great concern for reconnaissance or security. To the right was Lake Trasimene, and abruptly rising to the left were heavily vegetated hills. On that spring morning, fog was hanging, characteristically, above and nearby the lake. Even higher hills immediately to the east were obscured from the view of the advancing Romans. On these crests, waiting to spring the trap, were Hannibal's Gaulic and Numidian cavalries. Down they swept out of the fog and forest, falling on the surprised Romans from three sides, driving them to the lakeshore. More than three-fourths of the army was captured or killed, making this one of Rome's worst defeats. Hannibal, a master of using the terrain to advantage (as also demonstrated at Cannae), used fog, vegetation, and topography to mask the concentration of his cavalry, a tactic that surely contributed greatly to the Roman rout.

century. Similarly, in 1588 Spain's King Philip II's Armada was defeated as much by violent gales as by the English navy.[3]

Sometimes a short-term weather pattern, less spectacular than a devastating storm or long-paralyzing winter, can have a great influence on the tactics and outcome of battle. Overcast skies, where very low clouds form an opaque barrier between the upper atmosphere and ground, can protect troops from airborne observations or bombardment, nullifying the opposition's aerial operations, at least for a while. Fog can reduce visibility to virtually zero and completely mask troop movement from visual observation. While inclement conditions are commonly thought of as adverse, they also offer windows of opportunity to the commander who exploits them to advantage. Thus, knowledge of the weather and climate, including cloud and fog formation, has acted and will act as a force multiplier in battle.

Battle of the Bulge, 1944

In the more recent past, fog and overcast skies have significantly affected many military campaigns involving air operations. Two examples among many which clearly illustrate this are World War II's Battle of the Bulge (1944) and the siege at Khe Sanh in Vietnam (1968). However, the atmospheric conditions that caused equally poor visibility on both of these battlegrounds differed markedly, thus illustrating the diversity of apparently similar conditions on the battlefield. At the Bulge, low clouds and fog that persisted for days favored a large German force to concentrate and initiate their last great, yet desperate, offensive through the Ardennes in December of 1944.

On a rainy, foggy morning the rugged countryside of the Ardennes erupted as 25 German divisions crashed through the forest along a 135-km (85-mi) front near the Belgian-Luxembourg-German border, catching the thinly stretched

American forces totally by surprise. What followed was to become one of the most bitterly contested struggles on the World War II Western Front. By the end of the month this fight would involve more American troops than any other battle in U.S. history.

The Battle of the Bulge was fought in the Eifel and Ardennes regions of northwestern Europe. Twice before German armies had marched westward through this rugged and wooded terrain. In 1914 Bismarck's army moved from the Eifel and through the Ardennes to engage the unsuspecting French near the French-Belgian border. One of Marshal Ferdinand Foch's generals, Charles Lanrezac, said of this region, "If you go into that deathtrap of the Ardennes, you will never come out."[4] Again in 1940, over the objection of the German general staff, Hitler ordered his Wehrmacht to attack through the Ardennes, thus launching the awesome blitzkrieg against Belgium, Holland, and France that quickly produced stunning victories in the Low Countries and, a short time later, the surrender of France.

However, near the end of 1944, six months after D-day, Germany faced defeat on both its Eastern and Western Fronts. After breaking out of Normandy, the Allied armies had moved northeastward across France and into the Low Countries. Bernard Montgomery commanded English and Canadian troops on the north flank. Omar Bradley (with William Simpson's Ninth Army on the left, George S. Patton's Third Army on the right, and Courtney Hodges's First Army in the center) was responsible for the Ardennes/Eifel sector.

After the failed September effort at Arnhem known as Market-Garden (see chapter 7), the Allied offensive had markedly slowed. During early December General Hodges's troops were recovering from a bitter, costly, and unsuccessful November offensive near Aachen in the Huertgen Forest, one objective being to capture the Roer River dams to prevent the downstream flooding of the countryside. At its end the offensive in the Huertgen Forest had cost 33,000 American casualties, a harbinger of the struggle ahead. Meanwhile, to the immediate south, Maj. Gen. Troy Middleton's VIII Corps was thinly spread along the line of contact with no expectation of a German attack.

Having previously elected to advance along a broad front in Western Europe, Eisenhower was acutely aware of the overextended supply system and wanted his armies to catch their breath and refit, while allowing the logistical train to catch up before beginning the final assault across the Rhine into Germany. In addition, German resistance had stiffened along a defensive position called the "West Wall."

The weather, too, had deteriorated as autumn rains signaled the approach of winter. With a front so broad, all areas could not be defended equally during this brief pause before Christmas. Although the Ardennes was the most lightly held sector, and despite Hitler's great success here in 1940, none of the Allies expected a German offensive here.

During November and early December, German troops, equipment, and supplies were gathered in the Eifel for Hitler's counterattack. This region lies mostly in Germany, among the Rhine, Moselle, and Roer Rivers. Being a largely forested complex of steeply sloping hills, the area provides good concealment in any season. Two roughly parallel ridges extend the length of the Eifel. Near the center of the region is a high, northeast-trending ridge called the Schnee Eifel (summit elevation = 697 m [2286 ft]) that, together with the Our and Sauer Rivers, forms an imposing natural western boundary of Germany.[5] The region had no cities but a number of small villages were connected by an extensive road and a limited rail system.

Westward the Eifel blends imperceptibly into the Ardennes, a dissected plateau called the Hohes Venn forming the gradual transition. Like the Eifel, the Ardennes lacks precise definition but exhibits three physiographic tracts: the High Ardennes to the south, the Famenne Depression in the center, and the Low Ardennes in the north. From the high ground along the German border, the Ardennes' altitude and ruggedness gradually decrease toward the south and west, eventually giving way to the rolling landscape in the vicinity of Bastogne.

Thus, the area's widespread forests, rugged terrain, high drainage divides, and narrow, deeply incised gorges restricted most movement of a mechanized army largely to roads that followed winding river valleys.[6] As it turned out, the terrain also decentralized offensive command and control while favoring small-unit defensive operations.

One frontier corridor in the Ardennes is especially suitable for military maneuver. A long, narrow lowland (10 km [6 mi] wide), the Losheim Gap, begins at the northern terminus of the Schnee Eifel. Through this relatively open and historic gateway, German armies marched out of the east in 1914 and 1940.[7]

Apparently Hitler's reasoning for selecting the Ardennes for the 1944 attack involved the following:

- He believed (correctly) that the enemy front in the Ardennes sector was thinly manned.
- A blow there would strike near the seam between the British and Americans, which might complicate communications and create confusion, or military disharmony, among the Allies.
- The distance from the jumpoff line to a solid strategic objective (Antwerp) was not too great (approximately 160 km [100 mi]) and might be covered quickly, even in bad weather.
- The extent of the Ardennes was somewhat limited, thus an offensive here would allow the concentration of fewer divisions rather than an advance on a wider front.
- The terrain leading to the intended breakthrough sector was heavily wooded and offered concealment from Allied air observation during both the buildup and the assault.
- An attack that regained the initiative in this area could split enemy forces and eliminate the threat to the Ruhr.

On 6 September, in a special briefing for Hitler, Generaloberst Alfred Jodl, Chief of the German Operations Staff, identified three main problems to be solved if the offensive was to succeed. These were Allied superiority in the air, the need for security in concentrating the attack force, and the problem of marshaling supplies necessary to sustain a major push.[8]

These first two factors, it was hoped, could be accommodated by timing the offensive to coincide with a likely period of inclement flying weather. Hitler was aware of the usually poor weather conditions of late fall in the Ardennes and initially chose November as the month for his attack. The date 25 November was then selected, based on advice from Dr. Schuster, Hitler's staff meteorologist, in answer to the Führer's demand for a period of at least 10 days of continuous bad weather and poor visibility.[9] In addition, this date corresponded to a new phase of the moon, further restricting Allied night reconnaissance and visibility.[10]

It soon became apparent that the necessary force could not be assembled by 25 November, so the attack was delayed until mid-December. Nevertheless, an extended period of poor weather was still likely to prevail, although the probability for clear skies was increasing. But why did the Germans anticipate a period of poor weather with extended cloud cover during November and December? And, more important, what effect did these conditions have on the Battle of the Bulge?

During late fall the Ardennes and Eifel lie beneath the fluctuating boundary of two distinctly different weather regimes. The first, and most dominant, is associated with the presence of moist maritime air flowing inland off the Atlantic Ocean and North Sea. The second is related to the westward extension of cold and dry air from an anticyclone centered over Siberia that grows in size and effect as the winter progresses.

Recall that Western Europe is essentially a peninsula of Asia, narrowing east to west. Much of the time the principal climatic control for the region is the Atlantic Ocean, which moderates temperatures and provides moisture that generally makes precipitation abundant. Furthermore, the absence of a north-south mountain range allows the prevailing westerly winds to carry the maritime air far inland, resulting in a low west-to-east temperature gradient for a given latitude.

Three major pressure systems dominate the weather of northwestern Europe: the Icelandic low to the northwest that produces stormy conditions; the Azores high to the southwest that generally brings fair and mild weather, especially during the late spring and summer; and the alternating high- and low-pressure (winter and summer, respectively) centers that form over Asia.

The Icelandic low is especially intense during the winter months when the temperature difference between the polar and subtropical air is the greatest. The variation is due in part to the warmer than expected (for this latitude) ocean temperatures in the North Atlantic and along the coast of northern Europe. The warmth is in part derived from the North Atlantic Drift (an eastward extension of the Gulf Stream) that splits as it approaches the west coast of Europe. One branch travels north past Norway while the other turns southward toward Spain to form the Canary Current. The west coast of Norway is often 20°C (35°F) warmer than the air temperature over the land.[11]

These, and other factors, favor the development of a persistent low-pressure trough extending from southern Greenland, through Iceland, to northern Norway. Here the contrasting cold, dry air from Greenland, northern North America, and the Arctic meets the warm, tropical air moving northeast from the south. The interaction of these air masses frequently forms Atlantic storms, which then cross the European continent, bringing cloudy skies, frequent rain, swift changes in wind direction and temperature, and overall poor weather.[12]

Thus, during late fall the storm-generating effects of the Icelandic low tend to dominate the weather of the Ardennes, the result being winds that carry air inland over the warm coastal waters to produce the moist, cloudy climate that Hitler was relying upon. However, at this time of the year, and especially as the winter progresses, it is not uncommon for the western edge of a strong high-pressure system centered over Siberia to intrude into Europe, bringing bitterly cold temperatures and clear skies. Which pattern would dominate in December 1944? Or could both of them appear?

In making their forecast for Hitler, the German meteorologists anticipated that westerly air flow would prevail in late November, a reasonable assumption by any measure. But they also surely recognized that a delay would increase the chances that a high-pressure system would move in from the east, bringing clear skies and favorable flying weather for Allied aircraft. By postponing their attack by three weeks the Germans increased the chances that an essential ally, poor weather, would desert them.

As usual, during 1944 the late fall weather of the Ardennes was harsh, cold, and wet, the result of repeated storms coming off the Atlantic and North Sea. Rainfall, which averages 90 to 100 cm (35–40 in) a year, is most abundant in November and December when evaporation is least. As a result the soil is often saturated, making cross-country movement of heavy equipment virtually impossible.[13] As winter temperatures fall, snow becomes more common and bitter winds commonly sweep across the dissected plateau. Conduction cooling of moist air blowing over the cold ground favors advection fog and mist that frequently linger until midday, only to reappear in late afternoon. On the yearly average, Bastogne has 145 days of freezing weather and it is not unusual for a single winter storm to deposit 25–30 cm (10–12 in) of snow in a 24-hour period.[14] Under these inclement conditions, the largest battle in American history was to be fought.

THE GERMAN OFFENSIVE

Through November and early December, Hitler concentrated 30 divisions in the Eifel. While the Germans benefited by the prevalence of generally overcast

MAP 3.1. Map centered on the Ardennes showing the planned three-prong German offensive in the Battle of the Bulge and the battle-lines on 16 and 25 December 1944. (Map modified from diagram by A. K. Lobeck in Army Map Service, *Continental Europe—Strategic Maps and Tables,* pt. 1, *Central West Europe.*)

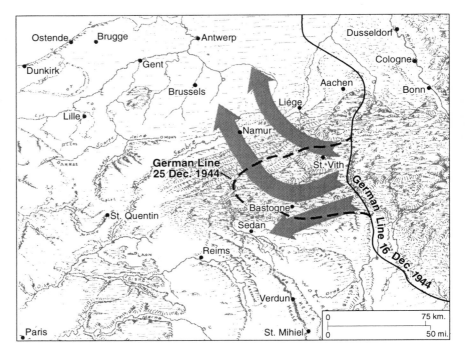

skies, they were also aided by what has been called the "most abysmal failure of battlefield intelligence in the history of the U.S. Army,"[15] a failure that allowed the undetected, massive concentration of German troops and equipment.

At 0530 hours on 16 December, three German armies attacked with complete and shattering surprise. On that morning, more than 200,000 German soldiers and 500 tanks went into action.[16] Their ultimate objective was Antwerp.[17] Their code-name for the Ardennes offensive was "Wacht am Rhein," intended to mislead the Allies into thinking that the German army was maneuvering toward a defensive position associated with the "West Wall."

The attack was launched along a 100-km (60-mi) front that extended from Monschau in the north, near Aachen, to Echternach in the south, near the juncture of the Our and Sauer Rivers (Map 3.1). The plan called for Generaloberst der Waffen–SS Sepp Dietrich's 6th Panzer Army to attack in the north along a front from Monschau to the Losheim Gap. The army was to move through the gap, cross the Meuse south of Liege, and then advance to Antwerp. General der Panzertruppen (Lt. Gen.) Hasso von Manteuffel's 5th Panzer Army was in the center. He was to attack through Saint-Vith, cross the Meuse, and then wheel northwest, protecting the 6th Army's southern flank. The primary mission of General der Panzertruppen (Lt. Gen.) Ernst Brandenberger's 7th Army on the left (south) was to protect the 5th Army's southern flank from Allied counterattack, primarily from Gen. George Patton's Third Army.

Facing Generalfeldmarschall Walther Model's attacking Army Group B were elements of six American divisions totaling about 75,000 soldiers. The most thinly stretched Allied division covered about 32 km (20 mi) of the front, an average of fewer than 400 soldiers per kilometer (or fewer than 650 per mi). Before the German attack the Ardennes was a relatively quiet sector, being manned by a combination of inexperienced and combat-weary troops, badly overextended for the length of front they were covering.

In the north, facing General Dietrich's nine divisions, were the U.S. 2nd and 99th Divisions, part of Maj. Gen. Leonard Gerow's V Corps. The experienced 2nd Division was in the process of passing through the 99th, which had been in the line only about a month, en route to attack the Roer dams. To the south was Major General Middleton's VII Corps, composed of the 106th, 28th, 9th, and 4th Divisions. The 28th and 4th had just arrived in the Ardennes for a "rest" after the difficult Huertgen Forest campaign. The seam between these two corps was in the Losheim Gap, certainly the most attractive avenue of approach through the Ardennes. Furthermore, on the day of the attack, this historic invasion route was lightly defended by just two squadrons of the 14th Cavalry Group, attached to the 106th Infantry Division.

The two corps of General Dietrich's 6th Panzer Army were designated to conduct the main thrust. The infantry divisions were to penetrate south of Monschau and form a cordon blocking Allied counterattacks from the north, while the I Panzer SS Corps' armored divisions would exploit the breakthrough. The timetable called for penetration and breakout to be accomplished by the end of the first day and the Meuse to be reached by the evening of the third. It was essential that the armored divisions not be detained during their advance to the Meuse. They would have to bypass strongly defended positions that could not be captured quickly, and they could not allow themselves to be deterred by open flanks. However, at the express orders of the Führer, Bastogne was to be taken.[18]

As planned, on 16 December, following an artillery barrage, the spearheading 6th Army attacked along a relatively narrow front. The terrain was not conducive to cross-country tank movement until the poorly drained, marshy Hohes Venn region, the objective for the second day of the offensive, was passed. As the Germans hoped, "Hitler's weather" was in full cooperation that very foggy morning, with abundant low clouds related to a frontal system that was approaching from the west (Map 3.2). Later in the day rain began to fall as a maritime polar air mass moved toward the area. With it came prolonged air flow from the Atlantic that provided abundant moisture which produced fog and dense, low clouds that persisted for several days.[19] To the relief of Hitler's staff meteorologist, his forecast, which was really an estimate based on probability, turned out to be correct, and Allied air superiority would not be a factor during the early days of this offensive (Map 3.3).

The attack in the north by 6th Army's XLVII Corps encountered stiff resistance from the U.S. 99th Division and made virtually no headway. The two Volksgrenadier divisions did not succeed in penetrating the American lines at Monschau and Hofen and experienced severe losses. Thus the northern shoulder of the German attack was neither secured nor expanded as the U.S.

MAP 3.2. Weather map for 16 December 1944 showing pressure centers and an occluded wave cyclone approaching the Ardennes.

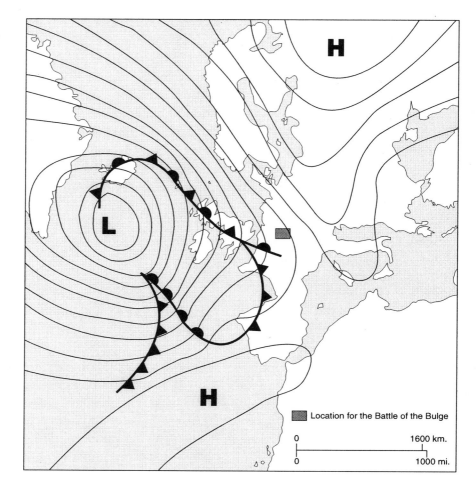

Location for the Battle of the Bulge

0   1600 km.

0   1000 mi.

2nd Division joined the 99th to successfully hold the line on Elsenborn Ridge, key high ground on the access route to Malmedy.

To the south General Lieutenant der Waffen–SS Hermann Priess's I SS Panzer Corps was more successful. The plan called for the II Panzer Corps to follow the I Panzer Corps in the breakout. The four SS Panzer divisions of these two corps were to spearhead the drive to Antwerp after crossing or going south of Elsenborn Ridge.

Meanwhile, the Germans broke through the American lines at Manderfeld on 16 December and essentially destroyed the badly overextended 14th Cavalry Group defending Losheim Gap. This allowed Kampfgruppe (Col.) Joachim Peiper's spearhead of the I SS Panzer Division to advance rapidly behind American lines near Malmedy. This local success, however, could not be reinforced, and by 18 December the main attack by the 6th Panzer Army was at a virtual standstill, due primarily to the strong resistance of the inexperienced 99th Division fighting in relatively isolated strongpoints that controlled key crossroads. Therefore, on 20 December the main German attack shifted to Manteuffel's more rapidly advancing 5th Panzer Army.

This center portion of the attack was experiencing some success, albeit less than anticipated. Both corps of Manteuffel's army, attacking toward the key road junctions of Saint-Vith and Bastogne, had broken through the lines of the American 106th Infantry Division and 14th Cavalry Group in the Schnee Eifel. Two regiments of the 106th Division, the 422nd and 423rd, were encircled and essentially destroyed. Both units surrendered by 20 December with 8000–9000 men either killed, wounded, or captured.

The 106th Infantry Division had left the United States less than two months previously and taken its place in a "quiet sector" of the Ardennes on 16 December, relieving the 2nd Infantry Division. Their front extended approximately 34 km (21 mi), including the Losheim Gap on its extreme northern flank, defended by the two squadrons of the attached 14th Cavalry Group.[20]

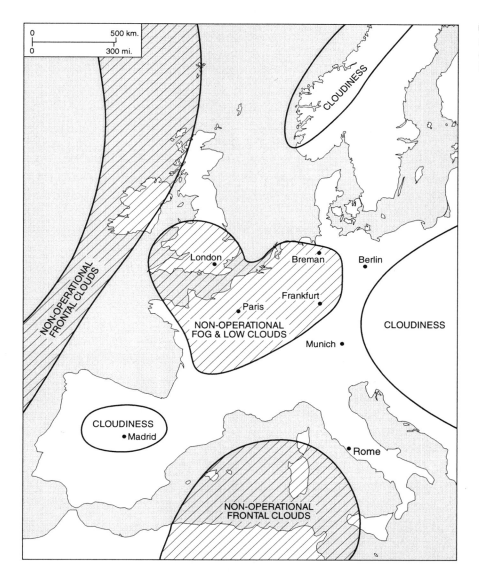

MAP 3.3. Average European weather conditions for 16–22 December 1944.

The German assault by the 18th Volksgrenadier Division quickly drove between the 106th Division and the 14th Cavalry.

A second breakthrough by Manteuffel's 5th Army on 16 December was opposite the 28th Infantry Division's 110th Infantry Regiment defending "Skyline Drive," the principal supply route paralleling the front. Three divisions of the XLVII Panzer Corps exploited the breakthrough and rushed toward Bastogne, arriving at the outskirts of the town on 19 December. Bastogne, the transportation hub of the Ardennes because of its seven crossroads, was key to the German plan. It was also to become the scene of some of the most heralded fighting of the Battle of the Bulge.

By the evening of 19 December, two major battles raged in the 5th Panzer Army area. Elements of the reinforcing American 7th and 9th Armored Divisions, as well as the remaining regiment of the 106th Infantry Division, were defending near Saint-Vith, site of another key Ardennes crossroads, against the LVIII SS Corps. To the south the other corps of Manteuffel's army, the XLVII Panzer, was at the outskirts of Bastogne, 30 km (19 mi) from the jumpoff point of 16 December, encircling the 101st Airborne Division and elements of the 9th and 10th Armored Divisions.

Farther south, Brandenberger's southern two division corps of the 7th Army, composed of four infantry divisions and clearly the weakest of the three attacking armies, was given a rather ill-defined containing mission while the northern LXXXV Corps was to move west and establish blocking positions south of Bastogne at Arlon.[21] Opposing Brandenberger along a 48-km (30-mi) front were elements of the American 9th Armored Division and 4th Infantry Division, as well as a regiment of the 28th Infantry Division.

Hampered by logistical problems and difficulty in bridging the Our and Sauer Rivers, the 7th Army made little progress across very rugged terrain. The German infantry was largely road-bound due to steep-walled valleys and very muddy off-road conditions. The Americans, defending in relatively isolated pockets of resistance, denied the attackers control of the key roads. The southern corps of the German 7th Army had particular difficulty, being fought to a virtual standstill by elements of the 4th Infantry Division and a reinforcing combat command of the 10th Armored Division. Only after three days of bitter fighting did they achieve the objectives they had planned to reach by nightfall of 16 December, the first day of the offensive. The 48-hour delay at the Sauer River, caused largely by inexperienced engineers and equipment shortages, forced the German infantry here to fight without heavy, follow-on assault weapon support.[22]

Similar bridging problems plagued Manteuffel's northern LXXXV Corps attempting to cross the Our River at Roth and Gentingen. It required nearly two days to construct a pair of bridges that finally allowed the 5th Parachute Division to begin its movement west.

It is important to note that as the offensive began, the weather was just about ideal from the German point of view. The clear skies of the evening of 15 December gave way to fog, drizzle, and low, heavy overcast conditions as a frontal system moved into the area from the west on the sixteenth. At this time the higher ground was covered by snow that had fallen on 13 December. As the front moved across the Ardennes, the base of a moist maritime air mass was chilled by the snow pack, decreasing its ability to retain moisture. Where the temperature fell below the dewpoint, fog formed. Contributing to this fog condition was evaporating rain, which added to the atmosphere's moisture content. This type of fog can form very rapidly and is associated with warm air riding up over cooler surface air. It usually develops above a shallow layer of cold air just ahead of an approaching warm front and is consequently often called frontal fog.

Thus fog, clouds, and drizzle concealed German movements as the attack began. At daybreak, 16 December, visibility was so poor that soldiers of the 14th Cavalry Group in the Losheim Gap did not detect the German infantry moving through their lines. The Germans, encountering no resistance, thought the Americans had withdrawn. Just north of this position, German infantry marched out of the fog in a column of twos into an American bivouac area and were thought to be friendly troops. The unit had nearly passed through the mess area when the Americans finally recognized and engaged them.[23]

The fog, drizzle, and overcast continued through 16 and 17 December. On the eighteenth, near Rocherath, the morning fog was especially heavy, visibility almost nil. The American infantry let the tanks of the 12th SS Panzer Division roll past, then tailed them with bazookas or turned to meet oncoming infantry at close quarters with grenades and even bayonets or knives. The first assault was beaten off, while a number of German tanks were crippled or destroyed by bazooka teams stalking successfully under cover of fog.[24]

Weather conditions improved somewhat during that afternoon with ceilings rising to 150–300 m (500–1000 ft) and visibility 3–10 km (2–6 mi). However, the fog and rain returned on 19, 20, and 21 December as the German offensive continued. Visibility on these three days was reported to be less than 100 m (100 yd).[25] While limited visibility certainly affects both combatants, the attacker generally benefits more. This was certainly the case in the Ardennes. The Americans found it difficult to adjust artillery fire against the moving, unseen enemy. Units found themselves suddenly engaged at close quarters; often they stumbled into one another's path with no warning.

It was in the air, however, that the notorious Ardennes winter weather had the greatest effect. The vastly superior Allied air forces could do virtually nothing from 16 to 22 December. German troops moved to the front, through the bridgeheads, and along the Ardennes roads with near impunity from aerial attack. Only during a brief period on 18 December were Allied planes over the battlefield a factor in the fighting. Without question, this absence of air power significantly contributed to the initial success of the German assault. If the

ACTUAL WEATHER CONDITIONS FOR TACTICAL AIR FORCE BASES AND NORTHERN BATTLE AREAS, BATTLE OF THE BULGE, DECEMBER 1944– JANUARY 1945. (ALL DISTANCES ARE APPROXIMATE.)

Date	Weather Summary
16 Dec 1944	Very low clouds and fog patches. Visibility poor. Light rain.
17 Dec	Overcast clouds, bases 60–300 m (200–1000 ft), with intermittent rain. Visibility 5–8 km (3–5 mi).
18 Dec	Overcast clouds, bases 90–180 m (300–600 ft), with light intermittent rain, becoming 150–300 m (500–1000 ft) broken during late afternoon. Visibility 3–10 km (2–6 mi). Also, fog patches in the southern sector.
19 Dec	Foggy conditions all day. Visibility less than 90 m (100 yd).
20 Dec	Foggy all day. Visibility less than 90 m (100 yd).
21 Dec	Foggy all day. Visibility less than 90 m (100 yd).
22 Dec	Overcast from 90–150 m (300–500 ft), with light intermittent rain and snow. Visibility 450–900 m (500–1000 yd), reduced to less than 90 m (100 yd) in precipitation.
23 Dec	Fog and stratus in morning, with visibility 450–900 m (500–1000 yd), improving in afternoon to scattered clouds with visibility 3–6 km (2–4 mi).
24 Dec	Clear. Visibility 5–8 km (3–5 mi).
25 Dec	Clear, except for fog patches in the morning. Visibility 900–1800 m (1000–2000 yd), becoming 3–6 km (2–4 mi) in the afternoon.
26 Dec	Clear visibility 2–5 km (1–3 mi), except 900 m (1000 yd) in fog patches.
27 Dec	Clear, except for ground fog. Visibility 450–1800 m (500–2000 yd) in fog, increasing to 3 km (2 mi) in afternoon.
28 Dec	Fog and stratus, bases of stratus 30–125 m (100–400 ft). Visibility 90–900 m (100–1000 yd).
29 Dec	Fog and stratus, bases of stratus 90–200 m (300–700 ft). Visibility 180–460 m (200–500 yd).
30 Dec	Broken to overcast clouds at 600–1500 m (2000–5000 ft), lowering to 150–300 m (500–1000 ft) in precipitation during afternoon. Visibility 900–1800 m (1000–2000 yd), reduced to 450–900 m (500–1000 yd) in patchy fog.
31 Dec	Broken clouds with snow showers. Visibility 5–8 km (3–5 mi), restricted to 900–1800 m (1000–2000 yd) in snow showers.
1 Jan 1945	Clear to scattered clouds with visibility 5–8 km (3–5 mi), except 2–3 km (1–2 mi) in patchy fog/haze.

Date	Weather Summary
2 Jan	Scattered to broken clouds 150–300 m (500–1000 ft) becoming overcast with light rain in the afternoon. Visibility 2–3 km (1–2 mi), except less than 2 km (1 mi) in fog and rain.
3 Jan	Foggy conditions in morning, 60–90 m (200–300 ft) overcast during afternoon. Visibility less than 45 m (50 yd) in morning, becoming 900–1400 m (1000–1500 yd) in afternoon.
4 Jan	Overcast with bases 30–150 m (100–500 ft) with snow. Visibility 2 km (1 mile), except 90 m (100 yd) in snow.
5 Jan	Broken clouds with snow showers. Visibility 5–8 km (3–5 mi), except less than 450 m (500 yd) in patchy fog.
6 Jan	Fog with visibility 45–180 m (50–200 yd), improving to 450–900 m (500–1000 yd) during afternoon.
7 Jan	Fog in morning with clouds 90–275 m (300–900 ft) becoming broken to overcast during the afternoon. Visibility 450–900 m (500–1000 yd) in fog, improving to 3–6 km (2–4 mi) during afternoon, except 2 km (1 mi) at 150–300 m (500–1000 ft) during snow showers.
8 Jan	Broken clouds to overcast at 150–300 m (500–1000 ft), with heavy snow showers. Visibility 3–6 km (2–4 mi), except 2 km (1 mi) in snow showers.
9 Jan	Overcast at 150–300 m (500–1000 ft), with visibility 2–3 km (1–2 mi) in snow showers.
10 Jan	Clouds and fog. Visibility 90–180 m (100–200 yd), locally less than 45 m (50 yd).
11 Jan	Fog with visibility 90–150 m (100–500 yd), improving to 450–1350 m (500–1500 yd) in the afternoon.
12 Jan	Overcast at 90–180 m (300–600 ft), with light snow. Clouds becoming scattered in afternoon. Visibility 450–900 m (500–1000 yd), except less than 450 m (500 yd) in local areas.
13 Jan	Foggy conditions with scattered clouds. Visibility 450–900 m (500–1000 yd), except less locally.
14 Jan	Clear to scattered clouds. Visibility 3–6 km (2–4 mi), restricted in local areas to 2–3 km (1–2 mi) in fog and haze.
15 Jan	Foggy with visibility less than 450 m (500 yd), becoming 700–1400 m (750–1500 yd) in afternoon.
16 Jan	Foggy in morning; becoming clear in afternoon. Visibility 150–1400 m (500–1500 yd), becoming 5–8 km (3–5 mi) during afternoon.

skies had been clear the bridgeheads and assembly areas on the Our, Sauer, and several other rivers would have made prime targets, as would the roadbound, and often congested, German columns (see Map 3.3 above).

The inability of the German forces to move cross-country was an important factor in their failure to meet their projected timetable. The snow and subsequent rain had turned open fields into quagmires for wheeled or tracked traffic. Tanks repeatedly bogged down, some sinking deeply when they tried to bypass congested roads or maneuver across open terrain.

The foul weather, said to be some of the worst this region had experienced in 50 years,[26] essentially eliminated the possibility of cross-country movement. The warmer temperatures and rain of the 16 December frontal system were soon replaced by colder arctic air from the northwest on 20 and 21 December. The ground at higher elevations began to freeze at some places, making stretches of the Ardennes' roads both slippery and muddy.

To the north Germany's 6th Panzer Army was becoming bogged down by the rain and mud. Farther south, the 5th Panzer Army was hampered in its swing around Bastogne by fog and snow. Snow fell continuously on the German supply roads leading from the Eifel. Meanwhile, at Bastogne, the now-encircled 101st Airborne Division, along with elements of the 9th and 10th Armored Divisions, battled the surrounding Germans to a standstill as Patton's Third Army, attacking northward, raced 200 km (125 mi) in seven days through terrible weather to relieve Brig. Gen. Anthony McAuliffe's Bastogne command on 26 December.

It was on the afternoon of 23 December that a major change in the weather commenced. Clouds became scattered and visibility increased markedly. By the next morning the skies were clear and a cold wind brought continually dropping temperatures. Obviously, the edge of a frigid Siberian high-pressure cell had finally drifted eastward across the battlefield. Allied aircraft immediately attacked German positions and began the Bastogne airlift. Desperately needed ammunition, food, and other supplies were delivered to the besieged "battered bastards of Bastogne" in the nick of time. As a measure of the changing weather's effect, tactical air commands flew 294 sorties to the battlefield on the twenty-third, followed by 2381 the next day, along with 2442 bomber missions. The German attack came to a halt. General Manteuffel later reflected:

> It was enemy air operations which tipped the scales. . . . The further clearing up of the weather on 24 December was decisive, since it allowed the enemy to attack from the air in great strength. They carpeted the roads and railways with bombs and succeeded in bringing the already inadequate German supply organization almost to a standstill.[27]

Fair skies continued for four days, 24–27 December, and Allied fighters and bombers continued their bombardment. Poor weather then returned on the twenty-eighth and related fog, combined with low stratus clouds, stopped Allied air operations. Visibility on the ground was reduced to less than 1000 m. But by then the point of the German offensive had been blunted and the shoulders contained.

On 3 January a powerful Allied counteroffensive was launched in cold, windy, and foggy weather over terrain covered by deep snow. The bad weather would continue for two weeks,[28] with Allied air power grounded all but three days. Even so, by 28 January the Bulge was completely cleared of German forces. Within a few days, centered on Christmas, both the tide of battle and the weather had completely changed, a relationship that was far from coincidental. Hitler's gamble had failed, and the final phase of the war in Europe was about to begin.

Khe Sanh, 1968

At Khe Sanh, protracted ground-hugging fog often enveloped and imperiled more than 6000 Marines at that isolated base, evoking haunting comparisons with the 1954 French humiliation at Dien Bien Phu (see chapter 11). One of the most publicized battles of the Vietnam Era, Khe Sanh is an enigma, its true significance still to be fully resolved. The base was established on a plateau near the Laotian and North Vietnamese borders, along Route 9, the main east-west road linking the Vietnamese coastal cities to Laotian towns along the Mekong (Maps 3.4 and 3.5). This small camp commanded the close attention of the press, other American forces in Vietnam, and then Pres. Lyndon Johnson for more than two months.

The outpost at Khe Sanh Combat Base was constructed by the French during their colonization efforts in Southeast Asia. From its beginning the fort was used as an obstacle to infiltration from Laos to the coastal cities in Quang Tri province. The area was quiet after the French defeat in 1954 until the late summer of 1962, when a Special Forces A detachment established a base camp there.

In 1966 Gen. William Westmoreland and Marine Lt. Gen. Lewis W. Walt decided to enlarge the garrison, upgrade the airstrip, and man Khe Sanh with Marines.[29] The objective was to stop the infiltration of North Vietnamese forces who were moving south via the Ho Chi Minh Trail, crossing the Laotian/Vietnam border into the northern provinces of South Vietnam, and proceeding eastward along winding mountain and jungle tracks. These lines of movement were largely obscured from air observation by rainforest canopies and dense elephant grass 4–5 m (12–16 ft) high.[30] But the thick undergrowth, including virtually impenetrable bamboo thickets, and rugged terrain severely restricted large-scale cross-country movement.

Khe Sanh is on a small plateau within the lengthy Annamese Cordillera, a rugged north-south trending upland that extends some 2700 km (1675 mi) from Laos to southern Vietnam. The plateau is triangular, roughly 5–6 km (3–4 mi) on a side. The combat base, approximately 450 m (1500 ft) above sea

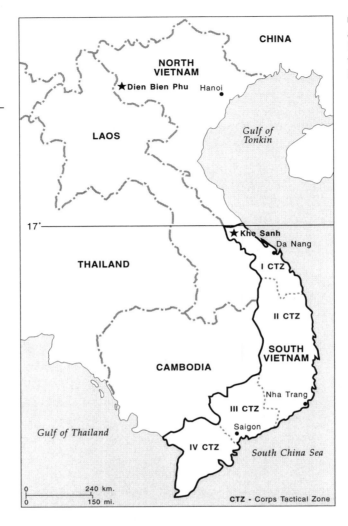

MAP 3.4. Southeast Asia in 1968 showing the relative location of Khe Sanh and Dien Bien Phu.

level, was positioned on the feature's eastern margin near a steeply sloping ravine that declines to the Rao Quan river.

In October 1966 U.S. Marines arrived, displacing the Special Forces detachment to the village of Lang Vei, some 5 km (3 mi) southwest. The base was nearly surrounded by hills that rose ominously above the Marines. Dong Tri Mountain, approximately 2 km (1.2 mi) north of the Khe Sanh Combat Base, was the highest summit in the region, rising to 1015 m (3329 ft), or more than 500 m (1600 ft) higher than the camp. Five lower hills—Hill 950, Hill 881 North, Hill 881 South, Hill 861, and Hill 558—commanded the approaches to the base from the west and north. The Rao Quan, a tributary of the Quang Tri, flowed past the base in the deep ravine, approximately 150 m (500 ft) below the dug-in Marines. The hilltops were key to the control of Khe Sanh and its airfield and were the scene of intense hand-to-hand struggles during the fight for the base.

The battle of Khe Sanh (21 January–18 April 1968) was one of the most prolonged and intensely contested of the Vietnam Era. For 2½ months, the remote Marine camp was virtually under continual fire by the North Vietnamese Army (NVA). From the onset, parallels were drawn to Dien Bien Phu, where in 1954 the French forces in Vietnam suffered their final defeat to Gen. Vo Nguyen Giap's Viet Minh Army after months of struggle (see chapter 11 for details). Certainly some similarities did exist. The surrounded Marines of the 3rd Division were essentially contained by North Vietnamese soldiers of the 304th, 325th, and 324th Divisions, the first of these being the unit that defeated the French at Dien Bien Phu.

A key feature of the combat base was an airstrip built by the French and upgraded by Navy Seabees from a 450-m (1500-ft) dirt runway to a 1200-m (3900-ft) aluminum-surfaced landing field. Nevertheless, erosion from heavy rains and the wear of cargo-loaded C-130 aircraft caused the runway to fail in August of 1967. Navy Seabees removed the metal sheeting, strengthened the base with a 15-cm (6-in) asphalt overlay, and replaced the aluminum planking atop the asphalt.[31]

Although now adequately constructed, the east end of the runway was immediately adjacent to the ravine, making it more vulnerable to attack. With the North Vietnamese in control of the surrounding forest and jungle, transportation and resupply over Highway 9 had long since stopped, with all bridges destroyed by the North Vietnamese and Viet Cong troops. The only lifeline for the forces within the camp was by air. The defenders at Khe Sanh were isolated, essentially hostages at their own base.

Enemy antiaircraft fire from guns surrounding the base was intense, but it

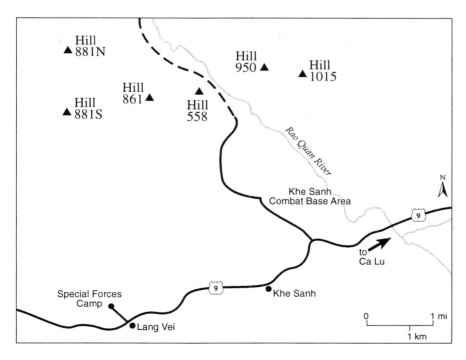

MAP 3.5. The Khe Sanh area, 1968.

was the weather that posed the greatest threat to the aerial lifeline. Low ceiling and poor visibility plagued pilots through January and February, the season of the northeast monsoons. The word "monsoon" is derived from the Arabic word *mausim,* meaning season, and has commonly come to denote a reversal of winds, a general onshore movement of air in the summer, and offshore flow in the winter. These regional winds are induced by differential pressures over continental and oceanic surfaces with the wind reversal caused by seasonal changes in temperature. Monsoons are present to some degree wherever low-latitude land masses are large enough to produce widespread temperature differences between land and adjoining ocean.

Traditionally the monsoons are interpreted to result from unequal seasonal heating of air over continents and oceans; that is, warmth-induced low pressure develops over land during the summer and attracts moist, moisture-bearing oceanic winds, with the situation being reversed during the winter. Actually the full explanation is much more complex and is closely related to upper-air-flow patterns. For whatever reason, as long as the seasonal circulation is from sea to land and a lifting process prevails, rainfall will be abundant. Conversely, when the winds are reversed, a distinct dry season may be expected.

The powerful East Asian monsoon that affects Vietnam is characterized by seasonally alternating winds, coming from the south and southwest from about May to October and north or northwest between November and April (Map 3.6). In one way or another, this system affects all of Southeast Asia, bringing most areas distinct wet and dry seasons respectively.

Two lifting processes work in concert to produce much of Vietnam's precipitation. In the first, moist maritime air drawn over the Southeast Asian land mass experiences convection induced by reradiation of solar energy from the earth's surface. The rising air cools through expansion, eventually resulting in condensation and precipitation, generally in the form of heavy rain.

The second lifting process, referred to as orographic, occurs where circulating moisture-laden air is forced upward as it passes over a topographic barrier, such as a mountain range or plateau. The related upward movement again causes cooling that may produce clouds, condensation, and precipitation that will persist as long as the circulation pattern prevails. The result is a long wet season over much of Southeast Asia. In fact the wettest areas on Earth are related to monsoons where warm, moist oceanic winds move upward and over major topographic features, notably in the foothills of the Himalaya.

It is in this fashion that the southern monsoon commonly brings large amounts of rain to most regions of Cambodia, Thailand, Burma, Laos, much of North Vietnam, and parts of South Vietnam. Curiously, however, the air passing over the Annamese Cordillera of southern Vietnam releases much of its moisture on the windward, or southwest, side of the range. As a result, the section of central Vietnam that lies east of the mountains near the South China Sea is then in a partial rain-shadow. Thus this part of today's Vietnam does not

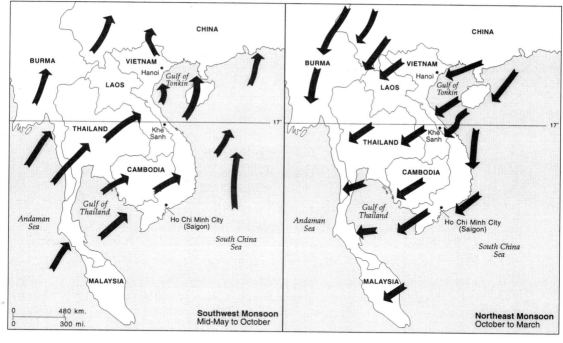

MAP 3.6. Direction of monsoon winds over southeast Asia.

receive as much rain during the months of May through August as its coastal counterparts to the south.

For most of Asia the reversed November-to-April north monsoon brings continental air and dry winds resulting in generally fair skies and little rain. For central coastal Vietnam, however, these same winds cross the South China Sea, picking up moisture before moving ashore and inland to ascend over the higher interior elevations. This advection and lifting then produces intermittent fog, drizzle, and rain in central Vietnam from November through March. Known as the *crachin,* it results in a climatic anomaly in that, unlike nearby areas in Southeast Asia, this region, which includes Khe Sanh, has no annual dry season. Instead, from November through early April, a prolonged period of widespread fog and drizzle exists along the coastal lowlands of Vietnam from the Gulf of Tonkin to the vicinity of Nha Trang (see Map 3.4 above). It is characterized by persistent low-level (600–1200-m or 2000–4000-ft thick) cloud formations accompanied by light drizzle that may persist for several days to weeks.[32]

While the north monsoon brings overcast skies and low ceiling to the Khe Sanh plateau, it is the accompanying fog that can most severely affect air operations there. For all practical purposes, it can be assumed that fog, like any cloud, will form when the relative humidity—the ratio of water vapor in the air to the maximum amount of water vapor that the air can hold at that temperature—is near 100 percent. The relative humidity can change either by an increase of the water-vapor content (by evaporation) or by a decrease in the saturation vapor pressure (by cooling). Since cool air can hold less moisture than warm air, any process that lowers air temperature near the ground can cause fog.

Regional and local environmental conditions at Khe Sanh were such that each of the four different types of fog could form at one time or another, and it was especially common, thick, and persistent during the winter monsoon. At night the temperature of the ground is decreased by the earth's continuing emission of long-wave radiation. As the land cools, the air in contact with the ground is also chilled, primarily by conductive processes. When air temperature drops below the dewpoint, radiation fog forms. Long, cloudless nights and a relatively calm atmosphere favor the development of this type of fog. When produced, it is generally less than 100 m (328 ft) thick and tends to hug the earth, so it is often referred to as ground fog.

Since different surfaces emit radiation at different rates, the nature of the ground material, vegetation cover, and terrain can be important in determining the likelihood, degree, and distribution of radiation fog. Hard, compacted, and barren surfaces give up heat at a much higher rate than thickly vegetated regions. Land cools more quickly than water, and metal runways, such as the one at Khe Sanh, favor radiation fog formation much more than the surrounding terrain blanketed by vegetation. This type of fog is also common in lowlands and valleys where cold air (being heavier than warm air) settles, or drains, from the higher and cooler surrounding hills to collect in the valleys or ravines.

Radiation fog generally begins to clear a few hours after sunrise because, as the earth warms, the near-surface temperature rises, thus increasing the air's capacity to hold moisture and favoring vaporization. Because the warming progresses upward, the fog "lifts" from below, clearing at the surface initially and working its way to the upper levels. Valleys or ravines protected from morning sun may hold fog until late morning or midday, and when incoming solar radiation is reflected by overlying clouds, the mist can persist for days. Most often, however, radiation fog is a diurnal (daily) phenomenon.

Advection fog forms where warm, moist air flows over a cooler surface. Here again, the heat-transfer process and resulting fog involve conduction, thus only the lowermost part of the atmosphere is affected. This type of condensation often develops in coastal regions where cold ocean currents chill the base of warm, moist air being drawn onto shore as a late afternoon sea breeze. (For example, the central coast of California, in the vicinity of Monterey and San Francisco, is known for such midsummer, midday fogs.) Advection fog need not be diurnal, and may persist for days, being characteristically thicker and less local than radiation fogs. At Khe Sanh the metal runway and cleared perimeter area provided the cooler surfaces to chill the base of moist winds of the northerly monsoon, often contributing to fog formation in the late afternoon and early evening.

A third process that can be responsible for fog generation is the up-slope movement of humid air. Through expansion, such rising air will cool at a rate of 10°C (18°F) for every 1000-m (3280-ft) increase in elevation until condensation occurs. Under certain conditions this process may lower air temperature below its dewpoint to form "up-slope fog." This type of fog can be thick and persistent. At Khe Sanh the plateau has a definite east-west-rising trend. Northeast monsoon winds, laden with moisture gathered from the South China Sea, flow onto the coast and funnel up the Quang Tri Valley. Moving inland, the elevation increases from sea level to 450 m (1475 ft) at the combat base, a sufficient elevation change to chill the incoming humid air, often to saturation.

A fourth process, evaporation, also contributed to occasional fog at the base. As rain falls a portion evaporates into the surrounding atmosphere. When the air near ground level is very humid, this additional water vapor can be sufficient to cause air saturation. During the rainy monsoon, the air is nearly always near a saturated condition, and it is the addition of this moisture that favors the formation of evaporation fog.

These four processes—radiation, advection, up-slope movement, and evaporation—all combined at Khe Sanh to produce conditions favorable for fog formation, a factor of major tactical importance as the siege progressed.

As long as the massive air support on call from bases all over Vietnam and Thailand could support Khe Sanh the Marines were believed to be secure. By 1968 the U.S. Command was quite familiar with the poor weather conditions to be expected during the northeast monsoon. Even so, decisionmakers thought that conditions would allow for adequate and effective air support. Although low ceilings and poor visibility had been predicted, planners were not prepared for the incredibly poor flying conditions experienced at Khe Sanh during January and February of 1968. General Westmoreland's staff was informed that ceilings below 600 m (2000 ft) and visibility less than 4 km (2.5 mi) could be expected at Khe Sanh on more than half the mornings from November through April. If typical, weather at midday was expected to improve, with average ceilings in the early afternoon rising to about 900 m (3000 ft).[33] Considering the data available, these forecasts for this northern province were reasonable. However, during February 1968 the conditions at Khe Sanh were far worse than average.

For any one day, the best weather during the siege lasted only six hours, when clouds were in a scattered to broken condition between 1000 and 2500 feet. Visibilities were never much better than five miles. In the early morning, afternoon, and late evening, weather and fog reduced visibility to less than a mile.[34]

It was the worst flying weather the Air Force had encountered in Vietnam to that time, and it threatened the Marines' lifeline. Gen. Rathvon McC. Tompkins, the Marine commander responsible for the combat base, remarked, "February 1968 made an old man out of me. Zero-zero, day after day."[35]

General Tompkins's "zero-zero," that is, zero ceiling and zero visibility, all but precluded base resupply and close air support. Gen. John P. McConnell, Air Force Chief of Staff, testified before Congress in 1968 that Khe Sanh was being supported by airdrop "because the weather is so bad you cannot land."

Throughout much of February, a persistent fog hung over the combat base. "No one was ready for the zero-zero conditions that nearly paralyzed airlift operations in February. The airstrip seemed particularly bedeviled by fog," concluded one official report. "On many a morning when visibility was excellent [elsewhere on the plateau], the runway remained shrouded in mist. . . . A deep ravine at the east end of the runway seemed responsible, channeling warm moist air from the lowlands onto the plateau where it encountered the cool air, became chilled, and created fog."[36] The ravine became known to the surrounded Marines as the "fog factory."[37]

At night fog also tended to form over the plateau, being induced by ground radiation from the runway and nearby cleared areas. As the cooler ground chilled, the overlying air condensation occurred and the heavier fog-filled air settled into low areas and nearby valleys.

With sunrise the metal-surfaced runway warmed more rapidly than the base. Differential heating between the airstrip surface and valley floor formed lower pressure located over the warmer plateau. As a result, air was drawn upward from the valley to form up-slope fog. Thus, instead of lifting by midmorning, fog could cover the base most of the day. General Westmoreland cited the "mists, low-lying fog, and drizzling rains of the 'crachin' from October through April" as posing "major problems for close air support and supply by air."[38]

The question that nagged General Westmoreland and President Johnson was whether Khe Sanh could be overrun by the much more numerous North Vietnamese forces. Under ordinary circumstances, with access to massive air and artillery support, Westmoreland was confident that the Marine garrison could hold. With air support grounded by the poor weather, however, it became a question of whether the North Vietnamese wanted to pay the price to take the camp. The parallels drawn to the French humiliation and defeat at Dien Bien Phu were inevitable and grew more compelling as the poor weather persisted.

The battle for Khe Sanh actually began in late April of 1967. Hills 881 North and 881 South became the battleground in April and May of that year when Marine patrols encountered an NVA battalion (see Map 3.5 above). After heavy fighting (the "Hill Fights"), the North Vietnamese withdrew. Khe Sanh was quiet until the enemy began to gather near the base in late 1967. U.S. Intelligence reports placed at least two enemy divisions in the vicinity by the first of the year. The elite 304th Division had moved from Laos to join the 325C Division.[39] Meanwhile, the strength of the 26th Marine Regiment garrisoning Khe Sanh was increased to three full battalions on 16 January 1968 under the command of Col. David E. Lownds. The 26th was joined by the 37th

WEATHER MODIFICATION IN VIETNAM

Weather, principally rain and poor visibility, was such a significant planning factor in Vietnam that the United States embarked on a weather modification program to try to induce precipitation and dissipate fog. The Air Weather Service was the operating agency for this highly classified project. Attempts were made in 1967 to extend the southwest monsoon season over the Ho Chi Minh Trail that snaked through Laos, Cambodia, and North Vietnam. Three WC-130 aircraft were assigned to the Air Weather Service in the attempt to increase rainfall on the main enemy line of communication. It was hoped that this, in turn, would reduce trafficability on the trails that supported the North Vietnamese logistical lifeline. Flying out of Udorn Air Force Base in Thailand, these WC-130s dispensed silver or lead iodide flares into the atmosphere. During 1967 and 1968, more than 1200 weather modification sorties were flown over Laos and North Vietnam and the A Shau Valley in South Vietnam. Using empirical and statistical techniques, it was determined that rainfall had increased by approximately 30 percent in the targeted areas. This would lead to the intuitive conclusion that enemy logistical supply was at least slowed by "cloud seeding." The cost of this operation was $3.6 million annually.

At Khe Sanh, in an effort to dissipate the fog, salt was dispersed from C-123 aircraft flying out of Da Nang. Fifteen missions were flown but the fog-clearing attempt was ineffective.

Eventually, in 1971, these weather modification attempts became public and congressional inquiries began. In August 1975, over Defense Department objections, the United States and the USSR submitted to the United Nations a joint draft treaty that called for the banning of "environmental warfare." Ironically, this "warfare" is more benign than conventional means. Supply lines interdicted by rain are certainly less lethal than those interdicted by steel. Nevertheless, there are serious environmental and perhaps judicial implications involved with weather modification. The feasibility of inducing precipitation over the battlefield was demonstrated in Vietnam. The potential for weather modification will certainly grow as battlefield automation increases and communication reliability becomes more critical.

Vietnamese Army Ranger Battalion and the 1st Battalion of the 9th U.S. Marines, bringing the strength at Khe Sanh base to approximately 6600. North Vietnamese strength in the area was estimated at over 40,000.[40]

General Westmoreland, in consultation with Lt. Gen. Robert E. Cushman, commander of the III Marine Amphibious Force, decided to defend Khe Sanh. The base commanded the main avenue of approach into eastern Quang Tri province. Westmoreland saw the critical importance of the little plateau: Khe Sanh could serve as a patrol base to block enemy infiltration from Laos, a base for operations to harass the enemy in Laos, an airstrip for reconnaissance planes surveying the Ho Chi Minh Trail, a western anchor for defenses south of the Demilitarized Zone, and an eventual jumping-off point for ground operations to cut the Ho Chi Minh Trail. "Khe Sanh commands the approach to Dang Ha and Quang Tri City," said Westmoreland. "Were we to relinquish the Khe Sanh area, the NVA would have an unobstructed invasion route into the two northernmost provinces."[41]

At approximately 0530 hours on 21 January 1968, the main battle for Khe Sanh began. Preregistered artillery pounded the camp from positions in Laos and the surrounding hills. The Marines were unable to locate and answer the enemy guns concealed by the morning fog. Damage was extensive. Blinding explosions ripped the base. Several helicopters were destroyed, the runway was damaged, fuel-storage areas went up in flames, and most important, the main ammunition dump, containing 98 percent of the camp's firepower (1500 tons), exploded.[42] The main attack was simultaneously directed against nearby Khe Sanh Village, which was soon captured while the outpost on Hill 861 was nearly overrun. As anticipation for an all-out, fog-shrouded attack by 40,000 enemy troops grew, so did the analogy with Dien Bien Phu.

After the opening assault on 21 January, the area became unexpectedly quiet, perhaps due in part to clear skies and aircraft swarming over the base. Supplies were rushed to Khe Sanh in expectation of a new North Vietnamese offensive. The attack soon came, but not at Khe Sanh.

On 30 January NVA and Viet Cong forces struck almost everywhere else in South Vietnam to begin the Tet offensive. Major population centers were the primary targets of approximately 62,000 enemy soldiers.[43] The threat to Khe Sanh began to seem like a feint to draw U.S. forces away from the cities. The pressure on Khe Sanh was not relaxed as daily bombardments continued. Even so, supplies poured into the base as favorable weather during this last week in January accommodated air resupply.

While the Marines prepared for an assault on the Khe Sanh Combat Base, at 0300 hours, 5 February, elements of the 325C NVA Division struck Hill 861A, which was defended by E Company, 2nd Battalion, 26th Marines. Bangalore torpedoes ripped holes in barbed wire. North Vietnamese forces moved through the gap and bitter fighting ensued. Fire support from Camp Carroll, the combat base, and adjacent hills, plus close air support, suppressed the enemy, enabling the Marines to counterattack, inflict heavy casualties, and drive the opposition from the hill.

With Khe Sanh surrounded by an enemy estimated at between 20,000 to 40,000, all eyes turned to the sky: the Americans for resupply, the North Vietnamese for the B-52 gunships and tactical air bombardments that rained death from above. For both, the weather held the key. The clear skies that followed the 21 January attack allowed continuous resupply to the base. For the next eight days, an average of 250 tons per day were delivered by C-130, C-123, and Caribou aircraft.[44] In addition, Phase II of Operation Niagara, a coordinated bombing of identified targets by B-52 and tactical aircraft, began on 22 January.[45] Phase I of this operation had begun on 5 January, designed to provide precise and consistently current information of enemy troop disposition and activity. Methods of intelligence collection ranged from sophisticated electronic and remote-sensing techniques (both ground and air) to the time-

DISASTER AT LANG VEI

The Special Forces A Team camp at Lang Vei, about 8 km (5 mi) west of Khe Sanh on Highway 9, was manned by 24 American Special Forces soldiers and approximately 500 Vietnamese and Montagnard tribesman. At 0042 hours, 7 February, nine Soviet-built PT-76 tanks, accompanied by infantry, tore through the wired perimeter and into the compound. The situation quickly became desperate for the defenders, and the team commander, Capt. Tom Willoughby, urgently called for help. Artillery and air support were called in on the camp itself, now in danger of being overrun. The captain then requested the Marines at Khe Sanh to activate the previously coordinated relief plan, calling for two companies of the 1/26th Marines to relieve and reinforce Lang Vei. The plan, agreed to by Gen. Robert E. Cushman, Marine Commander in I Corps, and Col. Jonathan Ladd, Commander of U.S. Army Special Forces in Vietnam, had been rehearsed two months earlier by Colonel Lownds's troops. He thought, however, that a relief force moving down Route 9 was likely to be ambushed and that a helicopter assault at night into a very uncertain situation was too hazardous. Colonel Ladd appealed to General Westmoreland to direct the Marines to send the relief force, but Westmoreland, "honoring the prerogative of the field commander on the scene, declined to intervene." A second desperate call for relief from Captain Willoughby at 0310 hours was again refused. The Lang Vei defenders were on their own.

Throughout the next day, the survivors, dodging enemy patrols, escaped to Old Lang Vei, approximately 0.8 km (0.5 mi) southeast on Route 9, where more than 500 displaced Laotian soldiers were garrisoned since their camp had been overrun by NVA forces weeks earlier. Finally, late in the afternoon of 8 February, 40 Vietnamese Civilian Irregular Defense Group (CIDG) troopers and 10 Green Berets were lifted into Old Lang Vei to evacuate the survivors. Of the CIDG defenders at Lang Vei, 316 were dead or missing. Ten of the 24 Americans were killed and another 11 wounded.

tested foot patrols and prisoner interrogation. Aerial resupply and bombardment became the heart of the Khe Sanh defense plan. Helicopter, fighter, and spotter aircraft swarmed around the base in layers while the cargo aircraft delivered the supplies needed to replenish the lost ammunition, sustain the base, evacuate casualties, and drop off replacements.

The B-52s were the only airborne system relatively immune to the local weather conditions. During the fight for Khe Sanh, the B-52s flew 2548 sorties and delivered 59,542 tons of bombs in support of the encircled Marines.[46] Each day formations of three B-52s took off from airbases in Guam, Okinawa, and Thailand every 90 minutes to maintain a near-constant presence over the area. They even provided close-in support, striking at planned targets within 1.2 km (0.75 mi) of the base perimeter.[47] These "arc-light" missions left the landscape around Khe Sanh a smoldering mass of mud, craters, and splinters, surreal in the morning and evening fog.

On an average day, 350 tactical aircraft, 60 B-52s, and 40 reconnaissance missions would be flown over Khe Sanh, all coordinated by Maj. Gen. William Momyer, General Westmoreland's deputy for air operations.[48]

Fog, low ceiling, and runway damage limited the airstrip's use. The weather was especially poor the first three weeks of February, preventing landings 40 percent of daylight hours. The North Vietnamese used the reduced visibility to position antiaircraft guns just off the runway approach and fire into the overcast sky as supply craft broke through the low ceiling. Several aircraft were hit on final approach while still completely wrapped in the fog.[49] Generally the ceiling rose slightly to several hundred feet during the late morning and early afternoon. It was then that North Vietnamese rocket and artillery fire tended to be heaviest, as enemy observers on the neighboring hills directed fire but remained hidden from spotter and strike aircraft by the low-lying clouds. As the afternoon fog settled over the base, the enemy artillery fire tapered off.[50]

In March, with clearing skies, enemy activity around Khe Sanh diminished and at 0700 hours on 1 April, Operation Pegasus, designed to relieve the siege of Khe Sanh, began. Combined Army, Marine, and Vietnamese forces moved along Highway 9 toward Khe Sanh, spearheaded by the 1st Cavalry Division. The force encountered light-to-moderate resistance as it pushed ahead against a withdrawing enemy. On 2 April, at 0800 hours, the relief of Khe Sanh was accomplished. Fighting continued until 14 April, and on 15 April Operation Pegasus officially ended.

The relief operation was timed to coincide with the breakdown of the northeast monsoon season. General Westmoreland, anxious to reestablish the ground link, ordered Highway 9 opened, commenting that "a study of weather in the region over the preceding ten years revealed that not until the first of April could I count on good weather for air mobile operations."[51] The meteorology, however, did not cooperate, and poor weather plagued the 1st Cavalry for the eight days until they reached the base. Seldom could Maj. Gen. John J. Tolson's (1st Cavalry Division Commander) helicopters lift off before 1300 hours because of the fog and low ceiling.[52] Nevertheless, as General Tolson said, this was the first time the cavalry had made an air assault as a division entity; every committed battalion came into combat by helicopter. In 15 days "the division entered the area of operations, drove off the enemy, relieved Khe Sanh, and were then extracted from the area only to assault again four days later into the heart of the North Vietnamese Army's bastion in the A Shau Valley."[53]

While the military significance of Khe Sanh may long be argued, there is no question that weather played a major role in its siege, defense, and eventual relief. Gen. Creighton Abrams wrote in 1968 that "never in the history of warfare have weather decisions played such an important role in operational planning as they did in Southeast Asia. Khe Sanh, the A Shau Valley, and Kham Doc are only a few of the many areas where weather has been a primary consideration in operational intelligence planning."[54]

The weather will continue to be of significant concern on battlefields of the future. Airmobile operations, a powerful weapon of today's arsenal, are especially vulnerable to fog and low ceiling and visibility. General Tolson com-

mented that in planning Operation Pegasus and the follow-on assault into the
A Shau Valley,

> Weather had been the key planning factor . . . from the beginning. The
> urgency to terminate Operation Pegasus in order to go into the A Shau
> Valley was based on inches of rain to be expected after . . . April [1968], not
> *ceilings* and *visibilities* which would prove to be so critical. In other words,
> the forecast monsoon rains (which did occur) never produced the terrible
> flying conditions of low ceilings and scud which preceded them in April.
> An air cavalry division can operate in and around the scattered monsoon
> storms and cope with the occasional heavy cloudbursts far better than it can
> operate in extremely low ceilings and fog. . . . The lesson learned, then, was
> that one must be very careful to pick the proper weather indices . . . for an
> airmobile operation. An inch of rain that falls in 30 minutes is not nearly so
> important as a tenth of an inch which falls as a light mist over 24 hours.[55]

These units become essentially immobile when confronted with adverse
weather. Similarly, obscuration drastically limits tactical air support, removing
the relative advantage of the side with established air superiority. Such considerations become especially important for regions where widespread low-level
clouds or fog persist because of seasonal or daily meteorological conditions.
No prolonged military operation has escaped the impact of poor weather, and
in innumerable situations it has served as a force multiplier that requires an
astute response if it is to be overcome by the disadvantaged.

CHAPTER 4

Invading Another Climate as Seasons Change

Napoleon and Hitler in Russia

For such is the Russian climate: the weather is always extreme, intemperate. It either parches or floods, burns or freezes the earth and its inhabitants, a treacherous climate whose heat weakened our bodies as if to soften them for the cold which was soon to attack them.

GEN. PHILIPPE-PAUL DE SEGUR, NAPOLEON'S AIDE-DE-CAMP

On 23 June 1812, more than 600,000 soldiers of France's Grande Armée began crossing the Niemen River into czarist Russia. It was the mightiest fighting force the world had ever seen, commanded by one of the most admired and feared military geniuses in history, Napoleon Bonaparte. Their goal: to engage and defeat the armies of Russia, which by all accounts were no match for the combined forces of the Napoleonic Empire, and to bring Czar Alexander I to terms before winter. Less than six months later fewer than 20,000 broken and defeated survivors of the Grand Armée returned to Poland.

On 22 June 1941, seven German field armies (totaling more than three million soldiers), spearheaded by four panzer groups, launched Operation Barbarossa against a surprised and unprepared Soviet Union. This was to be blitzkrieg, a war of swift-moving tanks, trucks, troops, and air power (although there were still 625,000 horses in the German invasion force). Like Napoleon's campaign, Operation Barbarossa was designed to be short, and to defeat the large but poorly led and trained Red Army before winter. Cold-weather equipment and clothing were not even in the inventory of Adolf Hitler's invasion forces. And like Napoleon, Hitler underestimated the enemy and overlooked the geographic challenge of Russia. His defeat came more slowly and cost an estimated 20 million Soviet and 6 million German dead (soldiers and civilians), compared with fewer than 1 million casualties on both sides in Napoleon's campaign. In the end the result for the aggressors was similar: the loss of an empire and the end of an era.

These two campaigns have been analyzed repeatedly to determine why they failed. Many factors may have contributed to their demise but one element, Russia's harsh continental climate, certainly played a major role. Some of the most intense accounts by French and German soldiers, as well as other witnesses, relate their terrible suffering in the extremely cold winter. And climatic problems in this part of the world go beyond frigid winters. Seemingly inconsistent with latitude, temperatures can rise to high levels in summer. Warm-season thunderstorms counterbalance fierce winter blizzards. The climate can parch and scorch, it can chill or freeze, or it can soak and flood. It is, as Gen. Philippe-Paul de Segur noted, a land of extremes.

It would, however, be inaccurate to claim that climate alone determined the defeat of Napoleon or Hitler. But these campaigns vividly illustrate the impact of seasonal changes on major operations, and, as such, they establish the potential importance of climatic considerations in military planning.

What Is Climate?

The climate of an area can be described as the long-term average of the changing weather, necessarily including its extremes and unusual phenomena. Temperature and precipitation are the two major elements by which climates are generally categorized. Considering the potential for variation, people live within a fairly narrow range of temperature, moisture, wind, and atmospheric pressure conditions, and human tolerance for climatic variation is quite limited. Temperatures more than 5 degrees above or below 21°C (70°F) can be uncomfortable for many people. Variations even less than 17°C (30°F) in either direction demand significant adjustments for people and equipment to function efficiently. Beyond these limits, human existence is tenuous without considerable artificial support.

Climates vary, primarily because of differences in latitude, altitude, and the atmospheric effects of land masses versus large bodies of water. Other factors include prevailing winds, moving air masses, ocean currents, and regional topography. In some combination these factors influence seasonal temperature and precipitation characteristics by which climates have traditionally been studied and classified.[1] Latitude and the huge size of Asia are the two most significant controls on the climate of Russia and account for its great seasonal variations in temperature, including notoriously cold winters. No place in the former Soviet Union is free from winter's cold. Even the small subtropical region on the Black Sea coast experiences killing frosts every year.[2]

A LAND OF EXTREMES

The Russian climate resembles Canada's. Being a northern country, its capital, Moscow, is at a latitude similar to that for the southern part of Hudson Bay (Map 4.1). Climates in the country vary considerably but every section has cold winters that are exceptionally severe in the northern forests and tundra. Eastern Europe's climate is transitional between two extremes for the middle latitudes.[3] To the west conditions are temperate, being influenced most by the moderating maritime influences of the Atlantic Ocean. Toward the east, however, the Asian land mass is increasingly influential (a characteristic referred to as "continentality"), resulting in large seasonal temperature variations.

In contrast to Russia, the climate of France and Germany is most influenced

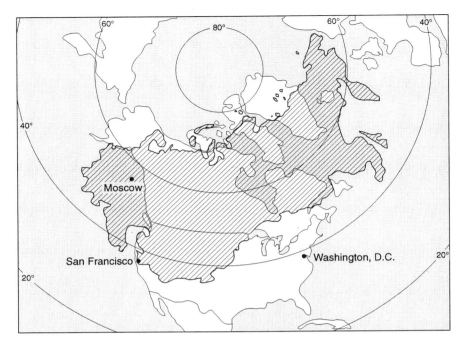

MAP 4.1. Size and latitude of the superposed former Soviet Union are two of the key factors controlling its climate. Note that the latitude of Moscow is higher than anywhere in the conterminous United States. (CIA Map 76114 5-69.)

by air that originates over the Atlantic Ocean and penetrates inland to lessen temperature extremes, both winter and summer. These milder conditions exist primarily because the ocean heats and cools more slowly than the terrain of Asia. That is, unlike land, water warms slowly while absorbing and storing large amounts of energy, only to gradually release it to overlying air year-round. Differences in albedo (reflectance) and specific heat, along with the ability of light energy to penetrate water, be absorbed at depth, and circulate with currents, account for the variance between temperature regimes for air originating over oceans and continental air masses that develop over large land areas.[4] As a result, in the upper-middle latitudes maritime air masses moving eastward bring milder temperatures where they cross the land. The result in Western Europe, which is essentially a narrowing extension of Asia with many peninsulas, is the largest area of Marine West Coast Climate on Earth.

In Russia, deep within the world's largest land mass, maritime effects are minimized, being replaced by continentality, which ensures a climate of extremes. During winter a giant Siberian high-pressure system dominates the weather, with temperatures in northeastern Siberia colder than anywhere outside of Antarctica and the highest mountains. This huge, dry continental air mass, far from the moderating influence of an ocean, spreads cold outward in all directions including European Russia (Map 4.2).[5] As a result, contrary to general expectations, temperatures decline more rapidly eastward than northward.

Though these differences might initially seem insignificant, a 9°C (15°F) drop in average monthly temperature can present severe problems for an unprepared military unit. Extreme cold is always a formidable hazard, often requiring extraordinary efforts for operations and survival (Map 4.3).[6] Additionally, the winters grow longer toward the east, as rapid cooling during the fall brings freezing temperatures early. Then, winter's snow cover reflects much of the sun's energy, helping to retard the arrival of spring.

During the late-arriving spring, the number of progressively longer, sunny days slowly increases average temperature. Concurrently the Siberian high is gradually reduced and eventually replaced by low pressure centered over south-central Asia. As a result, with the arrival of summer the average temperature across Eastern Europe and European Russia is consistently warm, exceeding 35°C (90°F) on many days (Maps 4.4 and 4.5).

Seasonal fluctuations in Asia, which are among the greatest on Earth, vividly

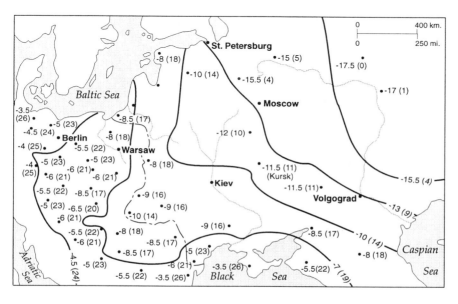

MAP 4.2. Average minimum temperatures (Celsius and Fahrenheit) for January. Isotherms show a distinct north-south alignment, revealing the increasing effects of continentality inland.

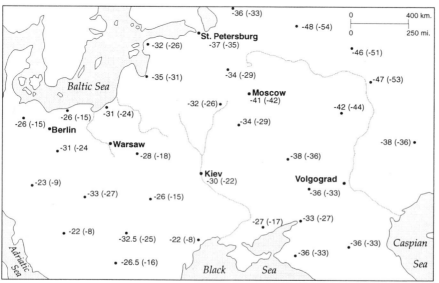

MAP 4.3. Absolute minimum winter temperatures (Celsius and Fahrenheit) showing a marked continental influence toward the east. German troops recorded the reading of −53°C (−63°F) west of Moscow in January 1942. (Department of the Army, "Historical Study: Effects of Climate on Combat in European Russia," Pamphlet No. 20-291, Washington, D.C.: GPO, February 1952.)

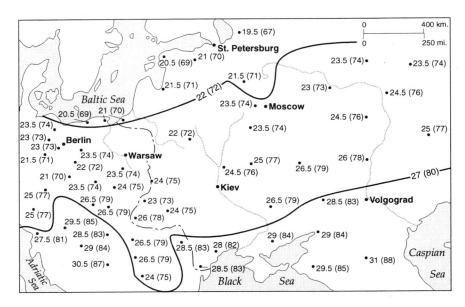

MAP 4.4. Average maximum temperatures (Celsius and Fahrenheit) for the warmest months. Note that there is not a major difference between Central Europe and the European section of the former Soviet Union.

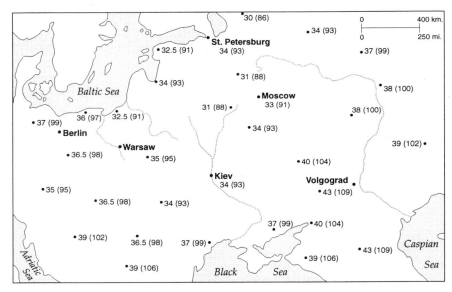

MAP 4.5. Maximum summer temperatures (Celsius and Fahrenheit).

demonstrate the magnitude of continental influences on the climatology of a place. For example, the world's largest known annual temperature range was recorded in Siberia, where the mercury has fallen as low as −71°C (−96°F) and risen as high as 33°C (92°F). To a lesser degree such seasonal extremes extend throughout Russia.

Although extreme heat and mud from occasional summer rains presented some severe problems, it was clearly the extreme cold that eventually presented the greatest natural threat to the invaders of Russia and the Soviet Union. Apparently, this notion never entered the minds of Napoleon and Hitler as their armies moved eastward from Western Europe's pleasant summertime climate. In only a few months these soldiers—inadequately equipped and unsheltered—would be enveloped in a winter environment that was commonly more lethal than the enemy.

RAIN AND SNOW, MUD AND DUST

After temperature, the second major climatic factor affecting movement is moisture in its various forms. The impact of rain, snow, and ice on trafficability and mobility can be critical. Problems accompanying mud, dust, and humidity must also be considered in the maintenance and proper functioning of weapons and equipment. Ballistics and visibility are often impeded. Mud and deep snow reduce shrapnel radii of artillery and mortars. Cloud cover and fog inhibit air operations. Certainly, health and morale can also be stressed under very wet or exceedingly dry conditions.

When compared with that of the eastern United States, annual precipitation in European Russia is low. Whereas 100–150 cm (40–60 in) are received along America's eastern seaboard, the Saint Petersburg area receives about 70 cm (27 in) annually, and the total gradually decreases eastward and southward to only 15–18 cm (6–7 in) around the lower Volga River (Volgograd) and the northern Caspian shoreline. Averages along the Baltic coast are similar to those for much of Central and Western Europe, but these areas do not have the marked decrease inland so characteristic of Russia. Distance from the major source of moisture, the Atlantic Ocean, is the decisive factor that controls this trend. Most of the moisture in European Russia and Ukraine falls as rain in summer, except for a small area along the Black Sea coast that has a winter maximum associated with storms moving northeast from the Mediterranean Sea.

Although there is more precipitation in summer, the seasonal variation is not drastic. Temperature, however, dictates the form of condensation and precipitation, as well as the rate of evaporation. During summer and early fall, high rates of evaporation cause rapid drying and dust, despite the larger amount of precipitation. This is particularly true in Ukraine, where fine-textured soils combine with somewhat drier conditions to produce dust that is easily stirred by wind, vehicles, horses, and humans, with sometimes serious effects on people and machines. However, not uncommon, brief but intense thunderstorms can quickly transform the dusty surface into an army's even more formidable adversary—mud.

When water saturates soil it displaces air that occupied spaces between individual particles. As a result, the ground loses its firmness and its ability to support weight. Instead of resting only on one another, particles are also touching a pressurized and lubricating liquid film. An object that sinks into the resulting mud is easily enveloped by the cohesion of the soil and water; withdrawal creates a vacuum and related suction. Thus any attempt at extrication requires added energy that may be exhausting and is sometimes impossible. Saturation also limits infiltration of additional precipitation that, in turn, forces water to collect on the surface, causing flooding or rapid runoff. Furthermore, water clinging to the innumerable soil particles with relatively large surface areas allows the soil to retain moisture for long periods so that drying of moist soil may be prolonged.

In European Russia the greatest problems with mud occur in spring and autumn.[7] Snow melt and thawing ground are the chief reasons for spring mud, whereas cooler temperatures in late fall reduce evaporation, permitting water to remain in the ground longer until winter temperatures freeze the surface and restore firmness.

Few people have difficulty picturing the snow-covered Russia of *Dr. Zhivago*. Although less precipitation falls in winter (an average of 25 cm [10 in] of rain-equivalent falls in Moscow from October through March), most of it is snow. Considering that 2.5 cm (1 in) of rain equals approximately 25–30 cm (10–12 in) of snow, more than 280 cm (110 in) of snow can be expected during an average winter.

Actual snow depth, however, does not reflect total possible snowfall over a cold season. Some of the winter precipitation is rain. Furthermore, the weight of succeeding snowfalls compacts the first-fallen snow and makes it more dense over time.[8] Melting from insolation and ground radiation also reduces overall snow depth. By March an average winter maximum of 50 cm (20 in) is expected near Moscow, a considerable amount when compared with that in areas near the western borders of the country (Map 4.6). Higher snowfall or colder temperatures result in greater accumulation and depth. Strong winds may prevent snow from remaining in some areas as drifts accumulate elsewhere.

Studies and experience have shown that wheeled vehicles and persons on foot cannot travel efficiently on flat terrain or roads when snow depth exceeds 30 cm (12 in), and both are practically immobilized in depths above 50 cm (20 in). Some tracked vehicles may continue to operate in depths of 135 cm (53 in), but their mobility is considerably hampered, especially on slopes.[9] Any vehicle that lacks clearance soon acts as a snowplow, piling the mass ever higher, until resistance is too great for it to overcome. Similarly, persons walking through deep snow soon become exhausted.

Decreasing temperatures account for greater snow depths and longer persistence eastward in Russia even though total precipitation (in rain-equivalent) actually decreases. Moscow averages 150 days of snow cover each year, much longer than areas in Central and Western Europe. Nearer Siberia, the number of days with continuous snow cover increases as average winter temperatures decrease. In the direction of Ukraine and Central Europe, the reverse is true. Areas around Warsaw, Poland, have fewer than 80 days of snow cover, and Berlin only 40. Snow and cold far different from conditions at home face any army from Western Europe operating in Russia during the winter.

Added to the cold and snow, a dreary blanket of clouds fills the Moscow sky about 80 percent of the time in winter. As cool, moist air from the North Atlantic moves inland it rises in a gentle slope when meeting colder, more dense air of the continent. The result is condensation of water vapor to form a

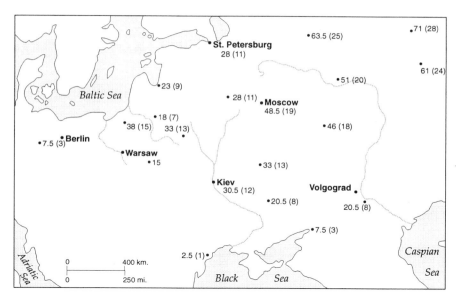

MAP 4.6. Average annual maximum snow depth (centimeters and inches) taking into account both melting and compaction. Snow usually reaches its greatest depth in Russia in March.

persistent cloud cover over much of Europe, including western Russia. Little daylight, long nights, and persistent cloud cover can even lead to emotional disorders collectively and locally referred to as "arctic hysteria." Though such mental states are not well understood, victims have been known to suffer from a variety of symptoms that include extreme timidity, passivity or fright, morbid depression, insomnia, suicidal tendencies, and claustrophobia.[10] Morale, such a vital factor in armies, certainly could become strained in a Russian winter.

The consequences, then, of high latitude and continentality in the former Soviet Union are many. Frequent mid-latitude cyclones often change conditions rapidly, making accurate weather prediction difficult. Seasonal variances are extreme, and major year-to-year changes are common. One winter might be unusually cold or have heavy snows, should atmospheric circulation patterns permit polar fronts to move south and mix with moist air from the Atlantic. Another cold season might be milder or drier. Prolonged, dry, hot summers may alternate with wetter years or prolonged winters to bring highly variable agricultural conditions that sometimes force reliance on imported grain shipments to feed the people.

Consider, then, what effect these same conditions could have on armies, particularly those not prepared for the challenge. In 1812 Napoleon's troops encountered a relatively early winter, so that even though the campaign ended by the middle of December, cold and snow took their toll. German forces faced a more severe test in 1941–42 when a prolonged wet fall created lasting problems with mud, only to be followed by one of the most severe winters in a century. In both cases, winter cold was far more damaging to invaders than the summer heat.

Military Consequences

NAPOLEON BONAPARTE—SIX MONTHS TO DISASTER, 1812

The 600,000-man Grande Armée of France had many reasons to be confident at the opening of the 1812 Russian campaign. Opposing czarist forces totaled only 180,000 in two armies separated by several days' march. Overall leadership in the Russian army was weak, and Emperor Alexander, though determined, had never led troops in battle. Napoleon expected to defeat the Russians early in the campaign;[11] he counted on Russian patriotism to ensure a decisive battle.

However, Charles XII of Sweden (see chapter 8) had been defeated by the Russians at Poltava in 1709, largely because his army was weakened from prolonged campaigning in a harsh environment far from its resources. Napoleon had studied Charles's campaign and was determined not to make the same mistakes of letting Russia's great expanses and harsh winter work against him. Logistically, this was his most carefully planned operation. A major system of advance supply depots was planned and the attacking forces carried 14 days' provisions, evidence that a prompt settlement was expected.[12]

Everything seemed to be in the emperor's favor but, genius that he was, he was not a fortuneteller. He could not prophesy that Czar Alexander would decide at the last moment to abandon plans for a coordinated defense near Russia's European border out of a well-founded fear that Napoleon would trap his armies. And, like many dominant leaders, Napoleon's confidence and pride precluded immediate recognition that the Russians would refuse to surrender. These factors, along with an inherent Russian understanding that their vast area and harsh climate were allies against any invader, led the Grande Armée into a situation where victory was impossible and escape unlikely.

Climate was formidable in Napoleon's campaign long before winter arrived. De Segur recorded that severe thunderstorms struck in June almost at the moment the offensive began crossing the Niemen River. The skies darkened and the rains came, followed by five days of poor weather, all of which was evidence for the passage of a large, mid-latitude summer cyclone. Since they had been using big, heavy wagons to carry supplies and equipment, much material was mired in the mud and left behind.[13]

Throughout the long march toward Moscow, Napoleon and his army faced many problems, the greatest being their inability to force the Russians into a decisive battle. This failure meant that a large battle force, not an occupying army, had to march eastward into an increasingly foreign physical and cultural environment. Summer heat took its toll in casualties. Drinkable water was scarce in the parched and dusty land. By the time the Grande Armée reached Vitebsk (Map 4.7) in August, the main force of 400,000 was reduced to 185,000. Of the original group, at least 100,000 were deserters or casualties; the remainder had been detached to the flanks or to secure lines of communication.[14] Although the troops were in Russia for two months, no major battle had been fought, and few of these casualties could be attributed to combat.

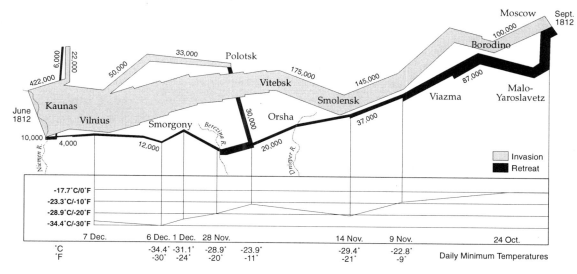

-17.7°C/0°F
-23.3°C/-10°F
-28.9°C/-20°F
-34.4°C/-30°F

	7 Dec.	6 Dec. 1 Dec. 28 Nov.		14 Nov.	9 Nov.	24 Oct.
°C		-34.4° -31.1° -28.9° -23.9°		-29.4°	-22.8°	
°F		-30° -24° -20° -11°		-21°	-9°	Daily Minimum Temperatures

0   80 km.
0   50 mi.

MAP 4.7. A modified English version of Joseph Minard's (1781–1870) famous 1861 cartogram showing lines of march, distances, attrition of Napoleon's army, and calendar dates associated with the Russian Campaign of 1812. Minimum temperatures during his retreat appear at the bottom. The crossing of Berezina River in late November where Napoleon lost an estimated 20,000 of 50,000 troops on hand illustrates the possible combined effect of climate and terrain on military operations. (*Tableaux Graphics et Cartes Figuratives de M. Minard, 1845–1869* [Paris: Bibliothèque de l'Ecole Nationale des Ponts et Chaussees, 1861].)

Napoleon considered bivouac at Vitebsk as his marshals pointed out the already weakened state of the army and the approaching "terrible winter."[15] But in this era, to feed and shelter such a large force through winter in a foreign land, at great distance from the principal supply bases, would have been most difficult even for the Grande Armée. Napoleon concluded that he could still achieve victory somewhere to the east before winter.

Autumn arrived with chill winds in early September on the eve of the Battle of Borodino, the only full-scale engagement of the entire campaign. De Segur went so far as to claim that a cold brought on by the weather robbed Napoleon of his genius for five days, and saved the Russians from destruction. "It is possible that without this ally [weather], Russia might have yielded to us on the fields of Moskva."[16]

Though de Segur might have oversimplified the issue, it was clear that, by then, climate and distance had significantly affected the physical and mental well-being of the intruders. Marching great distances in dust, heat, chill, and mud had weakened the soldiers, making them less aggressive and more susceptible to disease. Throughout Napoleon's advance the Russians inflicted many casualties in flanking and rear-guard actions. Each week witnessed a weakening of the Grande Armée and a relative strengthening of the Russians until Borodino, where the opposing forces were nearly equal in number and resolve. There, on 6 September, a great but indecisive battle ended with Russian withdrawal and continued French pursuit eastward. A few weeks later 100,000 French soldiers occupied an abandoned and burning Moscow with neither a major victory nor Russian surrender.

When Czar Alexander received the report of Moscow's occupation and learned of the French army's gloomy position he resolved to make no peace.

Plans to cut off the emperor's escape began with false negotiations that deceived Napoleon and kept him in the city as winter approached. Eventually recognizing the futility of his hopes and that it would be impossible to sustain his army in Moscow over winter, Napoleon ordered a retreat on 18 October with 15 rations of flour per soldier and wagons of booty that only slowed the rate of march. Unfortunately for the Grande Armée, winter weather arrived relatively early in Russia in 1812: The first hard frost came on 28 October, and snow followed on 6 November.[17]

The French troops initially marched south on the Kaluga road where, after a brief but bloody encounter with the Russians at Malo-Yaroslavetz, they turned north and then westward to retrace the original invasion route through Smolensk. Some have argued that Napoleon should have continued toward Kaluga to capture Russian supplies and then withdrawn along the more southerly path, one not despoiled by war, which could have provided more local provisions. Perhaps the greatest advantage of a southern route was that it was two days' (about 80 km or 50 mi) closer march to Smolensk, the earliest goal of the retreat. But winter along this route would have been no less harsh in Russia's continental climate, and foraging for such a large army would have been difficult along any line of march in a hostile land. Possibly the initial southward French move was simply a diversion, a chance to get the stalking Russians off their backs because depots and supply points were all located along the original invasion route. At any rate, Napoleon spent 10 days on the Kaluga maneuver, and winter was that much closer at hand.[18]

As the ground froze, the wagons moved more easily, but the accompanying cold was no comfort to the marching soldiers. Accounts of the retreat are a catalogue of horrors: famished troops among frozen corpses; reaching the depots at Smolensk only to find more starvation and disease among the wounded and stragglers; continuous harassment from pursuing Russian forces; and crossing the ice-clogged, but unfrozen, Berezina River (about 240 km [150 mi] west of Smolensk) in the face of the enemy, an action that alone depleted two-thirds of the remaining French force (for details of the river crossing see chapter 8). Possibly the only major consolation for the Grande Armée was that Napoleon escaped. Once past the Berezina River he departed for Paris, leaving Marshal Joachim Murat to oversee the final withdrawal.

Taking into account reserves and reinforcements during the six-month campaign, Carl von Clausewitz estimated that Napoleon left 552,000 dead, captured, or deserted in Russia. Thousands fell in countless skirmishes and smaller actions, but what cannot be ignored is that many more—hundreds of thousands—were lost to nonbattle causes related to the deteriorating weather. Nonbattle casualties (lack of shelter, inadequate medical care, and poor nourishment were principal causes) characterize warfare of this era. Additionally, large numbers of Napoleon's non-French forces (approximately half of the Grande Armée) deserted. All these factors were multiplied by the severity of Russian climatic conditions. Far from friendly soil, the Grande Armée could

FROM CHOSEN RESERVOIR TO THE SEA, KOREA 1950

It was December. Bitterly cold winds blew out of the northwest from Siberia, across Manchuria, and over the Korean peninsula in the wake of an outbreak of polar air. Members of the 1st Marine Division trudged slowly along a narrow mountain road toward the sea. At the end of November, it looked as if the northward-advancing United Nations Forces would be home by Christmas, until hordes of Chinese attacked across the Yalu River to engulf the exposed flanks and rear of the advancing UN armies. Suddenly, the Marines were fighting in the opposite direction (toward the south) against the Chinese, the icy winds, and drifting snow. Periodically, a truck lost traction on the slippery road and crashed down the side of the mountain, its occupants leaping to safety. With the 1st Marine Division were 181 survivors of the 1st Battalion, 32nd Infantry, U.S. Army. The battalion had been isolated farther north by the Chinese offensive and severely mauled in assault after assault during the long cold nights, waiting for relief that never came. A final attempt to break out as a unit had failed, and survivors had to escape in small groups at night, mostly over the frozen waters of Chosen Reservoir. Now they too moved toward the sea where ships would rescue them to fight another day. At a rest halt, a Marine slowly prodded with a spoon held between gloved, but numb, fingers to dislodge a frost-covered bean in his can. He held the single bean in his mouth to let it thaw. When asked what he wanted if he could be granted any wish, he spoke with great effort through jaws numbed by cold, "Give me tomorrow."

From David D. Duncan, "Retreat Hell," *This Is War* (New York: Harper and Brothers, 1951); and Russell A. Gugeler, *Combat Actions in Korea* (Washington, D.C.: Combat Forces Press, 1954), 62–86.

not adequately cope with the environment or replenish its losses. As a result, it eventually succumbed more to the elements than to the enemy.

Some of the greatest blows of winter came as the remnants of the Grande Armée neared the Russian frontier and anticipated safety. According to one source, on 6 December temperatures fell to $-38°C$ ($-36°F$); de Segur told of ice crystals that hung in the still air as birds fell dead to the earth. Most likely the ice was sublimated from the breath of soldiers and animals, remaining buoyant in near-surface air within a temperature inversion produced by the intense cold. De Segur estimated that between Berezina and Vilna as many as 40,000 soldiers perished from the cold within four days. A fresh division of 15,000 that had come south from Vilna to meet the main army was reduced to 3000 during the three-day march. Clausewitz more modestly estimated a French loss of 18,000 in the last 10 days of marching (approximately 300 km or 186 mi) to the Niemen River, including the majority of the fresh division sent from Vilna. There were no battles during this period.

The Russians did not fare much better. In the long, nearly continuous pursuit of the French, they too were exposed to the frigid weather. Though more accustomed to such conditions, at least a quarter of Count Peter zu Sayn-Wittgenstein's army of 40,000 was lost to cold in the weeks before reaching Vilna.[19] According to de Segur's estimates, Mikhail Kutusov's army of 100,000 had dwindled to 35,000, and of 10,000 reinforcements sent from the east, barely 1700 reached Vilna as a result of the elements.

Winter was an enemy to all, but in combination with hunger, disease, exhaustion, and decreasing morale, it took its greatest toll on the Grande Armée even though the campaign was over before the coldest weather arrived in January. Ironically, at the Berezina River, casualties would most probably have been lower had the temperatures been colder and the river frozen over. Then crossing the river would have been less a problem. On the other hand, fewer might have reached the river if conditions had been colder at the start of the retreat.

Although their army was destroyed, in the end the French soldiers apparently still believed that as individuals they were better than their opponents. "Theirs was the ghost of an army; but it was the ghost of the Grande Armée. They felt they had been defeated only by Nature."[20] They failed to recognize that the destruction by climate was much greater than that which could have been inflicted by the Russian foe.

OPERATION BARBAROSSA, 1941

At 0330 on Sunday, 22 June 1941, the roar of cannons marked the beginning of armed conflict between Germany and the Union of Soviet Socialist Republics. It was to become an enormous war in every respect: the largest and most elaborate invasion scheme ever attempted and more men, machinery, and horses than ever before committed (600,000 motorized vehicles in the German army alone). The battleground covered an area several times larger than all those of Western Europe. From north to south, the front extended for 3200 km (2000 mi), and the terrain presented a wide variety of challenges. The opposing armies fought across mountains, swamps, forests, steppes, deserts, and several wide rivers.[21] The savagery and ferocity far exceeded that of previous wars. Casualties on both sides, but especially among the Russians, were staggering.

To the Soviet citizen, it became the Great Patriotic War to save the Motherland and eventually to avenge her suffering. To the German citizen it began as a crusade against atheism and the Red Peril to rid the world of the Communist menace forever. It evolved into a desperate fight for survival, a struggle between life and death in which only one nation would survive. It was as near total war as the world has ever seen.

Powerfully superimposed upon this titanic struggle was a climate unmatched in its severity in any other major theater of war. If the frailty of armies within a climatic context was ever to be fully tested, Russia was the place to do it. Through parching hot summers and bitter winters, by war's end some 25,000,000 people died on and near the battlefields of the Eastern Front. To the invading Germans, who were neither accustomed to nor prepared for what they encountered, the climate was equally as foreign as the country and its people.

In retrospect the observer may be tempted to judge Adolf Hitler's Operation

Barbarossa as a failed mission from the start, an invasion that should never have been initiated and one that had no hope of success. Unlike Napoleon, Hitler did not enjoy numerical superiority. The vast terrain and the poor transportation network foretold countless logistical problems for such a huge operation. The Soviet Union was by far the world's largest country and the third most populous. Its human and natural resources were vastly greater than Germany's. Ahead loomed the infamous Russian winter through which the forces of both Charles XII and Napoleon had suffered defeat.

A number of factors, however, seemed to favor Hitler's daring strategy. Germany's blitzkrieg had quickly defeated all previous opponents, some more technologically advanced than the Soviet Union. Hitler viewed the Soviet Union as a tottering giant on clay feet that would quickly collapse when faced with a strong opponent. Russian military leadership had been severely depleted by Stalin's purges in the mid-1930s, the army was poorly trained and, it was believed, inadequately equipped. Citizens were not particularly satisfied with Stalin's ruthless dictatorship, and the racist Nazi regime was convinced that the Soviet people were part of an inferior culture.

As in their previous invasions elsewhere, German plans for Operation Barbarossa assumed a victory within weeks as their forces conducted vast encircling movements to trap and destroy the surprised Red Army (Map 4.8). The Germans did not seriously consider the long-term effects of Russia's size, climate, and resources because they expected the war to be over quickly. The whole campaign was predicated on the notion that the Red Army would make its stand near the border and certainly west of the Dnieper-Dvina River line, a tactic that would favor rapid envelopment and an early, decisive victory for Germany. Plans prepared for Operation Barbarossa repeatedly emphasized the need to conclude the campaign successfully before the arrival of the October muddy season and winter.[22]

Though most authorities now conclude that the Battle of Stalingrad (August 1942–January 1943) was the turning point of the Soviet Union's Great War, it is also clear that the Germans' chances for victory greatly diminished within five months of offset when they failed to capture Moscow. The German leadership underestimated the enemy in number, quality, and resolve; and the German army lacked adequate understanding of problems related to terrain conditions and transportation networks. Furthermore, Hitler did not agree with his generals on strategic objectives, and disagreements among military, economic, and political goals became major issues.[23] The plan to divide the offensive along the three major axes to seize Leningrad, Moscow, and the Donbas industrial region in the south stretched the Wehrmacht beyond its limits.

Still, Germany won great victories during the conflict's first four months. Entire Soviet armies were encircled and destroyed. Initially as planned, the German armored spearheads, supported by the Luftwaffe, advanced rapidly. Like the French and the Swedes of previous centuries, German power weakened over time as both distance and the width of the front increased. Finally,

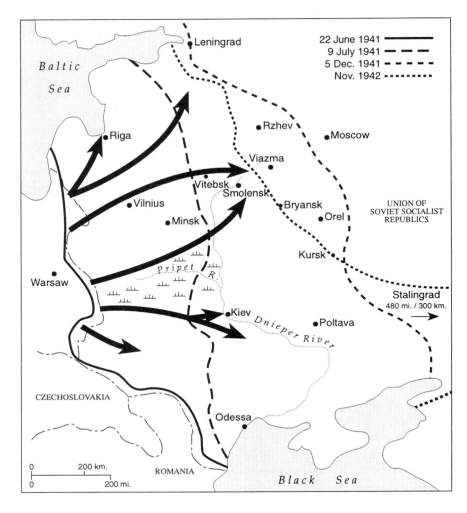

MAP 4.8. Main lines of thrust in Operation Barbarossa showing the front line for four dates early in the war. The map shows 1941 terminology.

Within the map:

Baltic Sea

Leningrad

22 June 1941
9 July 1941
5 Dec. 1941
Nov. 1942

Riga

Rzhev   Moscow

Viazma

Vitebsk
Smolensk

Vilnius   Bryansk

Minsk   Orel

UNION OF
SOVIET SOCIALIST
REPUBLICS

Kursk

Pripet R.

Warsaw

Stalingrad
480 mi. / 300 km.

Kiev   Poltava

Dnieper River

CZECHOSLOVAKIA

Odessa

0   200 km.
0   200 mi.

ROMANIA

Black Sea

near the gates of Moscow in early December 1941, the initial offensive ground to a halt. Then, in the harsh grip of winter, grand strategy gave way to a fight for survival against a fierce Soviet counteroffensive. The Germans fell back, but they held through the winter to launch a second major offensive in the summer of 1942, one that Army Chief of Staff Gen. Franz Halder and other prominent generals insisted was beyond the ability of the Wehrmacht after losses suffered the previous winter.[24] The pattern of 1941–42 repeated itself the following winter with a huge Soviet counteroffensive and decisive victory at Stalingrad. The Germans then staged one last large but ineffective offensive near Kursk in the summer of 1943 which was followed by an even larger Soviet counteroffensive that evolved into a two-year march to Berlin. The German gamble in June 1941 had been matched at Moscow in December 1941, put in jeopardy at Stalingrad in January 1943, and clearly lost by the beginning of 1944.

The campaign in Russia has been described from the German perspective as one great improvisation.[25] From the time the cannon first signaled the advance, weapons, equipment, and soldiers frequently did not function as expected in Russia's hostile environment. Many vehicles had air filters inadequate for the quantity of dust, which caused innumerable engine failures. Dust clouds from maneuvers also attracted Russian air attacks that became increasingly effective as the Germans gradually lost aerial superiority.[26] Water suitable for humans and horses was often hard to find, especially in the south. The summer heat stressed the troops as the fighting progressed.

Even so, summer was the best time to conduct blitzkrieg operations. Trafficability was good both on and off roads except during rainy spells that were generally brief. Localized swamplands were less a problem in summer, and with care some could even be traversed. These conditions strongly favored the more mechanized German army in their summer attacks of 1941 and 1942. Though the heat and dust caused problems, adjustments were quickly made. Filters and spare parts were improvised, maintenance was increased, water was conserved, and vaccinations helped minimize disease, which had so plagued the armies of Napoleon's day.[27]

By October 1941 the Germans had encircled and beaten Russian armies at Kiev in the south and at the twin battles of Bryansk and Vyazma in the center. Leningrad was under siege and the final offensive on Moscow was imminent. According to the original battle plan, the Red Army by then should have been defeated, yet it still fiercely resisted. German generals blamed bad weather and inadequate roads for much of their lack of complete success, but credit should also go to the Soviets and the Germans' failure to assess accurately their strength and resilience.[28]

With October came rain from more frequent mid-latitude wave cyclones, bogging the offensive down into what some German veterans described as the worst muddy season of either World War I or World War II. Cart trails and rural roads soon became impassable. Even gravel and some hard-surfaced roads were churned into muddy morasses. Cross-country movement by vehicles was impossible. In the end, only light horse-drawn wagons (*Panje* wagons) and a few tracked vehicles could move under their own power. Thousands of horses died from exhaustion, stuck hip deep in the muck. Even when artillery could be concentrated, rounds exploded with much reduced effect within the mud or did not detonate at all. Only limited objective attacks with infantry were possible. Gen. Heinz Guderian's 2nd Panzer Group lost 60 percent of its tanks to mud while operating to the east of Bryansk in the autumn of 1941. While lead elements advanced as far as Tule by late October, the bulk of the 2nd Panzer's infantry and armor was bogged down west of Orel and Kursk, 240–320 km (150–200 mi) behind. One major problem was that tracks on German tanks were too narrow, and ground clearance on all their vehicles too low, to cope well with the muddy conditions.[29] In every sector, equipment and vehicle losses were high and replacement difficult at best.

Spring and autumn mud was the biggest obstacle to resupply and maneuver on the Eastern Front, yet forward movement was essential to the 1941 German offensive. In southern areas the mud was associated with widespread, fine-textured, humus-rich, and water-retentive soils. In the center, the huge Pripet Marshes and thousands of unstable marshlands restricted or blocked movement as abundant rain and reduced evaporation saturated soils and raised the groundwater table. Construction of corduroy roads helped to some extent in the wooded areas but movement was painfully slow and sometimes impossible in the virtually treeless south. Red Army tanks with wider treads and trucks with higher clearance were much better suited for the mud. Even so, the Soviets purposely limited their offensives during muddy seasons, at least until the autumn of 1943. They knew they could not afford the increased strain on men, horses, and equipment.[30] As their strength grew, by late 1943 the Red Army was pressing its attack during all seasons but, like the Germans, they too were stopped by the mud on a number of occasions.[31]

During October 1941, Hitler considered a halt incomprehensible and ordered the Germans to fight on despite the mud. Nature—much more powerful than the German dictator—soon brought the entire German army to a stop. Attempts to advance required extra energy and resources. Scarce fuel was consumed at higher rates and many bogged-down vehicles ruined drivetrains and engines in the struggle to move on. Weapons were difficult to keep clean and routinely malfunctioned, and soldiers suffered a variety of debilitations (hypothermia, trench foot, low morale) in the cold dampness of October. Resupply was increasingly difficult on roads continually clogged with armor and trucks. The German army approaching Moscow could no longer break through the Soviet lines. Field Marshal Fedor von Bock, Commander of Army Group Center, cited mud as a principal reason for failure to capture the enemy's capital.[32] Meanwhile, the Russians had improved their defenses and, using strategic reserves, increased the size of their forces near Moscow in preparation for their first great counteroffensive.

The autumn mud ended with the ground-solidifying frosts of November. The clear, freezing weather again opened the roads, and the battle for Moscow recommenced with the overextended Germans meeting increasing resistance. By 1 December elements of Hitler's Army Group Center at last approached the northwest outskirts of the city, but were much weakened. By then the terrible winter weather of 1941–42 had arrived to help slow their advance. The 6th Panzer Division was a mere 15 km (9 mi) from Moscow's outskirts and 24 km (15 mi) from the Kremlin. However, night temperatures reached −40°C (−40°F) (worsened by wind chill), paralyzing troops equipped with summer uniforms and jackboots. Some soldiers were so numb they could no longer properly aim their rifles. Firing pins shattered, and recoil liquid thickened and froze in machine guns and artillery. Shells detonated with little effect in the deepening snow. German tanks that had recently escaped the mud became immobilized in thick snow and intense cold. Yet the Red Army initiated a major counteroffensive.[33]

Repeatedly throughout the long winter and into spring, the Russians at-

WINTER AT VALLEY FORGE, 1777

Three or four days bad weather would prove our destruction. What then is to become of the Army this winter?

GEORGE WASHINGTON, DECEMBER 1777

The fortunes of the Continental Army under Gen. George Washington reached a critically low level in the autumn and winter of 1777–78. Defeat at Brandywine Creek opened the door for the British occupation of Philadelphia in September. At Germantown in October, the Americans suffered another loss. Problems in the Commissary and Quartermaster Departments led to a virtual collapse of the supply system, and the flow of basic necessities slowed to a trickle. Under these circumstances, Washington was compelled to retire to winter quarters at Valley Forge, 30 km (18 mi) northwest of Philadelphia, where he hoped to keep the army from disintegrating.

The name Valley Forge conjures up images of soldiers huddled around fires, of suffering and death in bitter cold and snow. By comparison with European Russia, winters near Philadelphia are relatively mild, with average temperatures hovering above the freezing point in the coldest months. However, without adequate food, shelter, and clothing, even those types of winters can be disastrous. By the end of the winter of 1777–78, as many as 2500 soldiers, a quarter of the entire force at Valley Forge, had died from disease or exposure. Many deserted, and many more chose to leave at the end of their one-year enlistment. On 23 December, shortly after arrival in camp, Washington wrote to the President of Congress that 2898 men were unfit for duty because they had "bare feet and [were] otherwise naked." Another 2000 were too exhausted to perform duties because they remained by the fires all night to stay warm. Blankets were unavailable. Human frailty was evident in these conditions and, without proper clothing and equipment, the army's strength and numbers were rapidly depleted. In his frustration and desperation, General Washington begged Congress for assistance and charged some legislators with holding the uncaring notion that soldiers were "made of sticks and stones and equally insensible of frost and snow."

From Noel F. Busch, *Winter Quarters: George Washington and the Continental Army at Valley Forge* (New York: Liveright, 1974), 67; and George Washington, "Letter to the President of Congress," December 1777.

tacked the Germans who were fighting to maintain their lines and to survive the cold. But German efforts to halt the Russian winter offensive failed, in some cases resulting in encircled units. Near Volokolamsk, about 110 km (70 mi) west of Moscow, the 4th Armored Infantry Regiment moved to counterattack a Soviet breakthrough on 29 December. Lack of winter clothing forced long halts at every village to warm men and equipment, while deep snowdrifts hampered movement on the roads. It took the regiment two days to travel 20 km (12 mi). Eventually the German counterattack encircled the Soviet penetration at Volokolamsk, but then it had to be abandoned because the German force was without shelter and could not have survived the nighttime cold.

Desperate battles were fought by both armies simply to obtain shelter. Near

FIG. 4.1. Mud played a major role in bringing the Germans to a halt in 1941. In October these motorcycle troops found the going rough in Ukraine. (National Archives.)

Rzhev, west of Moscow, the temperature dropped to $-52°C$ ($-63°F$), according to German measurements, and the Russians fought fanatically to overrun a surrounded German unit in the village before dark. They failed and were so weakened by the overnight ordeal that the next morning they could not counter the German escape within 100 m (328 ft) of the Russian positions.[34]

The situation at Rzhev shows that the Red Army also suffered in the winter, but not to the same degree as the Germans. The Russians were better equipped for such cold, and many had lived and trained in it. Furthermore, the Soviets could replace losses from their huge population while the Germans could no longer afford casualties without a reduction in force. The 4th German Army alone suffered 14,236 casualties from frostbite between 1 January and 31 March 1942.[35] Wounds that might otherwise have healed became fatal for soldiers weakened by the cold.

The winter of 1941–42 forced the Germans to innovate in their preparations for the next cold season. They distributed handbooks explaining how to cope with the cold; oil on rifles was replaced by powder; fires were lit under vehicles to keep them from freezing; sentries were replaced every 15 minutes during periods of extreme cold; men were taught to use snow to build defenses and insulate themselves. (Extensive tests under laboratory conditions have confirmed the effectiveness of snow against direct weapons fire and established parameters for thickness and density of snow fortifications.)[36] In a dramatic example of improvisation, the Germans built a railway bridge directly on the frozen Dnieper River by cutting and stacking huge blocks of ice and then fusing them with water that quickly froze when poured over the structure.[37]

Although such improvisations helped Germany continue operations in Russia, including their summer offensives of 1942 and 1943, they could not help them regain the momentum of 1941. Germany's failure at the gate to Moscow in December 1941, its defeat at Stalingrad the following winter, and its debacle at Kursk in 1943 all signaled the growing strength of the Soviet Union. When Hitler complained in 1943 that the soldiers of that year were not of the same caliber as those of 1941, Gen. Walther Model reminded him that many of the

FIG. 4.2. German troops using horizontal ladders to cross swampy areas in the glaciated plains of Russia. (National Archives.)

FIG. 4.3. A German vehicle stuck in the snow northwest of Kaluga in January 1942. (National Archives.)

men of 1941 were dead, "scattered in graves all over Russia."[38] Even so, Hitler continued to underestimate the Red Army, and insisted that it was on the verge of collapse. By 1944, with its greatly increased industrial capacity and its seemingly endless supply of manpower, the tide had turned fully in favor of the Soviet Union. Instead of eliminating the USSR as a European power, the initiation of Operation Barbarossa in 1941 ended with fighting in the streets of Berlin in 1945.

Conclusions

A committee of former German generals and staff officers who had served on the Eastern Front concluded that "climate is a dynamic force in the Russian expanse; the key to successful military operations. He who recognizes and respects this force can overcome it; he who disregards or underestimates it is threatened with failure or destruction."[39]

Obviously climate alone did not defeat Napoleon or Hitler in Russia, but it certainly played a powerful role. In both cases the Russian climate dealt devastating blows to unprepared armies invading another climate through changing seasons, a deadly combination for hundreds of thousands of soldiers. Although Napoleon had forfeited victory in October, that failure was dwarfed by the much larger November–December loss of his whole army.

Mobility was certainly one of Napoleon's concerns but dust, mud, and deep snow were far more serious problems for an outnumbered German army that depended so heavily on machinery, modern weapons, and rapid movement. Operation Barbarossa and blitzkrieg each required quick success. However, as the war slid into the muddy fall and frigid winter of 1941 the initial German advantage rapidly evolved into a strategic, operational, and tactical nightmare. Would the fall of Moscow in December have made the difference in the war's outcome, or would the populace have held out in a prolonged siege, as did Leningrad? Gen. Guenther Blumentritt, formerly chief of staff of the 4th Army that made the main attack on Moscow in 1941, thought that even if the Wehrmacht had been prepared for winter warfare, Moscow probably would not have fallen in December, and even if it had, the Soviet Union would not have been defeated. Although the bulk of the German troops survived their first winter in the Soviet Union, they were already victims of an awesome alliance between Russian geography and an increasingly formidable enemy.

Time, space, and environment were all enemies to Napoleon and Hitler. Each also favored Russia. By avoiding defeat the Russians drained the resources and vitality of an initially stronger enemy. Day by day, the strength and vitality of the invaders were drained, losses increased, and expectations diminished. Failure to win a quick and decisive victory forced Napoleon's France and Hitler's Germany into prolonged campaigns in a hostile setting that neither could conquer.

Because mobility and firepower are increasingly necessary for successful

FIG. 4.4. German columns containing both mechanized and horse-drawn vehicles advance toward Moscow in early December 1941, but the worst was yet to come. (National Archives.)

CLIMATES FACED BY AMERICAN FORCES, WORLD WAR II

In fighting against Germany, American soldiers first moved eastward in the western part of the Sahara, a low-latitude desert with hot days and cold nights. As a measure of its dryness, in advancing 1600 km (1000 miles) toward Tunis these troops never crossed a permanently flowing river. U.S. forces next advanced north and east through Sicily and into Italy, both having a Mediterranean climate characterized by hot dry summers and raw, rainy winters. The Italian Campaign (September 1943 to May 1945) is the longest continual engagement in American military history; as a result soldiers here fought through several changes in the seasons. The June 1944 landings at Normandy brought troops into the humid Marine West Coast climate of the middle latitudes, and as they advanced eastward toward Germany the onset of winter brought increasing cold that became severe with the December 1944 Battle of the Bulge.

Meanwhile, U.S. forces in the Pacific fighting the Japanese also faced a wide array of climates. For some it was the always wet tropical rainforest. Others had to cope with the wet and dry seasons of the hot tropical monsoon climate. Still others fought in the cold, harsh Aleutian Islands or at high elevations in the tropics. And as the war neared its end fighting took place in the subtropical islands of Iwo Jima and Okinawa.

Whether fighting Germany or Japan, just about all of the U.S. World War II combat troops came from a climate quite different from that in which they fought.

---

tactics, effective training and appropriate equipment are paramount in coping with battle in varying weather and differing climates. Strategy and overall operations should also be adjusted accordingly. Soldiers need to be confident that they can fight and survive in unfamiliar and hostile natural environments. Many of the improvisations in the German cold-weather handbook of 1942 are still applicable and have been used in training manuals for modern armies. Much new information is available and research continues on environmental constraints to armies. The real value of the information and research, however, depends upon its application at the strategic, operational, and tactical level, including the training of individuals.

The bane of modern technology is that it insulates humans from environmental hardship. The average citizen in an industrialized country is less and less accustomed to prolonged exposure to the adversities of nature. This condition changes immediately on the battlefield. The German campaign in the Soviet Union proved that even modern war can be reduced to a primitive struggle for survival against the elements. When technology fails, as it often does in combat, improvisation becomes the norm. Training provides soldiers with the confidence to meet all sorts of challenges. Wise planning, including considerations of the natural environment, increases their chances for success. The magnitude of Napoleon's and Hitler's problems in Russia vividly demonstrates that a responsible leader can never ignore the potential effects of weather and climate on military operations.

CHAPTER 5

Forests and Jungles

The Wilderness and the Ia Drang Valley

Where is there a field where his [Lee's] numbers would be so magnified in effectiveness and Grant's so neutralized by the natural difficulties and terrors of the woods, for dense woods do have a terror?

MORRIS SCHAFF

Vegetation, the product of heat, moisture, nutrients, and light, has affected battles and their outcome for centuries in ways as variable as the diversity in the size, shape, and density of plant growth. Although highly variable in nature, vegetation responds specifically to environmental factors. The most important are sunlight, temperature regime, moisture availability, and soil, which make its structure orderly and its distribution predictable. The thick, green cover of a dense forest or lush jungle may offer effective and possibly vital concealment from an enemy soldier. Many times the safety from observation offered by abundant vegetation has permitted the marshaling of forces for surprise attacks and facilitated the transport of supplies. But this same growth can also create serious problems in coordinating operations and substantially reduce a commander's options for maneuver and control.

In 9 A.D., Germanic tribes near the present German city of Minden annihilated Roman legions under Varius as they attempted to move through the dense Teutoberg Forest during adverse weather. The Romans, trained to fight in formation in open areas, were attacked by the tribesmen who, using the woods for concealment and as a barrier to Roman deployment, wore away at the fabric of a marching column that included not only soldiers but Varius's supply trains and the families of the troops. In the Teutoberg Forest, the Romans bitterly learned the potential impact of vegetation in "unconventional" warfare.

Nearly 2000 years later, during the American Civil War, federalists and separatists battled for Vicksburg, a Confederate stronghold on the east bank of the Mississippi River and their last point of control on that important waterway. To attack the city from the east Gen. Ulysses S. Grant ordered some of his forces upstream (to the north) of Vicksburg to move westward across the mighty river, then southward on the Louisiana floodplain, past the city, to the village of Hard Times, approximately 24 km (15 mi) south of Vicksburg. Here, near Bruinsburg, he again shuttled his army across the Mississippi, this time to its eastern bank, with plans to move rapidly north and then west to attack the city.

Once east of the river, they encountered almost impenetrable lowland thickets, the product of a long growing season, abundant moisture, and rich but poorly drained bottomland soil. After struggling across this floodplain Union troops moved onto higher ground with numerous steep hills and deep ravines where rich loess soils of western Mississippi supported abundant vegetation. Fighting through thickets and across southern woodlands and fields, Grant's forces finally captured the town of Jackson and then turned west to initiate the siege of Vicksburg, which was to end with a great Union victory in early July 1863. This great 270-degree flanking maneuver required Grant's forces to abandon their supply lines, conduct two major river crossings, and traverse some of the region's densest vegetation.

Examples of vegetation's influence on decisionmaking during World War II are numerous but nowhere is it more apparent than in British tactics on the Malay Peninsula. The British believed that the dense jungle of northern and central Malaya would prohibit a southward land attack against Singapore. As a result, in 1941 all the emplaced defense guns in the city pointed toward the ocean. In sharp contrast, during December 1941 and January 1942, a large invading Japanese army successfully moved south on the peninsula. The British, having placed only limited forces astride the two principal military corridors, expected that these small blocking units would prevent any significant advance by the enemy. Unfortunately for the English, they did not understand that the Japanese were experienced in jungle warfare and were well prepared to move rapidly and stealthily in areas thought impenetrable by others. As a result the Japanese quickly outflanked the British and captured Singapore from the rear of its strongest defenses.

Nearly three years later and halfway around the world, Adolf Hitler launched a major early winter counteroffensive through the Ardennes (see chapter 3), a densely wooded upland between the Belgian border and the Rhine River that provided ideal cover and concealment. In the 1960s North Vietnamese and Viet Cong control of the forests and jungles of South Vietnam (just as the Viet Minh had when the French were trying to reestablish their empire in Indochina during the late 1940s and early 1950s), including the establishment of the Ho Chi Minh Trail, led to much of the most fierce fighting of the war. No matter the century, on both the operational and tactical scale, vegetation can be a critical factor in combat.

In this chapter the effect of second-growth, mid-latitude forests on two savage Civil War battles in Virginia are juxtaposed with one of the fiercest encounters in the Vietnam Era, the desperate 1965 fighting in the Ia Drang Valley. In the Wilderness of Virginia, an outnumbered Robert E. Lee twice hamstrung his Union adversaries by combining large-scale maneuvers with knowledge of the area's trails and vegetation. About 100 years later, in South

Vietnam, the regular forces of North Vietnam battled the U.S. 1st Air Cavalry Division in the first major test of air-mobile operations, tactically designed to counter the friction of the terrain and the cover and concealment offered by elephant grass, jungle, and triple-canopy forests of Southeast Asia.

Vegetation

Except for places with most extreme conditions, such as permanent ice and snow cover, ongoing volcanic action, or severest aridity, all land supports some form of vegetation. The factors that most influence the amounts of flora—plant life—in a given area are solar insolation, which affects photosynthesis and temperature, and the availability of liquid moisture. Because these two elements are also basic to long-term weather, the distribution of vegetation assemblages is closely related to the world's climatic regions (Map 5.1). However, variations in local environmental factors (i.e., soils, exposure, and topography) often influence and complicate distributions in vegetation and biomass.[1]

The annual temperature regime is especially important to plant characteristics and growth rates, with each species having an optimum temperature range for its survival. In humid middle-latitude areas with snowy winters, many types of trees adapt by shedding their leaves and becoming dormant during the cold season. This frigid weather slows growth in many plants; it also tends to reduce the number of species that will survive in the region. In warmer moist climates, growth may continue, more or less, year-round, and here the biomass tends to be relatively large and species more diverse.

It is important to note that the moisture evaporation rate, which is closely related to temperature, can have a controlling effect on vegetative types and distributions. That is, the warmer the temperature, the greater the amount of evaporation and plant transpiration. The extreme response exists in areas where the annual evaporation potential far exceeds yearly precipitation, the result being xerophytic plants especially adapted to the desert.[2]

Only in remote parts of the world does the earth's vegetation remain truly natural. At most places humans have destroyed the original vegetation for agriculture, grazing, or fuel, or have harvested it as a product. In some subsistence cultures, agriculture is based on shifting crop sites. The natural vegetation is destroyed, the residue burned, the area cultivated until nutrients have been depleted and the soil no longer productive. The land is then abandoned and the process repeated elsewhere. Once agriculture ceases, second-growth plants may become reestablished to begin a succession that may eventually evolve into a mature, natural, climax forest. During the initial stages, when plants are small and canopy sparse, a wide variety of native species have ample opportunity to grow into dense thickets. Later, as the branches and leaves of higher trees capture increasing amounts of sunlight, the lower, less hardy growth beneath the canopy diminishes.

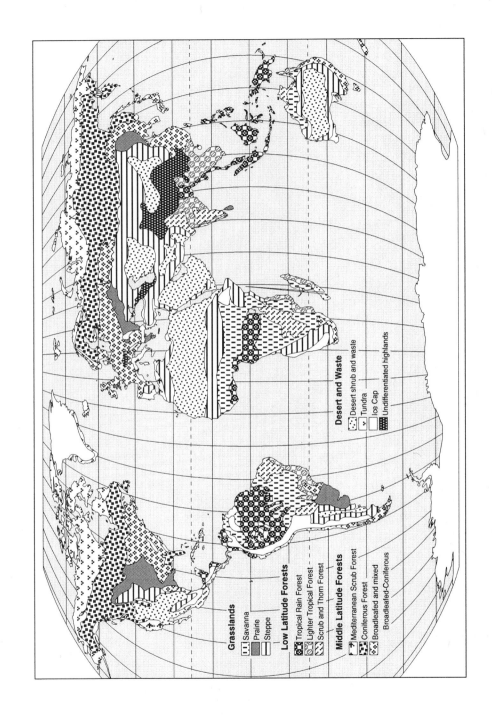

Grasslands
- Savanna
- Prairie
- Steppe

Low Latitude Forests
- Tropical Rain Forest
- Lighter Tropical Forest
- Scrub and Thorn Forest

Middle Latitude Forests
- Mediterranean Scrub Forest
- Coniferous Forest
- Broadleafed and mixed
- Broadleafed-Coniferous

Desert and Waste
- Desert shrub and waste
- Tundra
- Ice Cap
- Undifferentiated highlands

Warfare in forests has always been influenced by the size and distribution of the various species. Dense growth limits visibility, restricts small arms and artillery fire, and obstructs vehicles. Conversely, trees also provide cover and concealment, can enhance defensive positions in many ways, and may enable surprise attack and ambush. The two following battlegrounds, one in the middle latitudes, the other in the humid tropics, show how dissimilar forests, whether cut over or natural, can guide and shape military operations and tactics.

The Wilderness

At the time of the Civil War, the Wilderness was a dense vegetated thicket that evolved after humans decimated the original mid-latitude forest.[3] For more than 100 years furnaces smelting ore from the mines in this area had consumed charcoal derived from Wilderness timber. This irregularly shaped area west of Fredericksburg was approximately 23 km (14 mi) from east to west and 13 km (8 mi) from north to south. The natural successors to the harvested trees consisted of second- and third-growth pine and black oak plus a wide variety of less hardy species set among a thick scrub undergrowth. The upland soils of the region, being either highly acid clay or loamy, were low in both organic content and nutrients, characteristics that retarded the reestablishment of larger trees.

The thicket impeded cross-country movement while briars and thorny bushes tore at clothing and shoes. The density of the immature trees also brought early twilight and darkness. Moreover, plentiful rainfall had eroded the surface into dozens of ravines that in some places contained alluvial soil far richer than that on the adjacent uplands. As a result, these gully bottoms supported even denser vegetation.

Clearings were generally small, few in number, and next to widely scattered houses. Numerous narrow trails, used over the years by loggers, crisscrossed the region in a fashion obscure to the newcomer. The two most important crossroads were marked by the village of Chancellorsville to the east and Wilderness Tavern to the west (Map 5.2).

In such a forest, cavalry would not be able to race cross-country, supply wagons would be even more restricted, any artillery would have to be positioned in clearings to permit unobstructed firing, infantry movements would be slowed by the dense vegetation, and visual communication would be severely limited. The abundance of timber, much of it already felled for other uses, aided in the preparation of defensive breastworks, abatis (cut trees and branches arranged in an interlocking pattern), and barriers. Altogether these conditions presented potential disaster for an attacking force. Conversely, the dense vegetation of the Wilderness offered advantages to the defender and those who knew of its myriad trails and pathways. It was in this

MAP 5.1. World vegetation regions. (opposite)

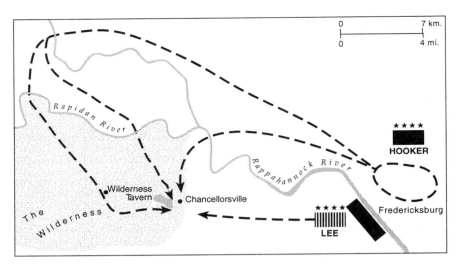

MAP 5.2. Hooker's initial maneuver and Lee's response leading to the first battle in the Wilderness near Chancellorsville in May 1863. The darker shading identifies the battle area shown in Map 5.3.

area, the Virginia Wilderness, that two of the greatest battles of the American Civil War were fought.

THE FIRST BATTLE OF THE WILDERNESS

In the spring of 1863, Gen. Joseph Hooker assumed command from Gen. Ambrose Burnside, who had directed the abortive attempt to seize Fredericksburg in December 1862 and the infamous Mud March of January 1863 (see chapter 2).[4] With a morale-stressed army in winter quarters at Falmouth just north of the Rappahannock River and Fredericksburg, Hooker effectively reorganized his forces and planned an offensive that was to commence in late April.[5] First moving westward on the north side of the Rappahannock River, Hooker then turned south to cross the Rappahannock River on 29 April. With a force of nearly 120,000, his objective was to attack the left flank of Lee's army of 60,000. With superior numbers, Hooker was well disposed for combat if the battle was joined in the open. Should this happen, the Union forces would have a distinct advantage over the Confederates.

On 30 April 1863 Union soldiers moved into the dense woods of the Wilderness to take up positions near Chancellorsville. Then, on 1 May, in an effort to engage Lee, advancing Union units, slowed and scattered by vegetation and terrain, met a large Confederate force sooner than expected. Upon learning that his forward elements met strong rebel resistance, Hooker ordered his troops to pull back to Chancellorsville and take up positions along two of the several roads leading into the settlement (Map 5.3). Once established, the Union left flank extended to the Rappahannock River, making it secure. The right flank, however, which was aligned westward along a turnpike, was "in the air," being unprotected by strong defensive positions or his cavalry. Across the lines and unable to see his opponent through the dense forest, Lee ordered scouts to determine Union deployment.

Recognizing that Hooker's right flank was exposed, at 0600 on 2 May Lee ordered Gen. "Stonewall" Jackson's corps to move first to the southwest and then northward along a series of trails and vegetation-canopied roadways to a position facing the right flank of the Union line. To Union observers Jackson's initial movement appeared to be a withdrawal from the line of battle. Within a short distance, however, the woods screened these troops while the Confederate cavalry skirmishers familiar with operations in the forest drew Union attention away from the maneuver. Hooker seems to have consequently concluded that his right (west) flank would not be threatened by the retrograde movement of Jackson's troops. To make matters worse, most Union cavalry were absent on a fruitless raid and the few remaining horsemen found movement in the forest so difficult that they had reduced their patrols. This left the Union general without these needed eyes. Hooker later reported "woods so dense that [Jackson] was able to mass a large force whose exact whereabouts neither patrols, reconnaissances, nor scouts ascertained."[6]

By late afternoon on 2 May Jackson's troops had completed their flanking maneuver. Then, at about 1800, with complete surprise, they smashed into the right flank of Union General Oliver Otis Howard's XI Corps. In the confusion of dusk and the thicket and undergrowth of the Wilderness, and operating without fear of heavy fire from artillery restricted by the forest, Jackson rolled up the right side of the Union line. Then with assistance from Confederate elements to the east, he knocked the XI Corps out of action and compressed three of Hooker's other corps into a 10-km^2 (4-sq-mi) forest area.

The battle continued throughout 3 and 4 May as Hooker's flanks folded northward toward the Rappahannock. Even in withdrawal the forest was a nemesis, restricting Union mobility and bringing death to many of his wounded who lay helplessly trapped by ground fires that blazed in the woods

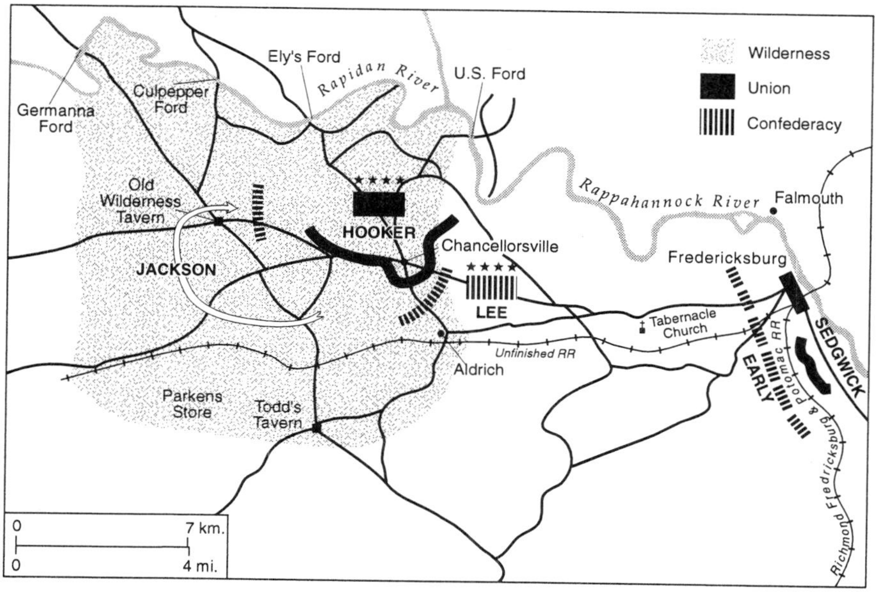

MAP 5.3. Union positions and Jackson's flanking movement during the May 1863 battle at Chancellorsville.

between the opposing front lines. On 5 and 6 May, Hooker withdrew north of the Rappahannock, ending the first Battle of the Wilderness.

The Confederate victory at Chancellorsville reflected in no small part the tactical abilities of General Lee and his subordinates and the Union leader's lack of aggressiveness. Nevertheless, the victory owed much to a dazzling flanking maneuver that took full advantage of the area's vegetative cover.

Gen. Winfield Scott Hancock, after the battle, reported that the Wilderness

was covered by a dense forest almost impenetrable by troops in line of battle, where maneuvering was an operation of extreme difficulty and uncertainty. The undergrowth was so heavy that it was scarcely possible to see more than 100 paces in any direction. The movements of the enemy could not be observed until the lines were almost in collision. Only the roar of the musketry disclosed the position of the combatants to those who were at any distance.[7]

Gen. Abner Doubleday, a Union division commander at the 1863 Chancellorsville battle, found that the "dense and almost impenetrable thickets . . . had a tendency to break up every organization that tried to pass through them. . . . It was worse than fighting in a dense fog."[8]

General Hooker had developed a masterful plan that he failed to execute. Success depended upon effective maneuver and coordinated attack by a large force, elements that were thwarted by a combination of poor leadership and the Wilderness. An initial vigorous and sustained advance would have taken Hooker's troops through the dense woodland and onto terrain that suited his plan, possibly turning the battle in his favor. Instead he let the vegetation "fix" his army, leading to Stonewall Jackson's heralded flanking operation and Robert E. Lee's greatest battle of the war. In the end, however, General Hooker did not credit the Confederates for his defeat. Instead he blamed his failure on the dense forest that deprived him of the ability to maneuver.[9]

THE SECOND BATTLE OF THE WILDERNESS

One year later, Union forces, then commanded by Gen. Ulysses Grant and led by Gen. George Meade, again crossed the upper Rappahannock to attack Confederate forces deployed south of the river west of Fredericksburg (Map 5.4). Since the fighting 12 months earlier, Lee and his command had ample opportunity to reconnoiter the area in and around Chancellorsville. Union units, on the other hand, had only a brief and unpleasant encounter with the Wilderness, and most of that was limited to the eastern part of the forest. Surely Lee would have favored another battle here, whereas Grant would have been wise to avoid it and seek combat on open ground.

After crossing the Rapidan River in force, Meade's forward corps paused at Chancellorsville and Wilderness Tavern so that supply trains could pull up behind the combat elements. The Union plan was to move out of this area the

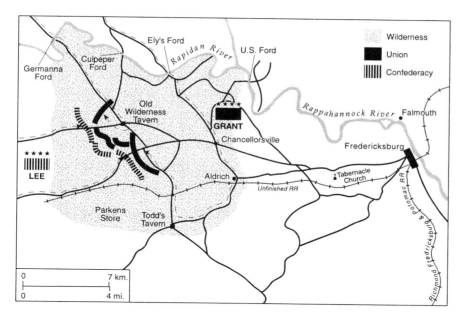

MAP 5.4. Initial positions in the second battle of the Wilderness, May 1864. Dashed lines show routes of approach by both sides.

following day and confront Lee farther to the south and east, in terrain more favorable for the Union forces.

Lee was not going to let that happen. Responding quickly he ordered Generals Richard Ewell and Ambrose Hill to place their soldiers in blocking positions astride the western approaches to the Wilderness, threatening the flanks of the advancing Union forces who, sensing the danger, prepared for attack. Although Gen. James Longstreet's corps had not yet arrived on the battlefield, putting the Confederate army at two-thirds strength, Lee ordered an advance toward an offense-minded Union force, neither side knowing its enemy's position until direct contact was made.

On 5, 6, and 7 May 1864, fierce fighting raged less than 8 km (5 mi) west of the 1863 battleground. With a force of 60,000 and outnumbered 2 to 1, Lee ordered continued engagement with Grant's forces, again hoping to use the woods and terrain to his advantage and repeat his victory of the year before. Soon after the battle began, the Southerners were driven back to defensive positions as the intensity of the fighting increased. In the process the effect of the dense vegetation and rough terrain became increasingly apparent. Units became separated, troops became lost, artillery was ineffective, communication and supply lines were broken. For two days the battle raged. Time and time again, the friction and confusion of the forest interrupted and directed the flow of battle as attacking forces unknowingly shifted direction while maneuvering. For example, on 5 May a Union division under Gen. James Samuel Wadsworth, while advancing through the woods, changed direction of march so as to expose its flank to Confederates who quickly attacked with devastating results.

Union plans to assault the center of the Confederate line on 6 May were disrupted when Burnside's corps failed to assume its position on schedule. Reportedly, the thickness of the forest caused what should have been a 2-hour (5 km or 3 mi) movement to take 18 hours, making the force unavailable at the time it was needed for the battle. Repeatedly, attacking forces of both the North and the South were disrupted by the nature of the environment. The cavalry had to dismount and therefore could offer little advantage or information. The ravines and narrow trails hid movement: three brigades of Longstreet's command marched unobserved along an unfinished railway around the flank of its Union opponents. In the first 48 hours of battle 4000 were dead, 16,000 wounded, and 6000 missing, divided almost equally between the opposing forces.

By the morning of 7 May, Grant recognized that further fighting where the "tangled forest inevitably disordered the attacking forces as they advanced, was not judicious."[10] But the Union army was not defeated and did not withdraw northward as other Northern forces had done in the past. Instead, Grant shifted units east and south toward Richmond to fight again at Spotsylvania only a few days later. This simple but costly tactic also permanently changed the war in the Eastern Theater. By remaining engaged and limiting Lee's mobility, Grant was converting a conflict previously dominated by Napoleonic maneuver into a death grip of attrition. He was also escaping the confinement of the Wilderness.

Air Mobility Meets the Rainforest and Jungle

By 1965, the conflict in Southeast Asia was escalating rapidly.[11] Large numbers of U.S. ground forces moved into coastal positions in support of the South Vietnamese military. Meanwhile, both the North Vietnamese Army (NVA) and the Viet Cong operated widely and effectively in the inland areas under the protective cover of tropical rainforest and jungle. North Vietnamese General Vo Nguyen Giap did not consider the arriving U.S. troops a formidable opponent:

> The organization as well as the composition and training of the American Army, generally speaking, are more or less unfit to help deal efficiently with our entire people's revolutionary war, not to mention great difficulties due to unaccustomed terrain and climate, and to the considerable needs in supply and logistics.[12]

The introduction of U.S. division-sized units into South Vietnam in 1965 soon tested Giap's view that the terrain would continue to offer his forces the same high level of security provided in battles against the Army of the Republic of Vietnam (ARVN). The first major U.S. unit deployed to South Vietnam was the 1st Air Cavalry Division. Equipped with more than 400 helicopters and fresh from two years of testing new concepts of air mobility in the United States, the division sought to apply those tactics to combat. Deployed in the

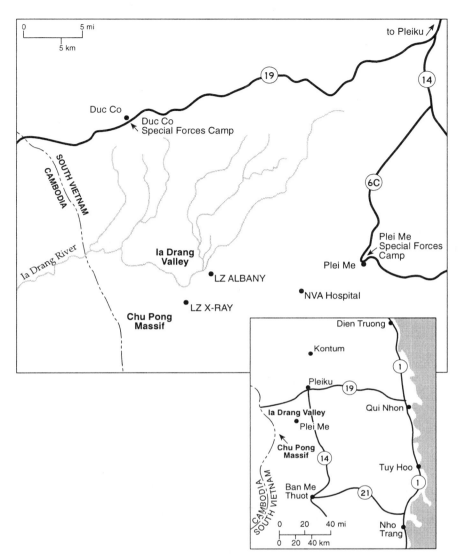

MAP 5.5. The Ia Drang Valley, 1965.

summer of 1965 to the central highlands of South Vietnam, the division stood ready to engage the enemy.

Meanwhile, as the dry monsoon arrived, the 32nd and 33rd regular-army NVA regiments crossed the Cambodian border and assembled on and near the Chu Pong massif that overlooks the Ia Drang Valley in the west-central highlands of what was then South Vietnam. From there, on 19 October 1965, elements of the NVA 33rd Regiment attacked a nearby ARVN special forces camp at Plei Me which had been established in an attempt to control enemy infiltration (Map 5.5). Immediately, and as the NVA expected, South Vietnamese forces in Pleiku were ordered to the rescue. While the 33rd continued its assault on the camp defenders, on 23 October the 32nd Regiment ambushed the relief force that was moving from Pleiku to Plei Me.

Unknown to the NVA, units of the 1st Cavalry were also standing by at Pleiku, and on 24 October these troops were moved to support both the relief column and the besieged camp. Carrying artillery by air to within range of Plei Me and the ambush site, the 1st Cavalry placed heavy fire on the attackers. Within hours the NVA withdrew into the jungle cover of the Ia Drang Valley to positions on and east of the Chu Pong massif.

Recognizing that the North Vietnamese force was substantial, Gen. William Westmoreland, commander of the U.S. Military Assistance Command, assigned Maj. Gen. Harry W. O. Kinnard, commander of the 1st Cavalry, the mission of searching the Ia Drang region to find, pin down, and destroy the NVA units in what was to become the first major test of the air cavalry concept.

Ia Drang Valley vegetation included dense jungle, mature rainforest, and widely scattered small grassland openings. Major General Kinnard saw it as

> devoid of roads indeed, even trails, it is cut by numerous streams, flowing generally from northeast to southwest, forming hundreds of small compartments . . . [and] nearly 80 per cent is covered with medium to heavy upland jungle, with trees towering as high as 100 feet above the jungle floor. Those areas devoid of jungle growth generally were overgrown with shoulder-high elephant grass, shrubs and bushes.[13]

For nearly two weeks, air cavalry and air-transported ground units crisscrossed the region, seeking signs of the NVA presence. They flew at treetop level, along river valleys, bobbing up and down from clearing to clearing looking for movement, signs of habitation, or an enemy response. Encountering such a high density of helicopters for the first time, some NVA soldiers fired at the low-flying aircraft, giving away their positions.

In response to this ground fire, air-transported infantry elements of the air cavalry squadron were inserted into clearings to find and make contact with the enemy. Other units, also brought in by air, took up ambush positions and waited for passing NVA soldiers. Initially, no major forces were encountered, but several small firefights did develop in the valley.

On 1 November air cavalry flying along a riverbed noticed activity in the adjacent treeline and put infantry forces on the ground. After a brief battle, Americans captured an NVA field hospital that had been moved forward to support the Plei Me operation. Information generated from prisoners, as well as other intelligence, indicated that the NVA was regrouping near the Chu Pong massif. Over the next two weeks, elements of the 1st Cavalry continued their search for the enemy. When air cavalry units encountered any signs of the NVA, they placed aerial rocket and machine-gun fire on the sighting, increasing the pressure on these withdrawing units. Except for an intense company-sized engagement on 6 November, these search missions drew only limited NVA response.[14]

Late during the morning of 14 November, as the search for the NVA moved farther to the west, soldiers of B Company, 1st Battalion, 7th Cavalry began a helicopter assault to LZ (landing zone) X-Ray, a 100 × 200-m (330 × 660-ft) clearing near the base of the Chu Pong. Although it was the largest of the few clearings in the area, X-Ray's limited size restricted the number of helicopters that could land simultaneously to eight. Abundant eye-height elephant grass in the clearing limited visibility. The opening also had a central cluster of trees, a number of man-sized termite mounds (often referred to as ant hills), and a shallow dry stream course. The best cover was in the dense vegetation surrounding LZ X-Ray and the nearby massif where, unknown to the Americans, three NVA battalions were positioned.

Initially unopposed, the U.S. soldiers rapidly cleared the area, took up defensive positions, and awaited the arrival of the successive flights that would bring the remainder of the battalion to the landing zone in 30-minute intervals, the time it took for a round-trip to Plei Me. However, NVA troops on the massif quickly moved toward the clearing and about noon the first shots were fired. By mid-afternoon the U.S. battalion of less than 500 men was opposed by several thousand enemy soldiers.

Through the day the fighting intensified. The battalion, plus about 100 reinforcements from a sister unit, the 2/7 Cavalry, faced an enemy on three sides. One platoon became separated to the north and was destroyed. Casualties mounted rapidly, the only American consolation being that the opponent's rate was probably several times greater. To support the troops, U.S. artillery airlifted to LZs east of X-Ray placed heavy fire on enemy positions, helicopters attacked with aerial rockets and machine guns, and Air Force fighter aircraft and B-52 bombers pounded the NVA. Those in the clearing were hampered in their effort to direct artillery and close air support because of the dense vegetation. While they could hear the explosions of incoming fires, they could not see where they hit, thus limiting their ability to report needed adjustments.

As daylight faded the firing continued, even though neither side could easily see the other because of the elephant grass, undergrowth, and increasing darkness. All through the night the two sides jabbed at each other, with several NVA attacks slowed by artillery and thrown back by close combat. Just after dawn on 15 November, the NVA initiated a fierce three-hour attack that penetrated the peripheral artillery zone and threatened to overrun several battalion positions, especially at the center. Finally, about 1000 hours, most of the NVA withdrew, bringing a lull to the battleground. At the same time, elements of the U.S. 2nd Battalion, 5th Cavalry, approached overland to reinforce the troops at LZ X-Ray.

As the day ended, the U.S. troops prepared for an NVA night attack. It came about 0430 on 16 November and was renewed at 0630 with a costly 15-minute frontal assault. When that attack failed, the bulk of the NVA withdrew to the Chu Pong massif, ending intense combat at LZ X-Ray. At that time there was good reason for declaring a U.S. victory, even though casualties amounted to

79 dead and 121 wounded (the NVA lost an estimated 1300 dead). The 1st Battalion had demonstrated that the air-mobile concept worked and that a greatly outnumbered unit could be combat-effective if firepower was employed as a force multiplier. Even more important, the battalion had not been defeated at LZ X-Ray. But, unbeknown to the Americans, fighting in the Ia Drang Valley was not over and the outcome of the next day's events would confuse the issue of victory forever.

As a third battalion (2nd Battalion, 7th Cavalry) moved into LZ X-Ray, the battle-weary 1st Battalion, 7th Cavalry (plus their earliest reinforcements) were removed by helicopters about noon. At 0900 on the following day (17 November) the two remaining battalions were ordered to leave the area before a B-52 bombing of the Chu Pong massif scheduled for 1100. Initially both units traveled overland along the same route but after about 3 km (2 mi) they separated, the 2/7th going toward LZ Albany. En route, this unit captured two of several NVA who were thought to be fleeing their unit. Unfortunately for the battalion, these men were not deserters, but instead were part of a large, nearby, well-organized force.

Around 1300 the 7th halted as it approached LZ Albany and most of its officers moved to the front to confer. Resting and strung out along 500 m (1600 ft) of the path, none was aware that the 8th Battalion of the 66th NVA Regiment had previously moved into ambush position along the trail. It was at this time that the NVA simultaneously attacked the entire column, quickly isolating some units. Through stealth and surprise, followed by close and intermingling combat, the NVA rendered U.S. artillery and air ineffective in fighting reminiscent of the most intense battles in Virginia's Wilderness. Again, as at LZ X-Ray, trees, shrubs, ant hills, deep grass, and brush became vital tactical factors. When the fighting stopped near dawn the following day, the U.S. battalion had 272 casualties including 151 dead.

The fighting had been very costly during those five days in the Ia Drang Valley: 305 American dead versus an estimated 3500 for the NVA. Fighting there also redefined the war rather than moving it toward an end. General Kinnard saw that through air mobility the 1st Cavalry was "freed from the tyranny of terrain." Years later, historian Gen. Dave Palmer labeled the fighting in the Ia Drang as "one of the war's rare decisive battles." Clearly air mobility had changed the way in which war would be fought in the jungle.[15]

However, George C. Herring, writing about the Ia Drang as a "First Battle," saw it as a deceptive victory for the Americans. While clearly proving that U.S. forces could by helicopter traverse broad areas of dense vegetation and rough terrain, it did not assure that, once on the ground, they could move into the jungle beyond the landing zone and destroy the enemy. Perhaps, as military analyst Harry Summers notes, it led the U.S. high command to believe that "we couldn't lose" and thus, erroneously conclude that if we could not be defeated that we would win the war.[16]

Conclusions

Despite dissimilarities of century, hemisphere, weaponry, strategy, tactics, terrain, and missions, combat in the Wilderness and the Ia Drang Valley have some intriguingly common attributes. Although these battles were not decisive, they did introduce a new and escalated phase into their respective conflicts. The Wilderness first brought on Lee's second invasion of the North and Gettysburg; one year later it moved the conflict more closely toward "total war." Ia Drang introduced large-scale air-mobile American combat involvement into Vietnam.

Although air-mobile operations proved effective for quickly moving light forces over hostile and difficult terrain, once combat began it quickly reverted to more traditional tactics. This is well illustrated in the overland (rather than helicopter) U.S. withdrawal of the 5th and 7th Cavalry from LZ X-Ray; once these battalions were restricted to ground maneuvers they were susceptible to some of the same kind of fighting that took place in the Wilderness more than 100 years before.

Possibly the greatest similarity between fighting in the Ia Drang Valley and the Wilderness was the effect of vegetation on battle. Although quite unlike in terms of biomes and species, the size, density, structure, and distribution of plants guided maneuvers and influenced tactics. Combat depends primarily upon training, strength, missions, commanders, and soldiers, yet in many situations environmental factors, often considered passive, become effectual elements. Vegetation did exactly that in Virginia's Wilderness and Vietnam's Ia Drang Valley.

CHAPTER 6

Terrains and Corridors

The American Civil War's Eastern Theater
and World War I Verdun

Terrain is not neutral—it either helps or hinders each of the
opposed forces.

U.S. ARMY *FIELD MANUAL 100-5* (1993)

Washington, D.C., and Richmond, Virginia, are on approximately the same
meridian, separated by about 160 km (100 mi) of low, rolling hills bisected by
the southeast-flowing Rappahannock River and its tributaries. Just east of both
cities is a low, poorly drained coastal plain with its Chesapeake Bay, by far the
largest and most deeply penetrating embayment on the Atlantic shoreline of
the United States. In contrast, toward the west, across the higher Piedmont, are
outliers and ranges of the rugged Blue Ridge Mountains. Even farther west is
the distinctively linear terrain of ridges and valleys bounded by the imposing
Allegheny Front of the Appalachian Plateau, which for many years discouraged movement farther westward.

In this varied, yet organized setting the eastern campaigns of the American
Civil War evolved. Here, each side was vitally concerned with protecting its
capital while conducting maneuvers designed to engage and defeat the enemy.
Within a 15-month period early in this conflict, both Washington and Richmond were threatened by large-scale flanking movements. The Union tried it
first with an amphibious operation by way of Chesapeake Bay, enabling them
to attack westward toward Richmond on land between the James and York
River embayments. After this operation failed, Gen. Robert E. Lee's Confederate forces twice invaded the North to threaten its security and lines of
communication.

The Union's operation produced the Peninsular Campaign and the Confederate invasions resulted in major battles between Antietam Creek and Sharpsburg in 1862 and at Gettysburg in 1863. The Union's Peninsular Campaign
shows how a land army can utilize a body of water as a route for offensive
purposes. In contrast, Lee's two invasions of the North illustrate that, when
used wisely, compartmentalized landscapes can increase mobility for both
offensive and defensive purposes, while also keeping an enemy off balance and
puzzled. Whether by water or land, the strategy and tactics of these wide-
ranging campaigns were guided as much by geography as by the generals
themselves.

In contrast, on the Western Front during most of World War I such wide-ranging maneuvers quickly became impossible. After 1914, quite unlike the Civil War, this battle line had no flanks; instead it extended without interruption from the English Channel to neutral Switzerland. Near its center at Verdun, fortified ridges and escarpments presented formidable barriers to movement in adjacent lowlands, minimizing the opportunities for obscured maneuver that Lee had executed so effectively in his invasions of the North. The result was that, as elsewhere during the Great War, most attacks were frontal and losses high. Over a 10-month period in 1916, the French and Germans suffered huge casualties along a 24-km (15-mi) front that, in the end, changed position very little. For General Lee the terrain of the Appalachians presented great opportunity. At Verdun it was part of a terrible problem.

Rock, Structure, and Topography

All continents have a foundation of very old, exceedingly thick, relatively resistant, highly deformed rocks that weigh somewhat less than oceanic and more deeply buried crustal material. As a result the less-dense rocks of the continents, being somewhat like a barely floating raft, tend to rise to a slightly higher altitude, often projecting their surfaces above sea level.[1] Current tectonic theory proposes that the continents are also associated with moving segments of the earth's crust, which are called plates. Six major and many smaller plates are now recognized. Because these are all experiencing relative movement, they must separate, collide, or slip past one another (see Map 6.1 and Figure 6.1).[2]

Where plates converge forcefully the compression deforms the crust accordingly. The collision is accompanied by breakage and slippage, which generates shock waves that produce earthquakes. At the same time, deep in the subsurface, vast masses of molten material form which may intrude upward and slowly solidify at depth or find a way to the surface to erupt as lava or volcanic ash and dust. Elsewhere, such as along California's San Andreas fault zone, these crustal plates slide side by side, also occasionally producing earthquakes. Finally, plates may separate from one another to produce rifts at the surface, along with deep, tensional ruptures that provide routes for huge amounts of molten material to rise from great depths and pour out on the earth's surface. Thus, a continent, riding along with a moving plate, may converge on one land mass, slide past another, and at the same time become more distant from a third.

Because the foundation rocks of all continents are very old and plate movements are believed to have taken place through most, if not all, of Earth's history, these parts of the crust may provide evidence for ancient episodes of crustal deformation that, once introduced into bedrock, remain until being either destroyed by erosion or altered beyond recognition. Thus, rocks of a continent may show both the effects of ongoing deformation and relict effects of geologically ancient tectonic events. Whether old or modern, these scars form distinct geologic patterns that can greatly influence the topography of the land.

Both old and new convergent plate boundaries tend to be curvilinear, forming great salients. The arc of the Aleutian Islands marks a currently active boundary between the American and the Pacific plates. Interestingly, the rock and structure of the Appalachian tectonic system in eastern North America reveal several similar salients, but these are all associated with a long-inactive, convergent plate boundary.

While severe crustal deformation creates mountain ranges and complex linear structural belts at the edge of plates, much more broadly based crustal uplift or subsidence can also occur to form arches, domes, troughs, and basins. These sorts of structures are commonly associated with the interior areas of continents and, like intensely deformed marginal salients, are also revealed by the regional geology of an area.[3]

All of this becomes more relevant to settings for certain battles when erosion is considered. Once elevated above sea level, whether at plate borders or within continents, weathering and erosion etch their effects into surface rocks. Because of differences in composition, rocks of the earth's crust vary greatly in resistance. With the passage of time, decay and wear can gradually carve out an erosional landscape in one place while depositing the removed material in catchment areas that often form relatively flat plains elsewhere. In rugged, deeply eroded areas, stronger rocks are generally associated with the ridges,

MAP 6.1. Major tectonic plates with their relative movements shown by arrows.

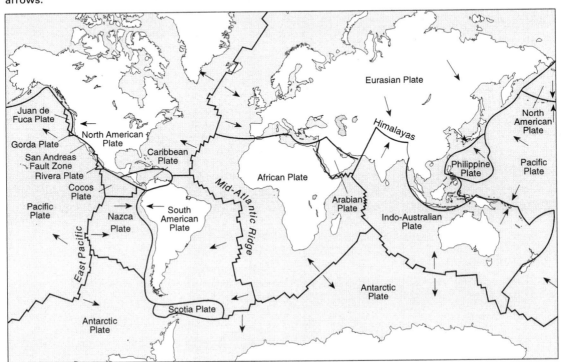

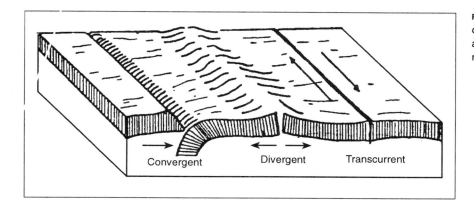

FIG. 6.1. Three types of plate boundaries and their possible relative movements.

Convergent   Divergent   Transcurrent

uplands, and mountains, while lowlands and valleys are usually developed on weaker, more easily eroded bedrock. As a result, the topography often reflects structures in the earth's crust.

Equally important is the fact that each of the erosional processes (including the effects of mass wasting, fluvial activity, glaciation, wind, waves, and groundwater) leaves distinct imprints on the land which are different from all others. Furthermore, their effect likely evolves with the passage of geologic time, leading to a gradual progression of "stages" in landscape development.

Traditionally the landforms of an area have been explained as the combined product of "structure, process, and stage": structure being rock type and its configuration; process involving all types of natural erosion and deposition; and stage referring to phases of progressive geomorphic development. Although this "formula" is simplistic and must be applied with care, it does provide an overview of genetic geomorphology and terrain development.

In this fashion, deformed weak and strong rocks were deeply eroded to form the compartments and barriers in the Appalachians in Virginia and the hills and lowlands around Verdun in northeast France. The first is a region of highly deformed rocks pressed tightly together and later torn apart through plate tectonics. In contrast, Verdun is on the east margin of an elliptical structural basin containing layers of broadly warped, stream-eroded, sedimentary strata. Knowledge of the rock, structure, geomorphic processes, and related topography at both places enhances understanding and analysis of geographic factors that influenced three early Civil War invasions and the incredibly costly attacks near Verdun in 1916.

The Eastern Theater of the American Civil War

The Civil War's Eastern Theater is composed of five physiographic provinces that form three barriers, and two corridors (or compartments). From east to west the provinces are the low, poorly drained Coastal Plain with its highly irregular coast; the adjacent and slightly higher and rolling Piedmont, which includes some areas underlain by Triassic rocks; the narrow but prominent

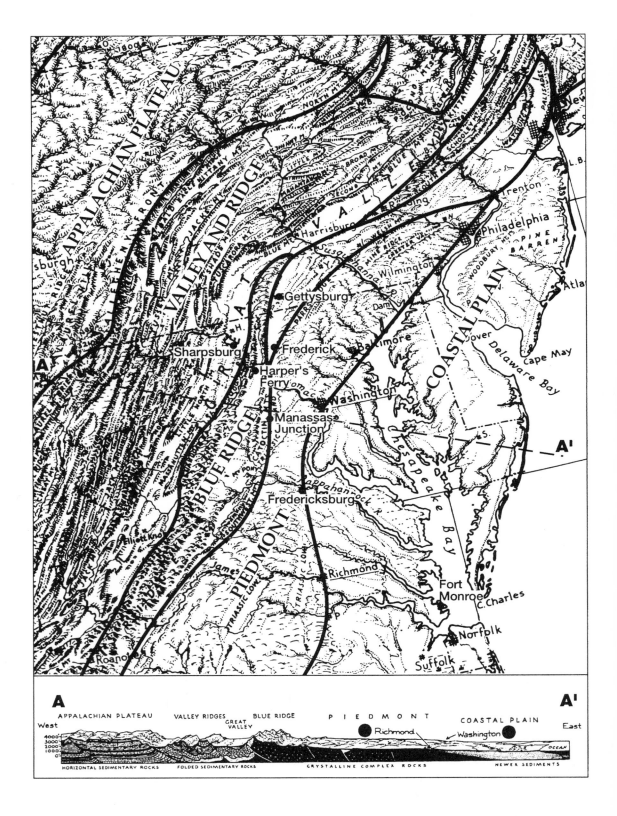

A
West

APPALACHIAN PLATEAU   VALLEY RIDGES   BLUE RIDGE   P I E D M O N T   COASTAL PLAIN

GREAT VALLEY

Richmond   Washington

A'
East

4000'
3000'
2000'
1000'
0'

OCEAN

HORIZONTAL SEDIMENTARY ROCKS   FOLDED SEDIMENTARY ROCKS   CRYSTALLINE COMPLEX ROCKS   NEWER SEDIMENTS

Blue Ridge Mountains; the Valley and Ridge Province with its fertile lowlands separated by rocky and forested linear ridges; and the distant and remote Appalachian Plateau (Map 6.2).

The northeast-southwest-trending Piedmont and Valley and Ridge Provinces form the corridors in which most of the battles were fought. The Blue Ridge Mountains form a central barrier that separates these two corridors even though it can be crossed at a number of places. The theater's western border is the high and long east-facing barrier of the Appalachian Plateau known as the Allegheny Front. And finally, the numerous embayments and inlets of the Coastal Plain form an eastern barrier to north-south movement; the only alternative would be amphibious operations.

Although this overall framework appears relatively simple, topographic details within each of these provinces were often of great importance in maneuver and battle. For that reason, knowledge of the physiography in the eastern United States adds greatly to an understanding of the Civil War's Eastern Theater.

GEOLOGY AND TERRAIN

The setting for battles in the east began to be determined hundreds of millions of years ago, when an ancestral North America, with a geography quite different in configuration and character than today's, was sliding toward a collision with its equally ancient counterparts that are now Europe and Africa.[4] These land masses eventually converged with others to form the largest known continent in Earth's history. Called Pangaea, for a time it combined all of the planet's continents into one enormous "world island."[5]

The old, thick, strong, and more stable rocks central to our continent's foundation were altered the least in this movement that added to Pangaea. But rocks along what was then the eastern and leading edge of North America, where the crust was somewhat thinner and weaker, were severely compressed and broken, the impact creating the distinct northeast-southwest-trending linear structures now revealed in the Appalachians. Some of these rocks were uplifted to form a complex mountain range that then may have been as high as today's Himalayas. Great ruptures, or thrust faults, along with giant upward- and downward-bending folds, accompanied this mountain-building as a belt possibly 300–500 km (200–300 mi) wide was eventually pressed into only one-third of its original width.

By 250 million years ago, this collision of continents had come to an end, but the tectonic effects remained permanently embedded in the rocks of eastern North America. At the same time this part of the continent had permanently risen above sea level, meaning that weathering, erosion, and off-shore deposition became the prime forces in forming the topography. Gradually, weaker rocks were removed to create lowlands, while areas underlain by more resistant material, although also deeply eroded, remained somewhat higher and more rugged. Erosion in regions of greatest uplift eventually uncovered rocks that originally formed deep in the crust where temperatures were once

MAP 6.2. Landforms and geology associated with the Eastern Theater of the American Civil War. (Base map ©1957 by Erwin Raisz; reprinted with permission of GEOPLUS, Danvers, Mass.) (opposite)

so high that the material was molten. Long since solidified, these rocks now form the igneous and metamorphic terrain of the Piedmont, immediately to the west of both Richmond and Washington. More resistant and higher than the Coastal Plain to the east, this area, after more than 200 million years of exposure to the elements, has gradually evolved into rolling hills such as those around the battlefields at Fredericksburg, Chancellorsville, the Wilderness, and Spotsylvania.[6]

Farther west a linear tract of ancient rocks even more resistant to erosion supports the Blue Ridge Mountains, which widen southward eventually to become the Great Smoky Mountains. Made up of especially strong material, these ranges are higher and more rugged than the adjacent Piedmont, thus presenting a nearly continuous obstruction to westward visibility and movement. Only one low-level water route, the gap formed by the Potomac River that contains Harpers Ferry, traverses the northern extent of these mountains. There are, however, other passes of different altitude and width, known as wind gaps, that provide westward access routes of varying difficulty.[7]

West of the Blue Ridge are folded and deeply eroded layers of thick sedimentary strata. The outcropping sandstone and conglomerate formations, being relatively strong, support numerous, linear, rocky ridges while the weaker limestones and shales underlie long, fertile valleys. The largest, easternmost, and most famous of these lowlands is the Great Valley, of which the Shenandoah Valley is an important part (see Map 6.2 above).[8]

Farthest west, for purposes of this discussion, are the relatively undisturbed and nearly flat-lying sedimentary formations that support the Appalachian Plateau with its high east-facing boundary escarpment. This remote, sparsely populated, generally infertile upland was so dissected and rugged that it proved impractical for large-scale military operations. Since for the most part the upland was strategically unimportant, combat in the province was limited largely to skirmishes and scattered cavalry raids, and these were generally near the valleys to the east.[9]

Another geologic realm, referred to as the Triassic Lowland, deserves discussion here even though it is relatively small and generally considered part of the Piedmont Province. It is treated separately because of its dissimilar geologic origin and its importance in the Civil War, most notably at Gettysburg. Its origin goes back about 180 million years. At that time the previously formed "world island" of Pangaea began to break apart, leading to the present arrangements of continents. At many places, including eastern North America, tensional forces produced numerous large wedges of down-faulted rocks, called grabens, as the land separated. As originally formed, these areas formed great linear lowlands, much like the present-day rift valleys of Africa, the low terrain occupied by the Dead Sea and the Red Sea, and the wide trough forming the southern part of Germany's Rhine River valley. As sediments accumulated in the lowlands, molten materials from deep in the earth's crust also moved upward along the weaker zones in the rock; most solidified along faults and between strata before reaching the surface. However, some of this magma flowed out as lava that quickly hardened to become especially resistant diabase and basalt. This episode of down-faulting, sedimentation, and volcanic activity continued for millions of years, tilting the rock strata as crustal movement progressed.

The accumulation of sediment and the faulting of these rocks occurred during the Triassic period of geologic time (225 to 190 m.y.) and represent the earliest events in the gradual separation of Africa and North America that formed the ever-widening Atlantic Ocean. The width of the ocean today shows how much divergence has taken place in the last 180 million years. Since that time both the Triassic rock and nearby older formations have been carried northwestward, along with the drift of the North American continent. Being above sea level much if not all of this time, the rocks have experienced much weathering and deep erosion. Where exposed at the surface, the diabase and basalt, being much stronger, formed residual hills or ridges. The relatively weak Triassic sediments (some with reptilian fossils of the Mesozoic era) were more easily eroded to form lower terrain. The relief along Bull Run near Manassas Junction, and the lower ground separating the slightly elevated Seminary Ridge from the even higher Cemetery Ridge near Gettysburg, directly reflect these differences in terrain associated with the most extensive Triassic lowland in the eastern United States.[10]

The last of the physiographic provinces to be considered here is the Atlantic Coastal Plain. Long-term erosion of all of the Appalachian provinces produced an enormous amount of sediment. Ever since the separation of North America from Eurasia and Africa, much of that material has been carried eastward by streams to be deposited and distributed along the continent's southeast shore. Over tens of millions of years these sediments have accumulated to form a thick wedge of relatively young rock that now supports the Coastal Plain. With a surface barely above sea level, this low, poorly drained, and partially inundated area is quite unlike any of the other provinces.

One last important and unusual aspect of the terrain involves the past and present river courses in and near Virginia. Although major geologic structures here extend northeast-southwest, the larger rivers of the Piedmont drain southeast, thus sharply cross-cutting geological trends. By their position, two of these rivers, the Rappahannock and Potomac, were of great tactical importance during the war. The larger of these, the Potomac, has its source far to the west of the Piedmont and Blue Ridge Mountains. Of its numerous tributaries the biggest by far is the Shenandoah River. In flowing southeastward toward Washington, the Potomac River does a curious thing. Instead of turning northward at Hagerstown, Maryland, to follow the Great Valley (thus avoiding the mountains) and drain toward Harrisburg, Pennsylvania, it continues southeast to cut, at Harpers Ferry, West Virginia, a spectacular gap through

several ranges of the Blue Ridge to form what Thomas Jefferson described in his Notes on Virginia as "perhaps one of the most stupendous scenes in nature . . . worth a voyage across the Atlantic" (Figure 6.2).

The Potomac's present-day course, like the origin of the Atlantic Ocean to which it eventually drains, was surely established long after the ancient continental collision that formed the crustal structures of the area. It is also apparent that the river could not have flowed southeastward if the higher Blue Ridge Mountains had always existed. In view of these two conditions, the only reasonable explanation for the river's present position is that, at some time in its long history, it must have flowed southeast on an inclined ancestral surface that lacked any important expression of the present-day Blue Ridge Mountains. Numerous tributaries must also have flowed southeastward on this same surface. Since then widespread uplift of the land resulted in down-cutting by all of these streams. Weaker strata were gradually eroded to a lower level while more resistant formations remained to support higher ground. Thus, the Blue Ridge gradually emerged, not as a product of intense structural deformation and mountain-building, but rather as a more resistant remnant, yet to be eroded to the level of the adjacent land.

The Potomac River, being the largest stream of the area, had the greatest available energy and thus eroded more rapidly, even on the most resistant bedrock. As this river cut downward into the resistant rocks of the Blue Ridge, one of its tributaries, the Shenandoah River, flowing over much weaker formations to the west, easily adjusted by both lowering its valley floor and expanding its drainage area through headward erosion (Map 6.3).

Meanwhile, the other smaller and less energetic tributaries, with eastward courses across the trend of the much more resistant rocks of the Blue Ridge Mountains, were not nearly as successful in cutting downward. In fact, one by one, the smallest first and progressing according to size and remoteness, the headwaters of each stream were, with time, pirated away by the ever-enlarging drainage basin of the more westward and lower Shenandoah River. As the upper headwater drainage for each stream was captured, further downward erosion of its notch through the Blue Ridge Mountains ceased.[11] Meanwhile the Shenandoah River continued to deepen, widen, and lengthen its valley.

It is in this fashion that ravines originally eroded by streams now remain as passes, or high openings, across ranges of the Blue Ridge. Geomorphologists call these wind gaps and include such familiar names as Swift Run, Manassas, Ashbys, and Snickers, all important at one time or another during the Civil War.[12] The gap at Harpers Ferry, however, was quite different and much more important than all the rest. It was occupied by the Potomac River and provided by far the easiest path through the Blue Ridge and into the fertile and tactically important Shenandoah Valley. Because of its site and situation, it was destined to become one of the most fought-over places in the Civil War.

Thus, from both Richmond and Washington, the geographical scenes to the east and to the west were diverse yet similar. In contrast, the military opportunities and problems presented to each side were quite different. For the North, with its superior naval force and the retention of its coastal outpost of Fort Monroe on the coast of Virginia in Confederate territory, Chesapeake Bay provided the potential for a flanking amphibious operation that could place Union troops east of the South's capital. On the other hand, the Confederates

MAP 6.3. Three-phase drainage capture by the Potomac River resulting in several gaps through the Blue Ridge Mountains. Earliest phase on the left, present drainage on the right. (Modified from a drawing by William J. Wayne in William Thornbury, *Principles of Geomorphology*, 2nd ed. [1969], 150.)

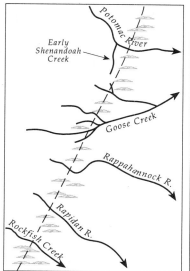
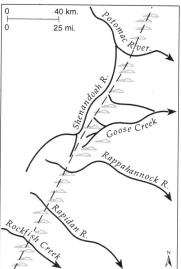
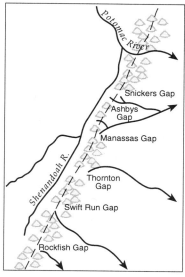

were well aware that the Piedmont narrows northward, bringing the Blue Ridge much closer to Washington than Richmond. By moving troops westward through mountain gaps and into the Shenandoah Valley, the South could use the Blue Ridge and its outliers to shield northeastward marches in the Great Valley which would, in turn, threaten the western and northern flanks of the Union capital. Conversely, a southwest Union march in that valley would lead not closer to Richmond but farther from the central area of the theater.

Though many battles of the east were fought in a relatively small tract within the Piedmont between Richmond and Washington, it is clear that lines of movement for the largest maneuvers early in the war were based, more than any other factor, on these major geographic characteristics. Early in the conflict the Union would take full advantage of the Coastal Plain and Chesapeake Bay to the east while the Confederates exploited the form and trend of the Appalachian topography.

PENINSULAR CAMPAIGN, MARCH 1862

Early in the summer of 1861, Confederate and Union forces faced each other both east and west of the Blue Ridge Mountains. Union demands for action led Gen. Irvin McDowell to initiate the first large-scale offensive maneuver, just west of Washington, D.C. When the Confederates learned of the Union plan, they moved reinforcements eastward from the Shenandoah Valley, first through the Blue Ridge at Ashbys Gap and then Thoroughfare Gap in the outlying Bull Run Mountains, to engage and defeat Northern forces near Manassas Junction at the first battle of Bull Run.

Shocked and bewildered, the Union badly needed an effective offensive if there was to be an early end to the war. The task was given to Gen. George B. McClellan, who by the following winter had devised his plan for victory. Instead of seeking out a mobile Confederate army in strong defensive positions along the Rappahannock River, he would use an amphibious flanking maneuver to capture their capital.

Recall that easternmost North America is underlain largely by relatively young, poorly consolidated sediments, in part derived from the long-term erosion of the Appalachian Mountains, which accumulated offshore in the gradually widening Atlantic Ocean.[13] Along the southeast edge of the continent, these deposits have been uplifted slightly above sea level to form the relatively wide Coastal Plain. Northward, however, the amount of uplift diminishes and is eventually replaced by subsidence.[14] As a result this plain gradually narrows and then disappears beneath the sea off the coast of New England.

Between these two extremes of uplift and subsidence is the irregular shoreline of Delaware, Maryland, and Virginia, formed from only partial submergence of the Coastal Plain. In the past, the Susquehanna River flowed within a lowland that is now Chesapeake Bay to meet the sea east of Norfolk, Virginia. In doing so, it received several large tributaries from the west, including the

James, Potomac, and Rappahannock Rivers. Later coastal submergence then drowned the lowermost sections of all these streams, eventually dismembering the lower part of the Susquehanna drainage system to create the great Chesapeake embayment.

With secession, coastal Virginia was within the northeastern boundary of the Confederate States of America, the declared border extending southward through the middle of the southern section of Chesapeake Bay. The Confederates, however, had never secured the eastern end of the large peninsula between estuaries of the York and James Rivers in Virginia. The peninsula led directly toward their inland capital, Richmond, and contained the Union-held coastal outpost of Fort Monroe. McClellan reasoned that by sailing south in Chesapeake Bay and landing troops unopposed at Fort Monroe, he would be on the enemy's flank. From there his army could invade the Confederacy and, after a mere 110-km (70-mi) advance across the Coastal Plain, attack Richmond from the east.

In early March 1862, large numbers of Union troops landed at Fort Monroe to initiate the offensive. Although delayed from early April to early May by Confederate defenses near Yorktown, they approached Richmond late that month. Fatefully, on 31 May Gen. Joseph E. Johnston, the Confederate commander who had moved his forces so effectively from the Shenandoah Valley to Manassas Junction and Bull Run 10 months earlier, was wounded while defending the city; his replacement was Robert E. Lee.

To counter this Federal offensive and relieve its threat on Richmond, Confederate plans evolved which used the topography of the Blue Ridge and Valley and Ridge Provinces to great advantage. Executed by Thomas Jonathan "Stonewall" Jackson, it became one of the most effective diversions of this or any other war. In long marches and swift encounters Jackson maneuvered widely in the Shenandoah Valley, using its intervening Massanutten Mountain and the numerous mountain gaps to preoccupy large numbers of Union troops. A geographical key to his effectiveness lay in the fact that, as mentioned previously, any northward movement in the valley brought his forces closer to Washington, thus threatening its security.

Jackson won many of his battles, and for three months his Valley Campaign harassed and confused Union leaders to the point that it prevented the major transfer of reinforcements to assist McClellan in his Peninsular Campaign.[15] For this and many other reasons, the Northern invasion across the Coastal Plain gradually lost momentum and eventually ground to a halt at the outskirts of Richmond. Lee then ordered Jackson to rejoin him, and the combined forces not only drove McClellan's troops from the edge of their capital, but also blunted and then thwarted the Union invasion.

With the Northern forces in the east separated into two major units (McClellan's army slowly withdrawing from coastal Virginia and the troops protecting the capital), Lee again moved decisively and effectively to set the scene for his first invasion of the North. Before Union troops were fully withdrawn to end the Peninsular Campaign, Lee quickly moved north across the Piedmont and once again went into battle near Manassas Junction (Second Bull Run), only 40 km (25 mi) west of Washington, D.C. The result was another Southern victory that in late August forced Union troops into defensive positions around their capital. The main Confederate force was now in northern Virginia, with the Union army to their east and an open path northward into central Maryland.

ANTIETAM/SHARPSBURG, SEPTEMBER 1862

Taking full advantage of the situation, early in September Lee directed his army of 55,000 northward across the Piedmont, to pass west of Washington, ford the Potomac, and move on to Frederick at the eastern edge of Catoctin Mountain, which is one of two ranges making up this part of the Blue Ridge Mountains (see Map 6.2 above). During this maneuver a copy of Lee's Special Order 191, dated 9 September, was misplaced in the field, only to be discovered by a Union soldier and eventually delivered to McClellan, now returned from the peninsula and directing the Union pursuit of Lee. If McClellan had moved quickly, the enemy's orders should have given him a great and possibly decisive advantage because they revealed that Lee had divided his army into three parts. One group was ordered to capture the water-gap town of Harpers Ferry with its important lines of communication and valuable supplies; another was farther north at Hagerstown; and the third was in and around Crampton's and Turner's Gaps of South Mountain, which in Maryland is the westernmost range of the Blue Ridge.

Upon learning that his orders were known to McClellan, Lee directed his forces at Hagerstown to move immediately southward. He also ordered troops at Harpers Ferry to join him after the water gap and town had been secured. All three groups were to come together near their midpoint at the town of Sharpsburg, which is a few kilometers west of South Mountain and 3–5 km (2–3 mi) east of a westward-bending meander of the Potomac River. Here, on 17 September 1862, with this large and confining river to his rear, Lee and his generals, Stonewall Jackson and James Longstreet, established defensive positions on slightly higher ground just east of Sharpsburg and prepared to meet the westward-advancing Union forces. Meanwhile, McClellan had delayed too long and lost any element of surprise offered by Lee's now-famous lost orders.

The Union's slow pursuit was at first the product of McClellan's cautiousness and later the result of an effective and hard-fought rear-guard action by Southern troops occupying Turner's Gap on South Mountain of the Blue Ridge.[16] However, this small force could only delay the Union advance which, once across the mountain, moved toward the south-flowing Antietam Creek, with its three bridges and numerous fords, to face the Confederate troops along a 5-km (3-mi) front. The right third of the Northern forces crossed upstream to take up more westerly positions, thus approaching their enemy from the northeast. Troops at the center and left occupied lower land nearer the creek.

Lee's soldiers were aligned in a crescent along slightly higher ground that formed a north-south trending drainage divide between the creek and the Potomac River.

Early on 17 September, the fighting began at the north end of the battle line and progressed southward during the day in a series of vicious but poorly coordinated and piecemeal attacks by the Union and costly yet effective Confederate counterattacks.[17] The stream-eroded landscape presented an immediate problem for the North. Antietam Creek joins the Potomac River about 3 km (2 mi) south of the battlefield. Because the Potomac is a powerful river, it has cut downward into the low, rolling hills immediately west of the Blue Ridge. As a result, the entry level of Antietam Creek has also been lowered. The lower confluence increases the creek's gradient, velocity, erosive power, and transport capabilities, allowing it to form a shallow yet steep-walled valley. This, in turn, permits incisement by the even smaller tributary creeks, creating numerous gullies that extend into parts of the battlefield.[18] These indentations offered concealment for some, but at the same time their depth, at many places deeper than the level of horizontal sight, impeded overall communication and troop coordination. In addition, Antietam Creek, with its steep banks and defensive advantage for the South, also markedly delayed a planned afternoon left-flank (south) Union attack.

In contrast, the Confederates were deployed in defensive positions with interior lines on higher ground with a better view of the battlefield than their enemy. Using all of this to advantage and practicing superior generalship, Lee adjusted his forces throughout the day to meet and stop one Union attack after another. By evening the casualties, totaling 22,000 for both sides, proved to be the greatest ever suffered in any one day by American troops.

Although the battle was not decisive, it did end Lee's first invasion of the North. On the following day Southern forces withdrew, largely unmolested, from their vulnerable position by fording the Potomac 3 km (2 mi) southwest of Sharpsburg and returning to the haven of the Shenandoah Valley.

It is conceivable that effective Union pursuit could have done great harm to Lee's Army of Northern Virginia, partially hemmed in as it was by a confining loop of the Potomac River. Such an event would surely have established McClellan as an effective general and favored future Union offensive operations in the Eastern Theater. On the other hand, outright Confederate success at Sharpsburg would have encouraged further antiwar sentiments in the North, possibly brought England into a more supportive position as a Southern ally, or convinced Maryland to secede and join the Confederacy. All three might even have produced the ultimate Southern objective, a war-ending truce.

But none of these things happened because neither side was victorious at Sharpsburg. Instead, Pres. Abraham Lincoln used Lee's withdrawal from Maryland as an opportune time to present his Emancipation Proclamation, which for the first time fully politicized the war's slavery issue. Early in November a concerned and frustrated Union president relieved McClellan of his command. Meanwhile Lee once again assembled most of his force in defensive positions on the south side of the most formidable geographic barrier between Washington and Richmond, the Rappahannock River.

Lincoln, having relieved McClellan, placed Gen. Ambrose Burnside in charge of the Army of the Potomac with orders to protect Washington and at the same time to defeat the South's Army of Northern Virginia. Burnside effectively shifted his troops southward toward Fredericksburg, Virginia, with plans to cross the Rappahannock River east of the town and attack before Lee was at full strength in established defensive positions. But here the river is both wide and tidal, characteristics that required materiel for the placement of six needed bridges. Because of the late arrival of flotation pontoons essential for bridging, Burnside's army lingered on the left (east) riverbank, thus losing any element of surprise. This delay gave the Confederates ample time to establish large numbers of troops in extremely favorable defensive positions in and near the city.

When the pontoons finally arrived, Burnside ordered an attack for 11 December. First the soldiers needed to, with no cover, cross a river more than 100 m (330 ft) wide. Those on the right flank then had to advance westward through Fredericksburg, providing an early version of urban warfare. Finally, in what amounted to a frontal attack, these troops met the largely concealed and well-protected corps of Longstreet positioned on a bluff overlooking the town and river. Each of these geographic elements was a major obstacle to the Union and strongly favored the Confederacy. After two days of fighting, the battle added to the series of formidable Eastern Theater defeats for the North.[19]

As Burnside regrouped his forces, he also planned a second attack against Lee and scheduled it for mid-January 1863. This, one of the few winter offensive operations in the Eastern Theater, ended in complete failure when a major storm with heavy rains resulted in the infamous "Mud March" described in chapter 2. For the remainder of the winter the two opposing armies bivouacked on opposite sides of the Rappahannock River near Fredericksburg.

Meanwhile an increasingly perplexed Lincoln replaced Burnside with Gen. Joseph Hooker. Hooker reorganized the demoralized Army of the Potomac and developed a promising plan for attack near Chancellorsville, about 11 km (7 mi) west of Fredericksburg, in early May of 1863. As described in chapter 5, however, the general's execution was poor while Lee's was superb, being highlighted by Stonewall Jackson's flanking maneuver, one of the most effective in military history. The result was another Union defeat.[20] As Hooker's troops retired northward toward Washington, Lee was presented with his second opportunity to invade the North. Thus began his march toward the most famous battle in American history.

GETTYSBURG, JULY 1863

Lee firmly believed that military success in the east was the key to victory for the Confederacy. Accordingly, during June 1863, in a series of carefully planned and deceptive maneuvers, he moved most of his Army of Northern

Virginia westward across the Piedmont, through Chester, Ashbys, and Snickers Gaps, and then northward in the Shenandoah Valley, crossing the Potomac River at two Maryland fords—one at Williamsport and the other on the familiar ground near Sharpsburg. Lee's troops then continued north within the Great Valley and into Pennsylvania behind the shield of the Blue Ridge Mountains immediately to the east.

From here the Confederate forces separated into groups that reached York, the outskirts of Harrisburg, Carlisle, and Chambersburg, Pennsylvania, thus threatening Baltimore and Philadelphia, as well as lines of communication with Washington, D.C. With Lee's troops in Pennsylvania, well to the north of both the Union capital and the Army of the Potomac, the situation became so frustrating for Lincoln that shortly after ordering pursuit, he replaced the army's leader, General Hooker, with Gen. George Meade, who was ordered to find and defeat Lee.

The pursuing Army of the Potomac knew neither Lee's location nor the deployment of his army. When the Confederate commander learned that Union troops were moving northward he again ordered his units to march toward a central point. In this fashion two great armies converged on Gettysburg, the South from the north and the North from the south.

Gettysburg is about midway between the east and west borders of a huge wedge of the earth's crust that subsided because of tensional faulting accompanying the separation of North America from Africa and the breakup of Pangaea that commenced some 180 million years ago. Being above sea level, streams transported sediments derived from adjacent uplands into the rift; with time these formed into weak shale and sandstone. In addition magma intruded through and between the sedimentary layers and then solidified into resistant diabase and basalt. As the continents drifted farther apart to form the ever-widening Atlantic Ocean, weathering and the elements slowly eroded the weaker sedimentary rocks to create a northeast-southwest-trending Triassic lowland that extends from northeasternmost New Jersey, southward through Pennsylvania into north-central Maryland. In most places this landscape is only slightly lower than the rolling hills of the Piedmont to the east but is well below the higher terrain associated with more-resistant Blue Ridge rocks to the west. However, in the limited areas where the resistant basalt is exposed, ridges of varying size exist, each proportional to the extent, or outcrop, of the stronger rock.[21]

The surficial igneous rock occurs in two forms near Gettysburg. The smaller of these are "dikes" of relatively thin, linear intrusions that cut through the sediments. These form the strong rocks that underlie Seminary Ridge and its northern continuation, the somewhat higher and wider Oak Ridge (Map 6.4 and Figure 6.3). Together these support a north-south-trending line of very low hills just west and northwest of the town. On the first day of the battle, 1 July, the Confederates captured Oak Ridge. On 2 and 3 July, their position was in large part on Seminary Ridge, facing Union troops to the east.

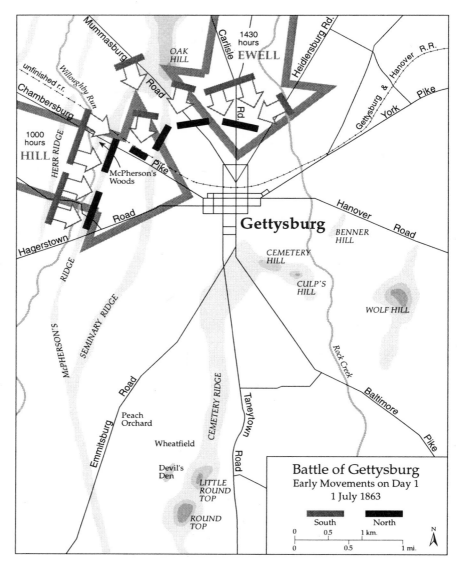

MAP 6.4. Gettysburg, nearby terrain, major roads in 1863, and Confederate attacks on 1 July 1863.

The second igneous formation, a "sill," is a mass of diabase and basalt sandwiched between sedimentary strata that dip gently westward. The outcrop of these rocks can be traced continually from Culp's Hill and Cemetery Hill in the north to Little Round Top and Round Top on the south, a distance of about 4 km (2.5 mi). The extent of this resistant rock is considerable because the unit is both thick and gently dipping. Accordingly, this sill makes Cemetery Ridge wider and higher than nearby ridges, with its related hills and "round tops" more prominent than other landforms on the battlefield. One small appendage of this resistant formation projects a short distance west from the southern part of the ridgeline to form the rocky Devil's Den.

Entrenchment in the outcrop area of the sill was difficult to impossible in most places because soils were generally thin or nonexistent, and the rock could not be excavated with hand tools. But because of greater altitude, locally rugged terrain, and numerous residual boulders and trees, the terrain offered good visibility and some concealment for many Federal soldiers.

On 1 July Confederate forces, advancing first eastward and later toward the south, attacked Union troops just outside of Gettysburg. During a long day of fighting, the outnumbered Northern soldiers gradually withdrew southeastward through the town to hold positions that evening on the advantageous defensive terrain of Cemetery Hill and Culps Hill. During the night large numbers of additional troops arrived for both sides. The Confederates reinforced troops around Gettysburg and occupied Seminary Ridge. Union forces were deployed in a curvilinear fashion along Cemetery Ridge and the two hills to the north. Thus it was the topography, developed on the resistant sill, that guided the eventual establishment of the North's now-famous fishhook position (Map 6.5). It also gave them the advantage of high ground and interior lines, reversing the pattern that so favored the Confederates at Sharpsburg.

During the next two days at Gettysburg, the Confederates attacked three different sections of the Union line without decisive success. Map 6.5 approximates Lee's battle plan for 2 July. Early that day the greatest pressure was against the Union right (north) at Culps Hill. In the afternoon the heaviest fighting developed at the Union's south flank, focusing on the Peach Orchard, the Wheat Field, and Devil's Den. The fighting here was markedly affected by a controversial maneuver ordered by Gen. Daniel Sickles and culminated in the crucial struggle for control of Little Round Top, which was occupied in force by Union troops only moments before the Southerners arrived (Map 6.6). If that situation had been reversed, it would surely have changed tactics by both sides and possibly the outcome of the battle.

Early on 3 July the Confederates renewed their offensive against the Union right with no success. Then, that afternoon, they made the major thrust against the enemy's center that came to be known as Pickett's Charge (Map 6.7). This attack was preceded by a long but generally ineffective artillery barrage. Once that ceased, about 12,000 Southern troops moved out of the woods along Seminary Ridge to align in full sight of the Northerners on Cemetery Ridge, about 1.4 km (.85 mi) to the east.

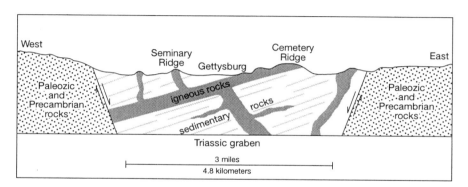

FIG. 6.3. East-west topographic and geologic cross-section of the Gettysburg area.

MAP 6.5. The Confederate battle plan for 2 July 1863.

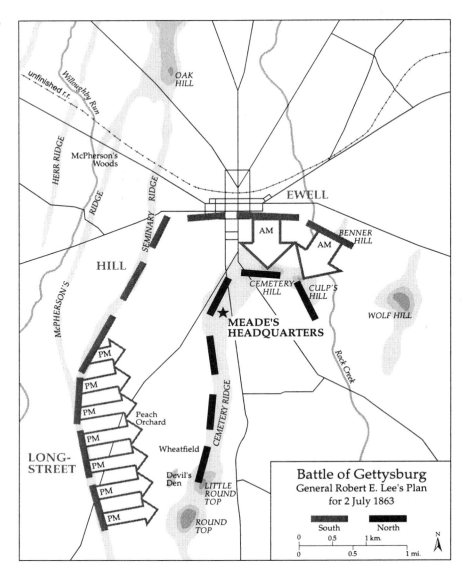

With no element of surprise the Confederates advanced toward their objective, which was penetration and separation at the Union center. Carrying needed equipment as well as their weapons and ammunition (for there was no plan for returning from the attack), the troops first marched forward over low, flat ground. Soon, however, they crossed the place where the inclined igneous sill began to control the topography to the east. At that point the land inclined gradually upward toward the enemy.

Initially Union fire was, by the deployment of both sides, frontal. As the Confederate advance continued, the rate of march increased as the troops moved over slightly increased slopes, both adding to their growing exhaustion. To make matters worse for the Southerners, the range and accuracy of Union rifle and artillery fire became more lethal as they moved with little or no cover into increasing crossfire from Cemetery Ridge. Few Confederates reached the Union line and about half of the attackers were killed or wounded. The charge, which lasted about 30 minutes, was the last major confrontation at Gettysburg.[22]

Once the fishhook defensive position was established, the terrain strongly favored the Union forces in just about every encounter. To reach its major objective, each Confederate attack eventually had to move forward over slopes that gradually steepened toward an enemy on higher ground in protected defensive positions. At Gettysburg the Southerners experienced something similar to that faced by Union soldiers at Fredericksburg only seven months before. Though the attack on the Union center during the final afternoon of battle reached well up on the west side of Cemetery Ridge, this "high tide of the Confederacy" never inundated its enemy on the Triassic sill.

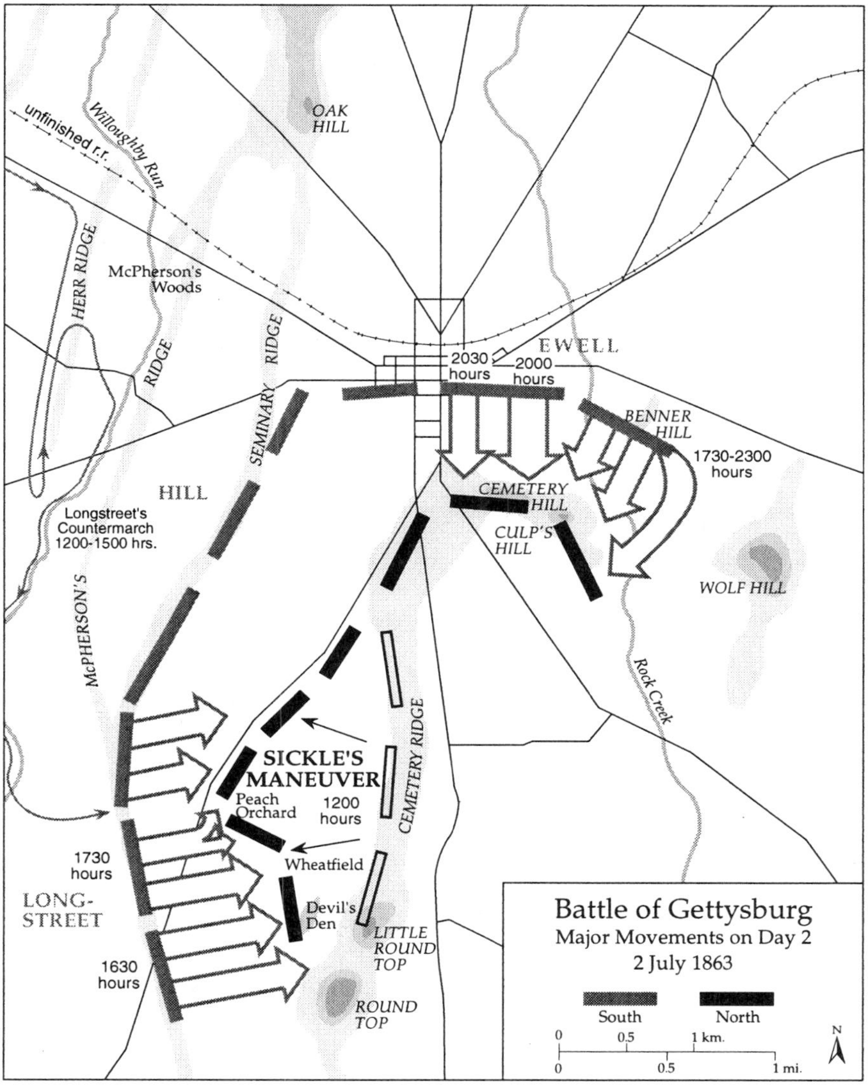

MAP 6.6. Overall troop configuration and major Confederate attacks on 2 July 1863.

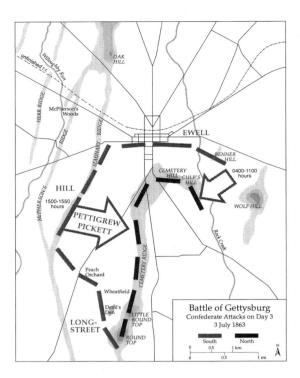

MAP 6.7. Overall troop configuration and major Confederate attacks on 3 July 1863.

As Pickett's remaining troops fell back, Lee's withdrawal from his second and last major invasion of the North began. There are many reasons that, within several weeks, Lee directed his finest battle at Chancellorsville and his worst here in Pennsylvania. In Virginia he knew the land, was in friendly territory, was aided by his two best generals (Jackson and Longstreet), and by using the vegetation and terrain to very best advantage converted his defensive position into several spectacularly successful offensive maneuvers. At Gettysburg he was in strange and foreign territory, faced a larger and better supplied enemy force, did not choose a favorable site for battle, and took the offensive on disadvantaged ground.

THE AFTERMATH

There was no Union counterattack late on the afternoon of 3 July at Gettysburg. Instead, as had happened many times before, both armies seemed satisfied to rest in place. The next day Lee's forces withdrew westward through two of the openings in the Blue Ridge Mountains of Pennsylvania, Fairfield and Cashtown Gaps. From there the army could follow the Great Valley southward into Virginia (see Map 6.2 above).

Initially the retreat was accompanied by rain. By the time Lee's forces reached the Potomac River it was too high to ford, and the only bridge in the area had been destroyed by Union cavalry. Unaligned and in some disarray, the Confederates were now especially vulnerable. But once again the slowly following Union force failed to mount a significant attack. Gradually the level of the river receded, and late on 13 July, Lee's army crossed the Potomac to the safety of Virginia.

The Army of Northern Virginia again took up positions south of the Rappahannock River and engaged the North in several battles that fall. During the following year, 1864, Lee again tried to use regional aspects of the terrain for wide-ranging maneuvers by sending Jubal Early's five divisions northward in the Shenandoah Valley and then eastward through South Mountain to the very outskirts of Washington, but the thrust was relatively small and had little lasting effect on the overall course of the war. In 1864 it was the Union, under the military leadership of Gen. Ulysses S. Grant, which assumed the initiative in the east.

To do so, Grant developed a set of coordinated plans for all of his armies. One force would move southward in the Shenandoah Valley, sweeping Confederate troops out of the area and destroying all resources there that would aid any Southern offensive, thus eliminating a major threat from that direction. Grant also moved troops inland along the south side of the James River toward Richmond to threaten the enemy's capital and right flank. Most important, he used his largest central force to battle Lee west of Fredericksburg. This was the second major battle in the Wilderness and was fought just about one year after the Confederate victory known as Chancellorsville (see chapter 5). When that battle was over, Grant refused to withdraw and rest his troops as others had previously done.[23] Instead he tenaciously pursued, keeping constant pressure on Lee's right flank by extending his position southward toward, and eventually past, Richmond. This tactic made it impossible for Lee to disengage without exposing his capital to attack. It also removed Lee's opportunity to use the geographic compartments leading to the North that had served him so well in the past.

Until the battle at Gettysburg, Lee used tactics and terrain so skillfully that much of the time military leaders of the larger and better equipped Union army in the East were confused, frustrated, or about to be relieved. After Gettysburg all of that changed. There are, of course, many reasons for the gradual demise of the Confederacy but one is especially clear. When Lee was prevented from using geography to best advantage, the hopes for Union capitulation because of a Confederate invasion in the Eastern Theater disappeared. The result was that Lee's initial stunning Napoleonic maneuvers were replaced by Grant's grinding war of annihilation, and, after Gettysburg, the tide of battle in the east shifted irreversibly against the South.

World War I Verdun

Unlike Sharpsburg and Gettysburg, little towns that only once were central to battle, the old city of Verdun-sur-Meuse in northeast France has been a military objective many times during the last two millennia. For more than three years during World War I, it was a vital location in the long and lethal Western

Front. Prewar German and French offensive war plans hinged on the city. The Schlieffen Plan dictated that several German armies, positioned well north of Verdun, should advance westward and then conduct a gradual south-turning movement. In doing so they would sweep through Belgium and into north-central France to capture Paris and defeat the French—the giant wheel-like movement turning on Verdun, but avoiding its strong defenses and rugged terrain. In contrast, the French Plan XVII called for an attack eastward from Verdun and Lorraine directly into Germany. In theory both plans were formidable, but the degree of success for either side depended upon adequate preparation, organization, execution, and resolve.

Following Count Alfred von Schlieffen's plan, the German army initiated World War I by advancing into north-central France and quickly reaching their most forward positions of the war, within 50 km (30 mi) of Paris. Verdun, near the eastern hinge of the giant maneuver, was nearly surrounded during the 1914 Battle of the Marne. Because of fierce French resistance, however, the city remained uncaptured, a symbolic stronghold that formed the foundation for a major salient anchoring the French right, which extended south through Lorraine to the border of neutral Switzerland.

Through the developing stalemate and trench warfare of 1915, the Verdun sector remained a conspicuous northeastern bulge near the center of the Western Front. Here in February 1916 the Germans began an offensive that led to one of the largest and most costly series of battles in the history of warfare. Although both the Germans and the French made enormous military efforts near Verdun that year, neither side achieved major gains or a significant advantage over the other. Physical geography was a vital factor contributing to the indecision.

GEOLOGY AND TERRAIN

The ancient geological foundation for Europe is centered on eastern Scandinavia and Finland. Like the Appalachians of North America, Europe has a marginal zone of severely deformed rocks produced during the convergence that formed the huge and ancient continent of Pangaea.[24] Here the resulting intensely deformed, ancient rocks are mainly in the west and south. In west-central Europe, their complex structures appear in a number of large residual crustal blocks (or massifs), including those of Brittany along the Atlantic coast, the Massif Central and Vosges Mountains of France, the Slate Mountains of the Ardennes, and the Schwarzwald (Black Forest) of Germany.

Among these rugged tracts is a circular lowland of younger and weaker rocks associated with a structural basin centered on Paris. This "Paris Basin" developed after the separation of North America and Europe as the Atlantic Ocean widened. With the intense compressional activity long past, parts of the ancestral European continental platform then experienced broad uplift at some places and wide subsidence elsewhere, such as in northeast France. As the surface here sank below sea level, marine strata accumulated in more or less horizontal layers. The amount of sedimentation approximated the rate of downward movement, thus keeping the water fairly shallow even though total subsidence eventually became considerable.

With continued basin development, the last-deposited horizontal strata were put down on underlying rock that had been previously warped. In this fashion, very gradually over more than 100 million years, a somewhat elongated east-west-trending structural basin formed through broad crustal warping.[25] Because this basin is centered on the site of Paris, all surrounding strata are slightly inclined toward the city. Furthermore, changes in depositional environments affected the strength and thickness of each tilted rock layer. Late in geologic time this area emerged from the sea to be weathered and eroded, processes that etched out concentric ridges supported by stronger formations and intervening lowlands underlain by weaker rocks.

CUESTAS

Verdun is near the east end of the Paris Basin, thus all underlying strata are inclined downward toward the west. Here, five resistant layers form corresponding uplands, each made more conspicuous by adjacent, poorly drained, river-eroded lowlands. Because all of the strata dip in the same direction, the uplands are both linear and distinctly asymmetrical, with a gentle western slope and a steep, east-facing escarpment of considerable height. Such features are known as cuestas.

From the position of the front early in 1916 (as shown on Map 6.8), it is immediately apparent that a German advance toward Verdun from the east would first cross low, often wet, and exposed ground, to then encounter the steepest side of the Meuse Cuesta, a high forested ridge that extends many kilometers in a north-south direction. Further westward movement would involve three more lowlands and cuestas. In contrast, the French ascent of the Meuse Cuesta from the west was over more gently inclined terrain, and their forces held the summit and had excellent observation sites. Obviously, any German attack from the east would be initially exposed, cross-compartment, uphill, and perilous if the cuestas were well defended.

In contrast, any German advance southward in the linear lowlands would move along, rather than across, compartments. But to be successful, such an attack would also require control of the flanking cuestas to eliminate French observation points and lessen threats from enemy flanking maneuvers and their artillery in the hills. Thus, whether the Germans were attacking from the east or north, the terrain around Verdun presented a formidable military problem for them. A closer examination illustrates the opportunities and difficulties in a landscape characterized by cuestas.

Many World War I battles in France took place within cuesta landscapes. Such features make up parts of the Somme, the Marne, and Lorraine, although their size and extent varies among these areas.[26] As the continuity, thickness, inclination, lithology, and strength of the underlying formations differ, so do

MAP 6.8. Landforms and geology of the Verdun area.

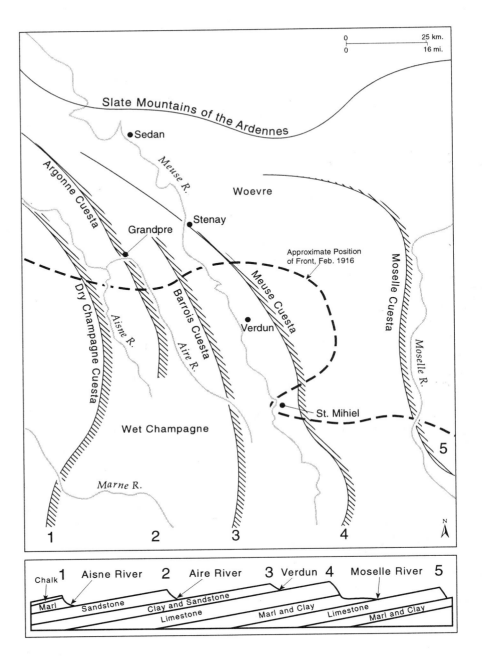

properties of these asymmetrical ridges. Where the strong rock layer is discontinuous, the cuesta is intermittent. Where the rock layer thins to disappear completely, the cuesta terminates. If the rock is both thick and strong, the upland is higher and more extensive. Where the inclination of resistant strata is relatively low, the terrain appears plateau-like, while more steeply dipping formations support a relatively narrow and possibly sharply crested ridge. The character of the bedrock also affects drainage and soil characteristics, which in turn influence both the natural vegetation and human use of the land.

Stream erosion is the most important process in cuesta formation. Larger rivers tend to extend along the trend of the adjacent lowlands developed on weaker strata. These are joined by tributaries draining the cuesta slopes. The streams flowing on the steep side of the cuesta tend to be shorter but have higher gradients, giving them more erosive power to form deep ravines containing rapids, cataracts, and possibly waterfalls. The rivers and creeks on the opposite, more gentle slope are slower-moving, less powerful, and generally not as entrenched, but because they drain a larger area and extend a greater distance, they can be expected to grow to a larger size.

Although their characteristics are somewhat different, both types of streams produce dissection lines that extend at right angles to the summit trend of the cuesta. Furthermore, headward erosion toward the summit area of the cuesta by these opposing stream systems may sharpen and scallop the crest of the feature, making it rugged and an even more formidable military objective, no matter which way it is approached. It was in such summit areas the French established some of their largest permanent fortifications—including the famous Fort Douaumont—for a time thought to be the strongest defensive positions around Verdun.[27]

Extensive forests remain on many of the French cuestas because their steep slopes and rocky soils make them unsuitable for crop agriculture. This is well illustrated by the coincidence of names for the Argonne Cuesta and the Argonne Forest, where many American troops fought and died near the end of World War I. Elsewhere, however, such as on the Dry Champagne, the cuesta may be underlain by porous chalk that permits rapid downward percolation of precipitation. Here the soil is thin, plants struggle for moisture, and the landscape has a barren look. Thus, the size, shape, and extent of cuestas may vary, but each in its own way presents characteristics, even if subtle, that may affect battle in different ways.

A second major element in a cuesta landscape is related to higher passes and lower gateways within these features. Some are the product of stream erosion, while others may be the result of variations in strata characteristics or structural control. Sedan, Stenay, Saint Mihiel, and Grandpre, as well as the centrally located Verdun, are all settlements established at openings through the cuestas east of Paris. Lines of movement and communication tend to concentrate on such points, often making them tactically important in military operations.

The lowlands between cuestas constitute a third important terrain characteristic. These areas are always underlain by weak rocks such as shale and marl. Such formations often weather into very small particles that, when wet, turn to mud, making movement of men and equipment difficult to impossible. Rivers that wind back and forth across the low, flat surface may also flood the land, making it especially difficult to cross when wet. Furthermore, being largely cleared for agriculture, this landscape generally presents little concealment during all seasons, making observation from adjacent hills a nearly continuous daytime threat. In this sort of landscape—hilly, forested cuestas and moist, open lowlands—the Germans began their great offensive against the French near Verdun.

THE 1916 OFFENSIVES

As World War I progressed, large flanking maneuvers, so common in earlier wars, became impossible on the Western Front. Trench warfare and frontal attack—accompanied by more effective artillery, the machine gun, and sometimes poison gas—became the order of the day. By 1916 German General Erich von Falkenhayn correctly concluded that the French would do everything possible to hold their strong point at Verdun. He also reasoned that a large-scale battle of attrition could produce a German victory at this pivotal site.

In relation to topography, the 1916 Western Front near Verdun may be divided into three sectors. The first is a westward-extending salient to the south, centered on Saint Mihiel. This was established by German advances early in the war but was not to be of special importance to their great offensive in 1916. The other two sectors, northeast and northwest of Verdun and separated by the Meuse River, were central to battle but presented somewhat different combat settings.

The Germans initiated their offensive on 21 February 1916 east of the Meuse River and then progressed southward and later southwestward obliquely onto the Meuse Cuesta.[28] This attack was preceded by one of the largest German artillery bombardments of the war. Furthermore, the advancing infantry was ordered to avoid strongpoints, to penetrate and encircle the enemy, and to gain as much ground as possible. This type of maneuver required that many troops move across the low, open terrain of the Woevre and then upward along the face of the escarpment to the crest of the Meuse Cuesta. Such tactics exposed troops to direct fire from soldiers in favorable defensive positions, and indirect enfilade from French artillery directly to the west across the Meuse River. As battle progressed in this sector, the German advance gradually became more westerly, shifting directly toward the cuesta's steep east-facing escarpment.

Years before, a number of French forts had been constructed on or near the crests of the cuestas near Verdun; these included Forts Douaumont and Vaux, both in this northeast sector. Although natural focal points on this battlefield because of their prominent positions, both had been disarmed of large artillery pieces by 1915 and were now sparsely manned.[29] Furthermore, French defensive works were limited largely to their forward positions, backed by only one major trenched secondary line.

The partially disarmed forts and the limited defensive positions would appear to favor a German offensive here and may have been factors in their military planning for 1916. But even before the German offensive got under way, it ran into geographic problems. First snow and poor visibility delayed the initial attack by many hours. Then a combination of circumstances slowed German movement and, as they advanced, made their logistical support increasingly difficult. These included cold, moisture, and mud; numerous cross-cutting gullies, ravines, and streams; steep slopes; and fierce French resistance guided by the motto, "They shall not pass." Close combat became common-place, and daily German advances on the rugged, forested cuesta were often measured in meters. Even so, Fort Douaumont, on the crest of the cuesta, was captured within a few days, adding to the German threat on Verdun.

To relieve pressure on the north (right) flank of their attacking troops and to spur the overall offensive, the Germans initiated another attack west of the Meuse River in early March. Here their forces advanced toward the south and south-southeast, directly along the trend of the Barrios Cuesta and its adjacent lowlands occupied by the Aire River to the west and the Meuse River to the east. Although the largely treeless lowlands offered greater visibility, the poor drainage, winding rivers with their tributaries, and nearby higher and forested cuestas all presented formidable tactical problems. Even so, German attacks here continued periodically through spring and into summer.

Because German forces held positions north, east, and south of Verdun, only one road, extending southwest from the city and known as the "Sacred Way," was available for the much-needed French reinforcements and supplies. A significant advance from any one of these directions could lead to the capture of Verdun. But by August, after a six-month offensive, the Germans finally ceased their attack within a few kilometers of the city. Their mission failed in the face of numerous strongholds, continuing tenacious French resistance, some of the roughest terrain in the area, Britain's newly initiated supportive Somme offensive far to the west, and their own measure of attrition.

Fighting was relatively subdued during September, but in October the French mounted a major counterattack. By December they had regained a small area on the cuesta northeast of the city which was tactically desirable for its defense. By the time the two 1916 offenses at Verdun had ended, the front had changed little, but casualties for both sides were huge, with estimates ranging from several hundred thousand upwards to nearly one million. The combination of tactics and terrain had taken an incredible toll. The fighting here, which started in 1914, would continue on through 1917 and much of 1918. Near the end, large numbers of American troops would also see combat in the plains, cuestas, and forests northwest of Verdun.[30]

Conclusion

The contrast between Lee's swift and far-ranging maneuvers early in the American Civil War and the dogged frontal attacks near Verdun illustrates how strategy and tactics may vary greatly between wars even though the shape of the battlefield may be somewhat similar in terms of corridors and barriers. Lee used the Blue Ridge Mountains and rocky, forested, linear ridges for cover and deception; the fertile valleys for movement and sustenance; and the connecting gaps for surprise and escape. At Verdun the wooded uplands became natural fortresses, threatening everything that moved in the nearby lowlands and valley passes. Topography for these two unlike campaigns was somewhat similar, yet the positioning of forces, military objectives, trend of the corridors, and form of the highlands combined to have opposite effects. During the first half of the American Civil War, Robert E. Lee used the terrain to great advantage for offensive maneuver. At Verdun, from the offset, the topography strongly favored defensive positions on cuestas that gravely threatened troops on the open, adjacent lowlands.

Known or not, all combat takes place in a physical setting that, as the fighting continues, in some way favors one side or the other. It is the fortunate soldier whose commander is wise enough to use the landscape to the very best advantage. Lee, with the option to maneuver, was often in that position before Gettysburg. At Verdun neither side developed and executed an effective offensive plan. Either way, the terrain could be as formidable in battle as the enemy.

CHAPTER 7

Troubled Waters

River Crossings at Arnhem and Remagen

We could not widen the corridor sufficiently quickly to reinforce
Arnhem by road.

FIELD MARSHALL BERNARD MONTGOMERY

Rivers, always important in war, are generally useful to the defense but often
obstacles to the offense. For both sides, rivers restrict maneuver and funnel
movement into crossing points that are generally known in advance by the
adversaries. Deep or shallow, broad or narrow, swift or slow, a river presents
the combatants with problems of depth, width, and velocity, as well as the
enemy. Even small streams, such as Antietam Creek near Sharpsburg, Maryland, may be apparent barriers in battle. But as rivers become larger, their
potential as a military obstacle increases at an exponential rate.

One relatively simple way to differentiate streams (and their sections) is
based on erosion versus deposition. For example, along the downstream parts
of large rivers, such as the lower Amazon and Mississippi, over geologic time
sedimentation forms a broad, flat-floored, marshy floodplain that is periodically flooded during times of high water. In contrast, steep gradient mountain
streams generally remove, rather than deposit, alluvium, eroding downward to
form narrow, steep-walled valleys. Either way, most streams are subject to
floods that not only widen and deepen the water, but also result in a marked
increase in stream velocity. Wide floodplains, steep-walled canyons, deep and
fast-moving water, unstable river bottom, and flooding are all factors that
must be considered in any plan to cross a river of significant size. Ironically,
environmental conditions can sometimes occur which will be an aid rather
than a challenge in crossing a river. One example is drought, which lowers both
water level and velocity of the stream. Another is extreme cold, which freezes
river ice to the point that it will support the passage of men, supplies, and
sometimes even heavy armor and large artillery.

Since the earliest days of warfare, astute military leaders have understood
that they should not be caught unprepared on the wrong side of a river.
Whether in forward or retrograde movements involving rivers, the potential
for disaster looms over any force that loses control of its crossing point. Over
2500 years ago Sun Tzu astutely wrote, "When an advancing enemy crosses
water do not meet him at the water's edge. It is advantageous to allow half of
his force to cross and then strike."[1]

During the American Revolutionary War, the British recognized that control of the Hudson River would prevent the colonials from easily crossing to support their allies, whether in New England or in the Middle Atlantic area. British efforts to limit communications by controlling the river led Gen. George Washington to establish a number of shoreline garrisons with cannon to threaten British ship patrols. And to limit navigation, as at West Point, chains were floated across the river at several places.

In the Civil War battle for Vicksburg, Gen. Ulysses S. Grant had to contend with both the Mississippi River and its nearly 160-km-wide (100-mi-wide) floodplain, consisting mainly of marshes, swamps, and backwater. All of these discouraged a direct thrust against the city from the west and north, and eventually forced him to attack from the uplands in the east after maneuvering 270 degrees around the city.

During World War II, in January of 1944, Allied forces suffered heavy casualties near Cassino, Italy, in their unsuccessful attempt to cross the well-defended, marshy, and steep-banked Rapido River. Several weeks later, they tried again, this time upstream from Cassino. However, by raising the water level at a downstream dam the Germans were able to turn the Rapido River floodplain into a quagmire for American troops. Well informed about water obstacles, the Germans also delayed Allied operations by flooding parts of the Ill River in France and the Roer River in Germany.

At the beginning of the Korean conflict in 1950, many South Korean soldiers retreating southward from the onslaught of the North Korean army faced death when premature detonation of a critical bridge isolated them on the north side of the Han River. Later in 1950 cold weather created ice floes on the Imjin River that, like battering rams, broke apart combat bridging vital for U.S. resupply. During the Viet Cong Tet offensive of 1968, destruction of important bridges across the Mekong River effectively cut off large areas of the delta from the rest of South Vietnam.

Great commanders from Julius Caesar to Dwight D. Eisenhower have recognized the importance of the Rhine River as a major barrier to military maneuver in Western Europe. The Treaty of Versailles following World War I reflected France's concern that the victors have some control over this militarily important waterway. By extending Allied occupation to the entire west edge of the Rhine, and by creating several bridgeheads on the east bank, the treaty provided the Allies with river-free access toward the heart of Germany should hostilities be renewed.

At no time, however, has the Rhine been more formidable and vital than for the Allies during their World War II offensives against Germany. The first of the two major river-crossing operations, Operation Market-Garden, occurred during September 1944. As a major joint Allied effort involving airborne, armor, and infantry, its mission was to leap across the Rhine by the simultaneous capture of bridges over three major rivers in the Netherlands and then secure their connecting highway. It ended as one of the largest Allied failures of the war. In sharp contrast, the second attempt to cross the Rhine was widely successful, in part because of the initial, fortuitous, and unexpected capture of the only remaining bridge at Remagen. The complete failure near Arnhem and the unexpected and startling success at Remagen illustrate the challenges and difficulties presented by rivers.

Rivers: Agents of Erosion and Deposition

Water running off the land quickly channels its energy to erode gullies, ravines, and valleys. Responding immediately and precisely to gravity, yet encountering friction at its channel perimeter, water flows turbulently, its velocity largely a function of relationships among volume, gradient (slope of the river bed), obstructions, and the cross-sectional dimensions of the stream channel. As each of these factors changes spatially and temporally, so does the nature of the river. The resulting variations are so great that different sections within one stream cannot be the same nor can two rivers be identical.

In moving from source to the sea, flowing water may, through chemical and physical processes, remove material from its banks and channel; transport material through solution, suspension, and traction; and deposit sediments as alluvium where the water slows.[2] In areas well above sea level, streams tend to erode deeply, carving valleys roughly proportional to their size, with steepness of adjacent slopes dependent upon the strength of the local bedrock. In arid areas the resulting erosional features tend to be more angular and, because of sparse vegetation, rock formations are commonly visible—possibly best illustrated by the Grand Canyon of the Colorado River.

In contrast, where long-term deposition occurs, rivers meander across wide, flat, floodplains underlain by alluvium. During widespread flooding the velocity of water overflowing the river's channel decreases, which in turn promotes quiet-water deposition of suspended clay- and silt-sized particles. Under natural conditions this process tends to be repeated by annual floods over many thousands of years to form floodplains, their extent being directly related to the total amount of alluvium deposited, the nature of the land through which the river flows, and volume-gradient-velocity relationships within the stream.[3] Similarly, where streams escape channel confines at the point where they enter larger water bodies, the decrease in velocity often favors the formation of deltas.

Centuries of deposition and channel modification on floodplains and deltas bring frequent changes as new river courses are created and older ones abandoned. All of this results in nearly flat, poorly drained areas marked by numerous marshes, bogs, and swamps. Even when not in flood, the level of both the rivers and the water table in these areas is only slightly lower than that of the land.

River volume depends upon the amount of precipitation runoff and ground-water springs (or seepage) within its drainage basin, minus evaporation, transpiration, and influent processes. In arid areas, where flow depends wholly on precipitation, streams are ephemeral, meaning that they exist only during and immediately after times of precipitation. Here large storms may produce flash floods in otherwise normally dry stream beds.

In contrast, rivers in humid areas may flow year-round; they are permanent due to a continual supply of seeping groundwater even though rainfall is intermittent. When both groundwater and surface runoff are abundant the resulting volume often exceeds the size of the channel, thus inundating the adjacent land unless it is protected by dikes or levees. Large, permanent rivers exist in a few arid regions, such as the Nile in Egypt and the Tigris-Euphrates in Iraq. In both cases the stream originates in a humid region and, by geographic arrangement, the excess water flows through a dry area in its course to the sea. Called "exotic rivers," these streams can be of enormous importance not only for use by the local population, but also economically, geopolitically, and militarily because supply comes from an area not controlled by the major users.[4]

Humans have used and altered riverine landscapes in a number of ways. Large floodplains and deltas are generally moist, low, level, and fertile. They favor agriculture and may support a dense agrarian population, sometimes for hundreds of centuries. The needs of these farmers were commonly served by ports and settlements, some of which evolved into major commercial or manufacturing centers. As property values increased, so did community effort to control stream flow and to prevent flooding. Ditches and canals were dug to expand agriculture, drain excess moisture from fields, or provide water for irrigation. Riverbanks were heightened with levees and roads were raised within marshes. Dams were constructed to control flooding. If navigation was important, channels were straightened and deepened to a specified depth, locks constructed where necessary, and dredging undertaken to insure and enhance port facilities. Bridging also became increasingly important, with their spacing within a region often roughly proportional to population distribution and the level of economic activity.

Fertile floodplains were often a vital source of food. Ports provided points of exit and entry. The stream itself was a natural avenue for movement. Some rivers served as effective boundaries. As humans densely populated and commonly modified riverine landscapes these areas also took on added military importance. Any army that must cross a river while engaged with a formidable enemy always faces a major military problem.[5]

Forces can cross a river in several ways: capture an existing bridge; build a bridge; wade a ford; motor, raft, float, or swim across; or fly over the river and land on the other side. The first is obviously preferable, but capturing an intact bridge in the face of a destructive enemy generally requires an extraordinary effort. The best way to do this is to secure both ends of the bridge at the same time, meaning that some troops must get to the far side of the river to carry out half of such an operation.

To build a bridge while in combat is one of the most formidable tasks for combat engineers. Whether by capture or construction, once a bridge is secured it is of enormous value because those crossing are less threatened by the changing nature of the river. But a bridge also funnels personnel and traffic into a precise and unchangeable location, making it an unmovable and ideal target for the enemy. Furthermore, if the bridge is destroyed before an adequate bridgehead is established, forces that have moved to the far side of the river are in immediate danger of being isolated.

The Rhine: Last Western Obstacle to Germany's Heartland

The ultimate target of Allied operations in Europe in 1944 was the heartland of Germany. Gen. Dwight D. Eisenhower believed that Adolf Hitler would not surrender until Allied ground forces had crossed the Rhine, had captured the industrially important Ruhr, and were moving through central Germany toward Berlin (Map 7.1). By late August 1944, Field Marshall Bernard Montgomery's 21st Army Group was headed across Belgium toward the Netherlands and the plains of northern Germany. At the same time, Gen. Omar Bradley's 12th Army Group, to the south of Montgomery, was advancing through the Ardennes toward central Germany. Farthest south, Gen. Jacob Devers's 6th Army Group was driving northeast through France toward southern Germany. To get to their objective each army had to cross the largest river in Western Europe.

The Rhine River originates as a high mountain stream in Switzerland and, fed by rainfall and melting snow and ice, flows swiftly out of the Alps. Receiving numerous tributaries, the northward-flowing river rapidly widens and deepens so that at no place north of Switzerland can it be easily forded, making it an effective natural boundary between Germany to the east and the Netherlands and France to the west. For example, during the "phony war," a phase of World War II between the German invasion of Poland in 1939 and their western offensive of 1940, soldiers on both sides of the river felt secure enough to move about freely in view of their enemy, believing that any cross-river attack would be obvious long before it occurred.[6]

Downstream (north) from Switzerland the Rhine has three distinctly different riverine environments. From Basel to Mainz, 275 km (170 mi), the river flows within a broad, flat-floored, steep-walled trench formed by a wide, down-faulted wedge of rocks weaker than those on either side.[7] Over the centuries the river has shifted back and forth across this lowland, to form a floodplain that is 3–5 km (2–3 mi) wide. By the early part of the twentieth century, engineering projects had greatly altered this part of the river. As a result it followed a more straightened course until turning northwest just downstream from its junction with the Main River near Mainz.

At Mainz the river curves westward into a completely different landscape—the Rhine gorge made famous by its rugged scenery, sloping vineyards, and the Lorelei. Here the river occupies most of the valley floor, leaving only meager marginal room for roads, railroads, and buildings. Cramped by this narrowness, villages extend upward along valley walls that rise 90–180 m (300–600 ft) above the river toward the surface of adjacent dissected plateaus spreading to the northeast and southwest.

After flowing past Remagen the Rhine gradually leaves the deep valley to traverse a belt of gently rolling terrain until it reaches the Ems River where its delta begins. Here it also nears the Maas River (Meuse River in France) that flows out of France. Together they have deposited the huge delta that makes up most of the Netherlands and nearby land in Germany. This low, flat area is crisscrossed by inter-river canals, agricultural drainage ditches, and embanked roads that

MAP 7.1. Approaches to the Rhine River. (Map modified from diagram by A. K. Lobeck in Army Map Service, *Continental Europe—Strategic Maps and Tables*, pt. 1, *Central West Europe*.)

BRIDGEHEADS OVER THE RHINE

As the authors of the 1918 World War I Armistice agreement began work, their desire to prevent a reoccurrence of German aggression became paramount. Recognizing the significance of the Rhine River as a barrier, the framers designated certain areas in Germany west of the Rhine River to remain as occupied territory free of German military presence. Thus, any future westward aggression by German troops would first require them to cross the Rhine, a move so obvious that it would immediately alert France to the attack. Control of the German territory west of the Rhine would not, however, provide the Allies with access to the east bank of the Rhine River if the Germans failed to abide by the Armistice. Any Allied offensive would require footholds on the east bank.

To ensure this advantage the Armistice also established four separate semicircular bridgeheads on the east bank of the Rhine River—Cologne, Mainz, Koblenz (Germany), and east of Strasbourg (France), each with a radius many kilometers in depth. When the Armistice went into effect, U.S. forces occupied the bridgehead at Koblenz, British forces at Cologne, and French forces at Mainz and Kehl. The Treaty of Versailles incorporated these bridgehead provisions and indicated that occupation of the Cologne bridgehead would end in 5 years, Koblenz in 10, and Mainz and Kehl in 15. Between 1923 and 1930, Allied troops were gradually withdrawn from all of these positions. Until abandoned, these bridgeheads served to discourage violations of the treaty. They also represented the Allies' awareness of the importance of physical features in military operations.

link the settlements and cities of the region. It was into this latter setting that British and Canadian forces advanced eastward in the late summer of 1944.

OPERATION MARKET-GARDEN, SEPTEMBER 1944

On 10 September 1944 General Eisenhower, on the recommendation of Field Marshall Montgomery, authorized a daring, deep penetration attack in Holland, code-named Market-Garden.[8] Early on the afternoon of 17 September paratroopers and gliders of the First Allied Airborne Army (including the U.S. 101st Airborne Division, the U.S. 82nd Airborne Division, and the British 1st Airborne Division) began to descend near the Dutch cities of Eindhoven, Nijmegen, and Arnhem (see Map 7.1 above, and Map 7.2). Their objective was to seize a line of bridges, the smallest over drainage ditches and canals, the three largest crossing the Rhine, Maas, and Waal Rivers. The airborne divisions were also to lay a ribbon of parachutists on the roadway from the Meuse-Escaut Canal to Arnhem that connected the bridges.

At the same time, some 25 km (15 mi) south of Eindhoven, elements of the British XXX Armored Corps began to advance north and east along the connecting highway. Their objective was to link up rapidly with the airborne forces holding the bridges: first with the American 101st, then the U.S. 82nd, and finally the most deeply inserted British 1st Airborne Division. The mission of the British XXX Armored Corps was absolutely essential to the success of the plan because it was the only way to secure the road, control the corridor,

MAP 7.2. Allied objectives for Market-Garden, September 1944.

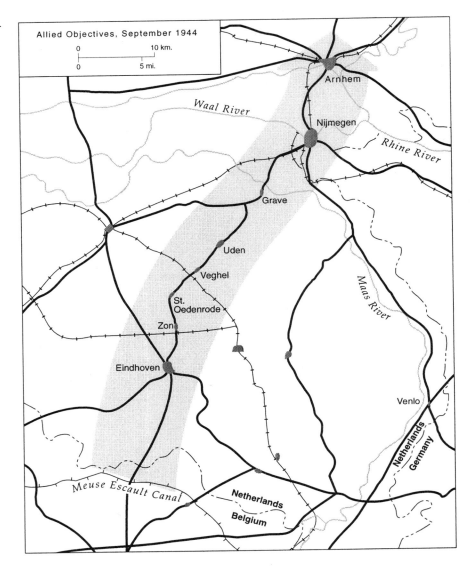

Allied Objectives, September 1944

0   10 km.

0   5 mi.

Arnhem

Waal River

Nijmegen

Rhine River

Grave

Uden

Veghel

Maas River

St. Oedenrode

Zon

Eindhoven

Venlo

Netherlands

Germany

Meuse Escault Canal

Netherlands

Belgium

and, most important, reach the three isolated airborne units who were to secure the major bridges.

All of the bridges were equally important but, because of increasing distance, the farthest one spanning the Rhine River at Arnhem would be the last and hardest to reinforce by land. If this bridge could be captured intact and the road to it secured, it would give the Allies their first bridgehead on the east side of the Rhine, which would in turn open the way for a deep thrust into Germany's heartland. But without all of the bridges and control of the road, Market-Garden would fail.

The British XXX Corps was to advance along the highway to link up with the U.S. 101st near Eindhoven on the evening of day 1 (D-day). On day 2 (D+1) the corps was to reach elements of the American 82nd holding bridges at Grave and Nijmegen. Finally, between the second and fourth day (D+3) and about

100 km (62 mi) from their starting point, the advancing armored troops were to reach the British 1st Airborne Division at Arnhem. Allied intelligence reported that the area was lightly protected, supporting the conclusion that the airborne troops could secure the bridges and that a rapid advance along the highway was possible. Some, however, disagreed, leading to Lt. Gen. Frederick Browning's now famous prophecy to Montgomery, "We might be going a bridge too far."[9]

At 0550 on 26 September, nine days after the attack began, the last remnants of the defeated British 1st Airborne Division withdrew to join the XXX Corps, still south of Arnhem, unable to reach the third bridge. Market-Garden had succeeded in crossing a number of rivers and canals between Belgium and the Rhine, but it failed to secure the vital final objective—a linkup with a bridge over the Rhine itself.

Without the Rhine crossing, a direct ground attack into Germany was not possible. The operation ended in failure and cost dearly. Nearly 12,000 Allied soldiers were killed, captured, or wounded. Postmortems of the battle have pointed out the indomitable spirit of the airborne soldiers who conducted their missions so bravely. Analysis also indicates that adverse, but not atypical, weather limited aerial resupply and contributed to the inability of the airborne forces to hold out in the face of strong enemy pressure. Furthermore, the Allies did not know that prior to Market-Garden a number of experienced and seasoned German army units had been posted near Arnhem for rest and recovery from combat in Normandy and following battles. Even so, in simplest terms, the operation failed because, in the face of strong and effective German resistance, the XXX Corps could not reach Arnhem to support the British 1st Airborne Division which had struggled so valiantly to hold that Rhine crossing.

Why did the XXX Corps fail to advance as scheduled on the route from Eindhoven to Nijmegen to Arnhem? An examination of the riverine landscape between the Meuse-Escaut Canal and the Rhine at Arnhem helps to answer this question. As the 1st British Airborne Corps recorded in its after-action report, "The country does not lend itself to quick advances."

The terrain from the Meuse-Escaut Canal to Arnhem, characteristic of deltas, is nearly flat. Natural drainage was minimal and mass vehicle movement over the moist, alluvial soil quickly became impossible because of water and mud. To overcome the problems of poor drainage and flooding, the highways connecting the scattered villages had been elevated as much as 2 m (7 ft) above the surrounding fields. The land itself was in carefully nurtured, productive farms. But the success of agriculture rested on the farmer's ability to drain the fields effectively. As a result, canals and ditches crisscrossed the region. Each field and roadway was commonly flanked by a ditch that served as a link in an elaborate runoff system. Most of the fields were open, numerous small orchards dotted the landscape, and tracts of woodland were interspersed throughout the area (Figure 7.1).

The main route for the British XXX Armored Corps was a two-lane, hard-

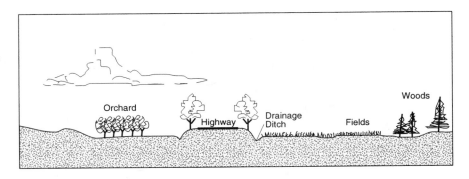

surfaced highway from Eindhoven through Zon, St. Oedenrode, Veghel, Uden, Grave, and Nijmegen to Arnhem (see Map 7.2 above). Between Eindhoven and Veghel, the route spanned two navigation canals whose widths varied from 30 to 90 m (100–300 ft). Between Grave and Nijmegen, the attackers faced the 245-m-wide (800-ft) Maas River and a 75-m-wide (250-ft) canal connecting the Maas and the Waal. At Nijmegen the width of the Waal varied from 75 to 110 m (245–365 ft), while the Rhine at Arnhem was 90 m (300 ft) from bank to bank.

The land between the Waal and the Rhine, or Nijmegen to Arnhem, became known as "the Island" and was dissected by an intricate drainage system that severely limited off-road mobility. The two main routes crossing the Island were so embanked that on a clear day traffic along the road was easily visible from more than 1000 meters. Here and there smaller roads paralleled the main route from Eindhoven to Arnhem, but none of these could support tanks and the heavy traffic required for Market-Garden, nor did they cross over the major waterways on the attack route.

On 17 September (D-day), strong German resistance stopped the XXX Corps in its breakout from the Meuse-Escault Canal 10 km (6 mi) short of Eindhoven. British tanks traveling the main roadways were easy targets for German 88mm guns positioned in nearby woods. Each German obstacle or resistance point required the advancing column to stop, bring in artillery and air support, and send out infantry to clear the road and drive away the enemy.

Lt. Gen. Brian Horrocks, in command of the XXX Armored Corps, recognizing that the Germans had prepared most bridges for demolition and that some might be destroyed before the airborne could seize them, had assembled more than 9000 engineers and 2300 vehicles to carry out an assault bridging mission.[10] Fortunately, the airborne forces were successful in gaining control over most crossings and lost only a small bridging span near the village of Zon which was quickly replaced by a Bailey assault bridge. Even with this good fortune, the first elements of the XXX Corps did not arrive at Nijmegen until noon on 19 September (D+2) and, together with the troops of the 82nd, could not gain control of the Waal River bridge until 1915 on 20 September (D+3).

For some reason, the German commander, Gen. Walther Model, did not order destruction of the bridge over the Waal River. (As a measure of vulnerability, had this demolition taken place, Market-Garden would have ground to a halt at that point, movement past Nijmegen being impossible.) Possibly the Germans did not destroy this bridge because, had they done so, their troops on the river's far side would have lost their most direct escape route, as well as all of their remaining heavy equipment.

The airborne and armored forces fighting between Eindhoven and Nijmegen were less successful in holding open that section of the highway. Periodically German forces counterattacked and established roadblocks that stopped all traffic (Map 7.3). Even when they were free to move, the continuing threat of attack made drivers especially cautious. Traffic jams were commonplace. Off-road maneuver for the airborne forces and the XXX Corps was extremely difficult. Once forced from the roadway, the Germans moved to nearby concealed positions and used their artillery and antitank weapons to delay and

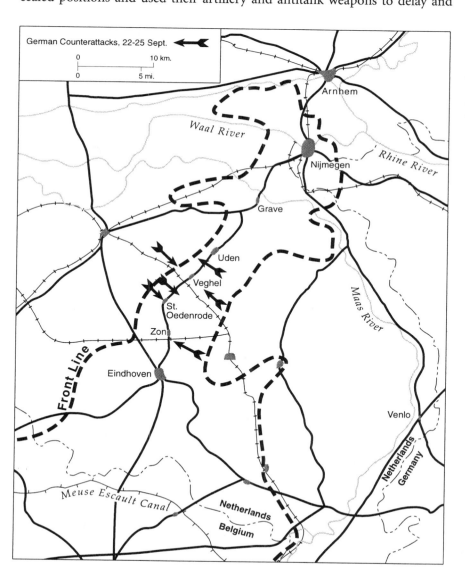

MAP 7.3. German counterattacks, 22–25 September, and Allied positions on 26 September shown as heavy dashed line.

disrupt movement on the elevated route. A disabled vehicle could block the road for hours, leaving Allied forces in the open much of the time. Meanwhile, German troops maintained advantageous positions in scattered woods and numerous ditches.

On 21 September, four days after Operation Market-Garden was launched, forward elements of the British XXX Corps, which had crossed the Nijmegen bridge the night before, finally began what they expected to be a quick 16-km (10-mi) trip to Arnhem to join their 1st Airborne colleagues who were holding the north end of the bridge over the Rhine. Moving on "Hell's Highway," the column suddenly ground to a halt. The elevated road between Nijmegen and Arnhem became a shooting gallery for the Germans. Montgomery found that "it was almost impossible to maneuver armored forces off the roads which generally ran about six feet above the surrounding country and had deep ditches on both sides."[11] British tanks were standing targets for enemy gunners. When the attack stalled, the armored forces of XXX Corps fell back to regroup.

Early on the morning of 22 September (D+5), tanks of the British XXX Corps, moving northwest from Nijmegen in a circuitous route toward Arnhem, finally managed to contact some troopers of the British 1st Airborne who had crossed to the south bank of the Rhine at Arnhem. At the same time, German forces were again interrupting the flow of traffic on the highway south of Veghel. During a critical 40-hour period on 23 and 24 September, the pathway from Eindhoven to Veghel was periodically shut off. Consequently reinforcements, so vital to the thin line of soldiers stretching from Nijmegen to Arnhem, never arrived. This phase of the operation virtually eliminated all hope of solidifying contact with the forces at Arnhem. On 25 September, facing superior German forces and threatened with annihilation, the British 1st Airborne was ordered to withdraw south of the Rhine. By 0550 on 26 September (D+9), the withdrawal had been completed and Market-Garden ended.

The XXX British Armored Corps had concentrated all of its effort on a single two-lane road that extended, with little or no concealment, for about 100 km (62 mi) directly into enemy territory. Success depended on a rapid and sustained advance over an unblocked, two-vehicle-wide front and on the adjacent terrain being secured from German attack. Neither happened.

The attackers had been forced not only to deal with crossing three major rivers and their tributaries, but also to conduct armored operations within an area of marshes, swamps, and low agricultural land crisscrossed by innumerable drainage canals. Cross-country movement of heavy equipment was impossible and even an individual soldier could not go far in any direction without encountering a drainage feature. Major military movement in this area required that a main road, along with all of its bridges intact, be captured and secured.

After the battle the Dutch convincingly pointed out to the British the cause of the difficulty the XXX Corps encountered in moving between Nijmegen and Arnhem. In previously studying the problem in their Command and Staff

College, the Dutch concluded that success depended upon avoiding the elevated roadway near the Island, crossing the Rhine east of Arnhem, and then advancing into the city along the river's north bank. In the opinion of the Dutch military, the direct route between Nijmegen and Arnhem would not support a large-scale attack.[12] Unfortunately the British had not consulted the Dutch before Market-Garden.

Much information on the physical setting for this battle was readily available to those planning the attack, the basic tool of the terrain analyst being the topographic map. A 1943 edition of the Holland 1:100,000 Series topographic sheet clearly showed the elevated highways, numerous drainage ditches, and generally flat land. These difficulties are confirmed in the 21st Army Group Staff preattack route analysis (prepared from information provided by the XXX Corps), which notes the extreme restrictions of the terrain.

A tactical study of the terrain, prepared on 11 September by the 82nd Airborne Division's intelligence officer, described the land around Nijmegen as "flat, damp, and intensely cultivated ground of clay consistency. The woods where found are thick. . . . Roads are all embanked, and vehicular movement off of them in the low portions is impossible."[13] Members of the 1st Airborne's staff were aware of these terrain characteristics and considered the operation risky and the distances too great.[14]

Before the assault the British XXX Corps commander questioned the wisdom of advancing on such a narrow front. Lieutenant General Horrocks, recreating his briefing to his officers before the battle, said, "The country is very difficult, woody and marshy . . . only possibility is to blast our way down the road . . . country between here and Nijmegen, low lying, intersected by canals and large rivers."[15] It is questionable to what degree Montgomery was briefed regarding related terrain difficulties. As a seasoned leader of many armored operations, with a reputation for always being concerned for the welfare of his soldiers, Montgomery was doubtless aware of the restrictions and risks for forces moving along limited-access routes, which all argued against the operation. On the other hand, his intelligence reports indicated that the area was lightly defended and he may have been confident that a surprise airborne attack would overcome the disadvantages of the terrain and permit quick control of the vital corridor.

Montgomery also surely recognized the great advantage that would result from a successful thrust across the lower Rhine and into Germany in the early fall of 1944. Such a move might convert operations in Western Europe from Eisenhower's "broad front" policy, which distributed men and supplies more widely, into a well-supported, power-concentrated, left flank attack that Montgomery had repeatedly professed. A British advance might also have countered, at least to some degree, the V2 rocket attacks against his homeland. A major factor could also have been related to increasing English geopolitical concerns regarding a defeated Germany. Among all of this, pride and confidence in his leadership and his army may have been decisive factors. Whatever the explanation for devising Market-Garden, the plan was basically Montgomery's. In the end, a highly experienced, calculating, and generally cautious general, as shown in North Africa, Sicily, and Normandy, and possibly traceable to lessons learned in the Somme and Flanders during World War I, devised and commanded one of the most daring and audacious Allied military efforts of World War II and failed.

The British Airborne Corps' after-action report, commenting on landing zones, stated, "Had there been time to tap all sources of information and to check that information thoroughly, alternative and possibly better detailed plans might have been made." Possibly a more reflective assessment came from General Montgomery himself in 1949 when he said, "What is possible will depend first on geography, secondly on transportation in the widest sense, and thirdly on administration. Really very simple issues, but geography I think comes first."[16]

The Ludendorf Railway Bridge at Remagen, March 1945

The crossing of the Rhine and the establishing of bridge heads on the opposite bank was looked upon as a major operation, second in magnitude and importance only to the crossing of the English Channel and the landing on the beaches of Normandy.

U.S. ARMY HISTORICAL REPORT[17]

APPROACH TO THE RHINE

The failure of Operation Market-Garden stopped, for a time, Allied efforts to cross the Rhine River. Montgomery's forces refitted and then, after slowly clearing the Scheldt estuary and opening the port of Antwerp, battled toward the Meuse River and the Rhineland north of Aachen. By 4 March 1945 the British had secured most of the Rhine's west bank in their sector.

After costly November battles in the Huertgen Forest, and recovery from the fierce December 1944 Battle of the Bulge, Bradley's 12th Army Group resumed its advance across Belgium, Luxembourg, and northeast France toward the Siegfried Line, the West Wall, and the Rhine. With the First Army in the north and the Third in the south, Bradley's forces approached the Rhine gorge in late February.

Farther to the south, and under the command of Devers, the U.S. 6th Army Group and U.S. Seventh Army, along with French support, battled eastward across Lorraine and the Alsace toward the upper (southern) part of the Rhine Valley. Thus, in 1945 during the winter, three Allied commanders approached the same river in anticipation of the greatest crossing operation in military history (see Map 7.1 above).

The Rhine River, however, presented a different setting to each leader. Montgomery's forces would, as they had intended in Operation Market-Garden, cross the Rhine in its down-river deltaic section that forms the lowlands of the Netherlands and the open region that borders Germany. Once across, his troops could move on the Ruhr from the north. They would also have a more open path toward Berlin than the Americans had to the south. Bradley, on the other hand, would have to cross the middle stretch of the Rhine which, for the most part, occupies a narrow valley deeply entrenched between the Hunsruck and Eifel uplands to the west and the Westerwald and related low, dissected plateaus on the east. Farthest to the south, and upstream from Bradley, Devers was advancing across eastern France toward a steep-walled but wide and flat-bottomed Rhine Valley that extends, uninterrupted, from Worms southward to the Swiss border.

Thus, within a single theater of war, Eisenhower's broad-front tactics had brought several Allied armies close to the west bank of the most formidable river in Western Europe. Once there, each would be in a riverine environment unlike that of the other two and all would simultaneously make a forced crossing to a resisting enemy in formidable defensive positions. Furthermore, the three army commanders—Montgomery, Bradley, and Devers—must have recognized that failure to establish a bridgehead east of the river would significantly diminish their mission in the defeat of Germany. In total, these operations and their possible implications were nothing less than extraordinary.

Doctrinally, a deliberate river crossing requires that the attacker secure a firm foothold on the far shore through an assault by troops that have crossed in small boats and are possibly supported by airborne forces. Once across the river, their mission is to establish a bridgehead that is then expanded by reinforcements and rafting of support vehicles and equipment. Once the bridgehead is extended far enough to prevent direct enemy observation of the crossing site, appropriate bridges are extended across the river.

For the Rhine crossings, the Allies assembled a massive force of more than 70,000 engineer, transportation, and naval personnel. The engineers were responsible for the assault boats, rafts, and bridges. The transportation organizations offered tugboats. The naval contingents brought LCPs (landing craft for personnel) and LCMs (landing craft for equipment) along with various other naval craft. All of this equipment had to be transported overland to the river's west bank.

Troops in powered storm boats were to make the initial assault crossings. Once they reduced enemy small-arms fire and secured landing sites, army rafts and naval landing craft would ferry equipment to the far shore. Floating field bridges would be installed when the advancing troops could protect the construction teams.

Like Normandy, this detailed planning for bridging the Rhine began long before the fact. However, the many variables in warfare prevented planners from predicting exactly how and when events would unfold. The Allies bombed 22 of the existing Rhine River bridges to prevent the eastward withdrawal of German forces and the resupply of German forces on the west bank. Eisenhower anticipated that as the Allies attacked, the Germans would demolish all remaining bridges to prevent them from falling into Allied hands.

Faced with the likelihood that all bridges would have been destroyed and that a multitude of unknown or unavoidable problems would appear, Eisenhower's engineers prepared for simultaneous and deliberate Allied crossings of the Rhine along a broad front. Recognizing also that a chance for an unplanned assault-crossing might occur, Eisenhower gave special instructions that every unit should always be ready to exploit such an opportunity. Thus the operations leading to the Rhine crossings were to be developed "with greatest flexibility to take advantage of any opportunities which events might offer."[18] If, as expected, no quick, unplanned crossing was made, the entire west bank of the Rhine would be secured first. Then and only then would deliberate crossings begin.

Initially the assault would be a two-pronged effort. The main crossing, by the British, would be in the delta country north and south of Wesel. A secondary crossing was to be made by the Americans either in the vicinity of Frankfurt, where the Main River joins the Rhine, or in the Rhine gorge near Koblenz. In early March, U.S. and British army engineers massed in wooded areas near the Rhine to prepare for the crossings.

Final Allied plans called for Montgomery's 21st Army Group to, in a joint airborne and ground operation (Operation Plunder), seize four crossings in the plains north of Cologne on 23 March. Shortly thereafter, Bradley's 12th Army Group, led by Patton's Third Army as well as elements of the U.S. First Army, was to attempt crossings along the middle Rhine between Koblenz and Oppenheim. The enemy situation permitting, the Seventh Army would cross near Worms. Intelligence indicated that all of these operations would be costly in casualties and equipment.

In the area of the British 21st Army Group, where the land is low, flat, and open, the river's width varies from 275 to 450 m (900–1500 ft). Velocity was relatively low and depth more than adequate for the use of landing craft. Because this area offered scant concealment, it was likely that Allied troops here would be subjected to well-directed artillery fire.

In the U.S. Third Army's sector, the most northerly crossings would take place in the river's rugged gorge. Even farther to the south, Devers's troops would face a wide river meandering on the Rhine graben. Here, too, troops approaching the river would have limited cover and be subject to the enemy's observation.

By 26 March all of the assault crossings had been accomplished but only at the cost of the lives of many gallant soldiers. The toll, however, was far less than anticipated, largely because of an incredibly opportune event two weeks earlier at Remagen, south of Bonn. Here, in an unplanned operation unanticipated by all, elements of the U.S. 9th Division captured the only remaining bridge across the Rhine, permitting the rapid establishment of a major bridgehead that immediately threatened the flanks and rear of German units defending the east bank of the Rhine. This in turn drew enemy reinforcements into the area, reducing their response to numerous Allied crossings elsewhere. The U.S. troops'

RHINE FLOOD PREDICTION SERVICE

When planning began for the proposed spring 1945 Rhine River crossings, it became apparent that one of the most critical factors would be the level, or stage, of the Rhine River during the assault. Heavy spring rains combined with snow melt from the Alps could raise the river's surface as much as 8 m (25 ft) and flood nearby low ground. Since such an occurrence would greatly affect any tactical plan, maximum advance notice of such an event was essential. At worst, the assault would have to be postponed. At best, should the river flood, the bridges being erected would have to be lengthened and strengthened. If this was the case, given limits on available material would require that the total number of spans be reduced. In addition, the Allies knew that the demolition of several dams on the upper Rhine could change the river level abruptly, an event that would markedly complicate the crossing operations.

To ensure that adequate hydrologic information was available for the commanders and their engineers, the Supreme Headquarters Allied Expeditionary Force (SHAEF) created the Rhine River Flood Prediction Service. The unit was established with the 21st Weather Squadron of the Army Air Corps using people formerly assigned to the U.S. Army Corps of Engineers Headquarters in Washington, D.C., its Waterways Experiment Station in Vicksburg, Mississippi (which for years had been studying the Mississippi River), and the U.S. Weather Bureau. Until the west bank of the Rhine was occupied, this group analyzed records and historical data concerning the river, while continuously monitoring weather conditions throughout the watershed.

As Allied units approached the Rhine they sent daily reports to the service for use in its predictions. Precipitation was monitored throughout Allied territory and estimated for regions the enemy held. The winter of 1944–45 brought heavy rains to some regions of the Rhine basin, and the highest floods in years occurred in several reaches. By early spring of 1944, the Service believed it could accurately predict the river's stages several days before any crossing and could forecast any weather-related flooding. It would also be able to predict the impacts from demolition of one or more of the upriver dams.

As fate would have it, the actual assault crossings of the Rhine, including the one at Remagen, took place when the level of the Rhine River was extremely low for the season. Also, the Germans did not destroy the upstream dams. Nevertheless, the Rhine River Flood Prediction Service provided information that immeasurably aided day-to-day riverine operations.

---

fortuitous, opportunistic, and courageous capture of the Ludendorf railroad bridge at Remagen in March 1945 is now legendary. In numbers involved, outcome, and impact, it is also a startling contrast to Operation Market-Garden.

REMAGEN

Elements of the 27th Armored Infantry Battalion approached Remagen on 7 March 1944. The town, in the upstream section of the Rhine gorge, is on the river's west bank, at the base of a steep slope forming the edge of the Eifel. Directly across the Rhine are the cliffs and high hills that form the precipitous western boundary of the Westerwald plateau.

FIG. 7.2. The Ludendorf Railway Bridge (at Remagen) in the distance and the recently established pontoon treadway bridge across the Rhine River (constructed by the U.S. First Army Engineers) in the foreground as they appeared on 11 March 1945. On 17 March the damaged railroad bridge collapsed, but by then the Allies had established a firm position on the east side of the Rhine River. (U.S. Army photo.)

To their amazement, the first U.S. scouts to view the entrenched river from the high ground immediately to the west saw that the 300-m-long (1000-ft) railway bridge was still intact. After some delay involving questions of command and existing orders, U.S. soldiers advanced through the town to approach the west end of the bridge, surely thinking that at any second the Germans would destroy it.[19] Then, heroic members of a vanguard platoon fought their way onto the bridge as defending Germans tried to set off demolition charges already in place. From span to span, the Americans fought their way across the bridge before the retreating Germans could detonate most of the charges. More than 640 kg (1400 lb) of explosives were removed from the bridge as it was being captured and secured. One major explosion did severely damage several important structural supports, but the bridge did not collapse.

Because of the rugged terrain, difficult approach, steep riverbanks, and swift current, the Allies had not considered Remagen as a major crossing site. This also became apparent to the Germans as they observed U.S. operations on the west side of the river. Thus, with the bridge prepared for destruction and the chance of an Allied crossing at Remagen unlikely, the Germans must have thought the east bank of the Rhine across from Remagen reasonably secure. Aside from the patrolling security forces and the bridge demolition teams, the

Germans had few troops in this area, their forces being concentrated at obvious crossing points elsewhere. Destruction of the bridge would, however, be delayed as long as possible to permit withdrawing German units to cross to the east side of the river.

The Americans' unexpected arrival and aggressive action changed the situation immediately. With the capture of the bridge, Germany lost control of the Rhine's east bank near Remagen. Within 24 hours, 28,000 troops poured across the bridge to establish firm positions east of the river. Day and night, German swimmers, boats, aircraft, artillery, and rockets targeted the bridge, and occasional hits threatened its already questionable stability. Meanwhile, with the Remagen sector increasingly secure, elements of the First Army erected two tactical bridges across the Rhine; these were in operation by 11 March. When the tactical bridges opened, the Ludendorf bridge was closed for repairs. Six days later, on 17 March, the bridge collapsed, taking with it 28 engineer soldiers, all losing their life in the Rhine.[20]

The numerous other planned crossings of the Rhine took place about on schedule.[21] Between 2200 on 22 March and 2400 on 28 March, 24 floating tactical bridges—whose length totaled 8000 m (over 26,000 ft)—traversed the Rhine in the largest river crossing operation ever. More than 100,000 British

FIG. 7.3. Pontoon bridge across the Rhine River at Remagen, 19 March 1945. From the water level on the distant shore it is clear that the river was low. (U.S. Army photo.)

and U.S. engineers participated in the effort. Yet casualties were surprisingly low. The 21st Army Group (including the U.S. Ninth Army) encountered light opposition; the U.S. XVI Corps suffered only 38 killed in action in their operation. At Oppenheim the Third Army's attack was virtually unopposed, although the crossing of the river south of Remagen did draw moderate casualties. In traversing the Rhine the Seventh Army lost only 18 people.

Numerous casualties were, however, inflicted at the Remagen site during the seizure of the bridge and subsequent construction of the float bridges, but far fewer than anticipated for such a crossing. As the U.S. V and VII Corps fought hard to seize and hold the bridgehead, they captured 11,200 Germans while losing 863 men in action between 7 and 14 March.

Although successful, the crossings were certainly not easy and far from simple. During the assault phase, the swift current of the Rhine swamped many of the storm boats and infantry support rafts. Underpowered tugs and power boats frequently were unable to push rafts against the current and went out of control. In some cases these runaways damaged, and even severed, tactical bridges downstream, disrupting construction or halting crossings on completed bridges. In the gorge, where enemy fire was heavy and the current swift, the crossings were made all the more difficult. Numerous assault craft, needed to carry combat soldiers to the river's far side, never returned for subsequent loads, having been sunk, disabled on the river, or abandoned on the distant shore.[22]

Many months of careful planning and expert preparation had gone into the successful Rhine crossings. But much credit must also go to the courage of soldiers who charged and captured the bridge at Remagen. Gen. Albert von Kesselring, the German commander who saw his reserves rush to the initial bridgehead in "a magnetic way," thus preventing reinforcement in other areas where the main attack was expected, spoke of the "Crime of Remagen. It broke the front along the Rhine."[23] Hermann Goering believed that Remagen "made a long Rhine defense impossible." Eisenhower saw it as "one of those rare and fleeting opportunities which occasionally occur in war and which, if grasped, have incalculable effects in determining future success."[24]

Conclusions

Although involving the same river, Market-Garden and the capture of the Ludendorf Railway Bridge differed in just about every way. The first evolved from a daring, multifaceted mission that might bring spectacular results. Yet the plan can also be justifiably questioned: It directed three airborne units to capture a series of bridges intact, each then to be reinforced within four days by an armored corps ordered to penetrate about 100 km (62 mi) along a single road in the face of a determined enemy. Although success might place a British army east of the Rhine on the North German Plain in late 1944, it did not guarantee that a deep thrust into Germany would follow. With Montgomery far forward from the Americans, it might instead have brought a German counterattack designed to trap his troops on the wrong side of the river. Germany's ability to conduct such a maneuver must be considered possible because Market-Garden predated Hitler's formidable December Ardennes offensive (Battle of the Bulge) by about three months. Moreover, the whole operation would have been thwarted if any one of the major bridges had been destroyed on the first day of the battle.

But such thoughts are speculative. The irony of Market-Garden is that all three of the major bridges were, at one time or another, captured intact by Allied airborne forces and were fully suitable for the planned river crossings. It was the inability of the XXX British Armored Corps to advance the full distance along a single connecting road that doomed Market-Garden. Thus in the end it was the riverine environment rather than the rivers that stopped the Allies.

Although months of planning preceded the 1945 Rhine crossings, none of it focused primarily on the Remagen area in the Rhine gorge. Here, in an unplanned operation, relatively few soldiers crossed the largest river in Western Europe and established a bridgehead before the enemy could respond in force. This fortuitous event accelerated Germany's certain defeat and surely lessened river-crossing casualties. The striking contrast in the failure and aftermath of Operation Market-Garden and the successful capture of the Ludendorf Railway Bridge is not unique nor should it be surprising. Just as no two rivers are alike, neither are any two military operations in a riverine environment.

CHAPTER 8

Glaciers Shape the Land

Alpine Fighting and the Road to Moscow

Those who wage war in mountains should never pass through defiles
without first making themselves masters of the heights.
MAURICE DE SAXE, *MES REVERIES*

During the Pleistocene Epoch, also known as the Ice Age, huge glaciers repeatedly gouged, scraped, and reshaped the earth's surface in a fashion unlike any
of the other geomorphic processes. In the higher mountains of the middle
latitudes, alpine glaciers sharpened peaks, deepened valleys, and steepened
slopes to transform the landscape into some of the most spectacular topography on the planet. Such prominent and formidable barriers can serve as natural boundaries between nations. However, if the adjacent countries go to war
these most rugged of all mountains can become the most demanding and
hazardous battleground imaginable.

In contrast, during the Pleistocene other glaciers of continental proportions
repeatedly covered northwest Eurasia and northern North America to alter the
surface in a very different fashion. These huge ice-caps covered everything,
scraped and beveled even the highest land, excavated extensive rock basins to
be later occupied by the Great Lakes and the Baltic Sea, transported enormous
amounts of soil and rock to new locations, and disrupted or obliterated preexisting drainage lines. Deglaciation then revealed a new array of topographic
features, including exposed polished and scratched bedrock in some places,
and piles of debris forming rolling and hummocky hills elsewhere. The jum-
bled nature of glacial deposits resulted in tens of thousands of new lakes,
numerous rivers that in places seem to wander aimlessly across the land, and
vast swampy areas. And, because this debris is unconsolidated and often thick,
it provides an ideal parent material for soil that, with changes in moisture
content, can easily support military traffic one day and arrest it the next.

Many important military engagements have been fought on glaciated landscapes with topographic characteristics so influential that they guided maneuver and affected combat tactics, sometimes even figuring in the outcome of
battle. The sites for forts at Montreal and Quebec, pivotal outposts in the
struggle between England and France in North America, derive much of their
prominence from glacial erosion of weaker rocks in the Saint Lawrence lowland and steepening of the edge of adjacent and more resistant hills. Bunker
Hill and adjacent Breed's Hill, so important in the American Revolutionary

FORMATION OF THE BALTIC SEA AND THE GREAT LAKES

Map 8.2 (below) shows the extent of Northern Hemisphere Pleistocene ice sheets. There are some striking parallels between the direction of ice flow, glacially scoured terrain, and resulting deposits in European Russia and the Great Lakes area of the United States. The real and mirrored images of the areas (Map 8.1) illustrate the similarities.

Both ice sheets rested upon and scoured very old, resistant rocks to the north (Laurentian and Scandinavian shields), eroded adjacent, less-resistant formations to form deep basins now filled with water (Baltic Sea and Great Lakes), and deposited an enormous amount of material to the south, some forming a low but conspicuous continental drainage divide with several especially convenient portage points (Fort Dearborn, Fort Wayne, Minsk, Smolensk). (William D. Thornbury, *Regional Geomorphology of the United States* [New York: Wiley and Sons, 1965], 217–45.)

The point here is that, when viewed geographically, places that may seem completely different from one another can actually have striking similarities that can enhance understanding of the land. In this instance the controlling process has been glaciation but similar parallels can be found for just about all aspects of physical geography. And many of these relate to warfare.

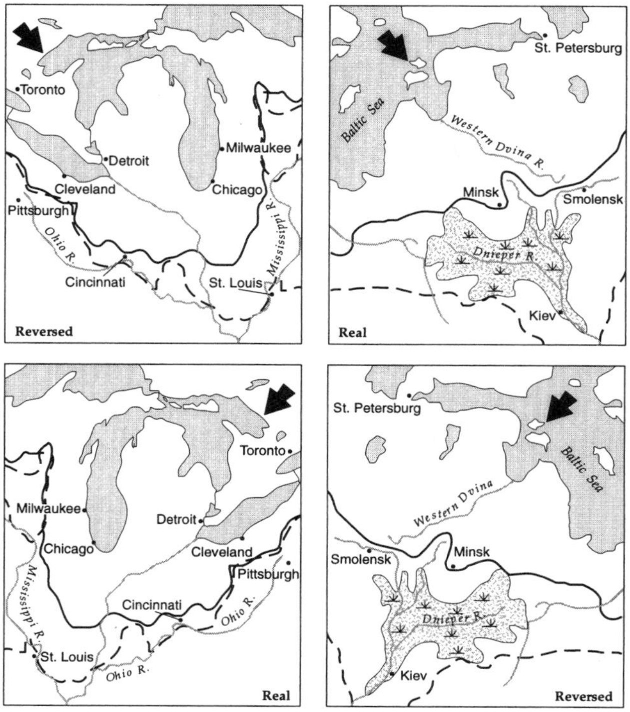

MAP 8.1. Real and reversed maps of the Great Lakes area and northwest Eurasia. The black arrows show the general direction of glacial movement. Dashed lines approximate the maximum position of ice during the Pleistocene, while the solid line shows the outermost ice margin during the last stage of glaciation. (Based on maps by Prof. Norman Meek, Department of Geography, California State University–San Bernardino, who first used real and reversed maps to illustrate geomorphic similarities of the Great Lakes area and European Russia. Thanks are extended for permission to reproduce his work here.)

War, were sculpted into their elongated and streamlined forms by action at the base of a vast glacier that also formed Cape Cod, Nantucket, and Martha's Vineyard nearby. The battle on western Long Island at Brooklyn Heights, involving troops of Gen. George Washington, was fought on low, rolling hills of debris deposited at the margin of the same ice sheet about 17,500 years ago. The famous elevated but flat plain, where West Point cadets conduct their celebrated parades, is underlain by glacially derived sediments deposited between the ancient, resistant rock of the Hudson Highlands and the edge of a shrinking glacier occupying the Hudson River gorge. The Great Lakes, of special importance in the War of 1812, occupy giant basins excavated largely by glacial action. Several military outposts, such as Fort Wayne and Fort Dearborn, were situated near portages across glacially formed hills that separate headwaters of two great drainage systems, one extending to the Atlantic Ocean and the other to the Gulf of Mexico.

Mountain glaciers reshaped the landscape to create high passes such as those that are associated with Hannibal's famous alpine maneuver of the Punic Wars (218 B.C.). Equally impressive are the movements of Napoleon Bonaparte's forces over the same mountains, followed by his effective tactical movements during the last half of 1796 near the glacially eroded basin of Lake Garda, culminating in the January 1797 battle at Rivoli. More recent fighting in similar terrain, such as the glaciated areas along the India-Pakistan border and in Afghanistan, shows that conflict continues in high-mountain areas.

Although each of these battlefields was in some way shaped by moving ice, the impact of continental and mountain glaciation on military activity is especially well shown by two extensive and dramatic lines of battle. The first is the incredibly rugged alpine landscape that formed part of the setting for the World War I conflict between Austria-Hungary and Italy. The other is the low but conspicuous terrain produced by continental glaciation that profoundly influenced lines of march eastward toward Moscow by three invading armies during three successive centuries. But where did these glaciers come from? When did they form? And why did the same process have totally different impacts on mountains and plains?

The Great Ice Age (Pleistocene Epoch)

Although the earth has experienced extensive glaciation many times in its long history, by far the most obvious effects are related to the last Ice Age, known as the Pleistocene Epoch. Probably beginning about two million years ago, there was a succession of glacial events separated by ice-free episodes with climates similar to that of today. Thus the notion that glaciation is the result of a cooling planet is disproved by ancient ice ages and the numerous warmer phases (including the present one) during the Pleistocene.

Numerous theories address these separate glacial events, such as variations in solar insolation, progressive changes in the earth's rotation and revolution properties, differences in the amount of carbon dioxide in the atmosphere, movement of the continents, alterations in ocean currents, the influence of high-altitude land barriers on atmospheric circulation, and combinations thereof. Be that as it may, the exact cause of repeated and widespread glaciation during the last two million years remains a complex and unsolved problem. Until that question is resolved there is no way of determining if, in fact, the Pleistocene has ended.[1]

Even so, the evidence is overwhelming that ice-caps of continental proportions repeatedly formed on northern North America and northwestern Eurasia; at the same time mountain glaciers in both the Northern and Southern Hemispheres were both more numerous and extensive. The Eurasian and North American ice-caps covered millions of square kilometers, were many hundreds of meters thick, remained centered on the upper-middle latitudes (rather than the North Pole), and markedly increased the earth's reflectance, or albedo (Map 8.2).[2] The last of these great glaciers reached their greatest extent around 22,000 B.P. (years before present), but by 6000 years ago they had both completely disappeared.[3]

The effects of repeated glaciation were not limited to areas overrun by ice. Because evaporation of ocean water was the basic moisture source for ice that accumulated on the land, sea level fluctuated worldwide, during glacial times dropping more than 100 m (330 ft) to expose a land bridge between Asia and North America and to connect the British Isles with Europe. During warmest episodes, when even the ice-caps of Antarctica and Greenland melted to smaller sizes than they are today, sea level was at least 30 m (100 ft) higher, enough to submerge part of the Capitol Building in Washington, D.C. Such changes altered the shorelines of all continents and islands. The extent and boundaries of the world's climatic zones, with their differing temperature and precipitation patterns, also shifted as the size of the glaciers repeatedly grew and shrank. This, too, affected the distribution of plants and animals, forcing massive displacement and repopulation as the ice margin advanced and retreated with climatic change. The fluctuating conditions also influenced the habitat and way of life of early humans.

If a continent-sized glacier ever returns to North America it would very likely cover New England, New York, the Great Lakes states, the northern Great Plains, much of Alaska, and just about all of Canada. Now-fertile farmland near the ice would become cold forests and tundra. Millions of people would be displaced, forcing inordinate changes on others. Ecological, political, social, and economic effects would be staggering. And a similar scene would be occurring in Eurasia. If glaciation does reoccur it is likely to be many thousands of years in the future. The important thing now is to understand how glaciers that disappeared so long ago have left their distinct mark on today's landscapes.

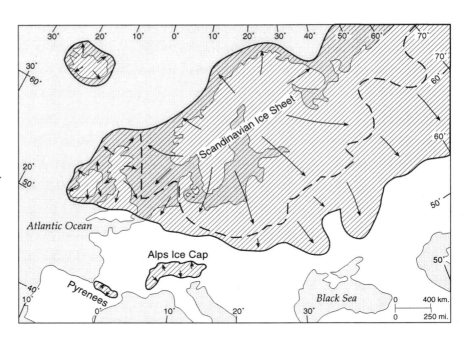

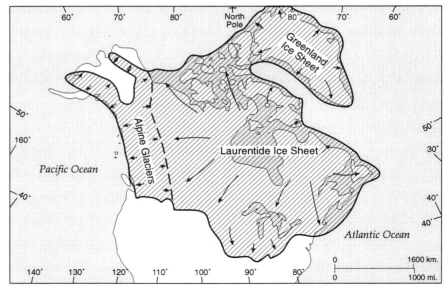

MAP 8.2. Maximum extent of the Scandinavian and North American ice sheets during the Pleistocene are shown by the solid line. The dashed line on the European map represents the outermost position of ice during the last major glacial advance. The dashed line on the North American map marks the approximate border between the continental ice-cap and alpine glaciers from the Cordilleran Ranges. Note the large difference in scale between maps. (Modified from several sources including Richard F. Flint, *Glacial and Quaternary Geology* [New York: Wiley and Sons, 1964], and Alan Strahler and Arthur Strahler, *Introducing Physical Geography* [New York: Wiley and Sons, 1994].)

Glaciers and Their Effects

Ice is a curious substance. Unlike all other relatively abundant solids at the earth's surface, it commonly has a temperature very near its melting point. Such a condition facilitates movement and recrystallization of molecules, or metamorphism. Most ice is derived from the modification of new "powder" snow into a denser "corn," and then "firn" (granular material). Further compaction forms ice.[4] As the ice thickens, internal compressive forces may become great enough to cause the movement that defines a glacier.[5]

The size and location of today's glaciers range from those occupying small, high-altitude rock basins to the ice-caps of Antarctica and Greenland. Altogether, these glaciers cover about 10 percent of the earth's land area (or 3 percent of the earth's total surface). During the Pleistocene glacial episodes the coverage increased to about 30 percent of the land, to affect a considerable amount of the world's terrain directly.

Colder, more brittle glaciers in polar regions or at higher elevations flow less readily and are generally frozen to underlying rocks, thus reducing their sliding movement and related basal abrasion. In contrast, more temperate mid-latitude glaciers—so extensive during the Great Ice Age—deformed much more easily and were not everywhere frozen to underlying material, conditions that increased their potential velocity. As a result, at some places the base of this ice could move directly over the underlying surface. But ice near its melting point is softer than a fingernail, meaning that it cannot abrade underlying bedrock. How, then, do these glaciers scour the land and make their distinct marks on the earth's surface?

Pressure, melting, and refreezing at the bottom of a temperate glacier result in the widespread incorporation of underlying rock fragments into the base of the ice. Once this debris is embedded and carried along with the glacier, it acts as an abrasive to polish, scratch, and groove the bedrock surface. As a result, with deglaciation, many areas of severe glacial erosion have smoothed bedrock now at or very near the surface.

Somewhat like flowing water, a glacier can both transport and deposit eroded material. But unlike the turbulence in streams, the ice in glaciers generally remains layered as it moves. Thus, much of the incorporated debris remains in one horizon within the glacier, all of it moving along at the same velocity as the ice. Large boulders are transported with the same efficiency as adjacent microscopic-sized particles without significant sorting by size, shape, or weight. All of this material can act as a powerful abrasive wherever it grinds over underlying bedrock. Eventually this eroded debris, called drift, is deposited either beneath the glacier or at its melting edge.

Sediment deposited directly from the ice forms a heterogeneous mixture of particles ranging in size from mud-making clay and silt to cobbles and large boulders; it is called till. As the glacier melts, other released debris may be carried away and deposited by streams that, during the warmer season, are numerous on and near the edge of the ice. The resulting glaciofluvial sediment is a water-sorted, stratified sand and gravel formation. Because of their smaller size, most clay and silt clasts remain suspended by stream turbulence, eventually to be deposited as layers of glaciolacustrine mud if the water becomes ponded. Finally, if the deglaciated landscape is barren of vegetation, wind can move exposed sand grains to form dunes, and smaller silt-sized particles may become airborne, eventually being deposited as loess, a form of windblown dust (Figure 8.1).[6]

Stratigraphic relationships among these different types of sediments are

FIG. 8.1. Diagrams showing the formation and resulting nature of landforms produced along the margin of a continental glacier. (Modified from a drawing by Arthur Strahler.)

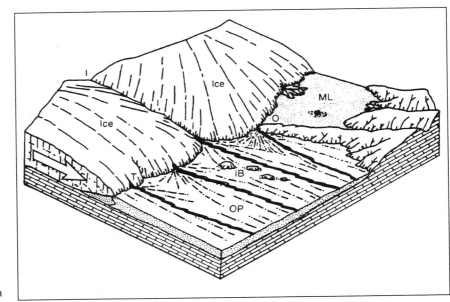

I = Interlobate areas
IB = Ice block
ML = Marginal lake
O = Lake outlet
OP = Outwash plain

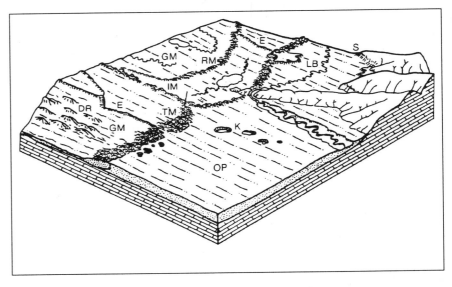

DR = Drumlins
E = Esker
GM = Ground moraine
IM = Interlobate moraine
K = Kettle
LB = Lake bottom
OP = Outwash plain
RM = Recessional moraine
S = Shorelines
TM = Terminal moraine

generally complex, partly due to the ever-changing nature and position of the ice margin and runoff waters. The complexity is well illustrated by variations in the behavior of the glacier's terminus. Although ice constantly moves toward the edge of an active glacier, its margin may act independently, advancing, retreating, or remaining in the same position. This is because movement of the glacier's edge is determined by the balance between the rate of ice advance and the amount of peripheral melting.

A glacier advances when the ice moves forward faster than its margin melts. Conversely, whenever the rate of marginal melting exceeds forward movement of the ice, the glacier's terminus retreats to release and uncover till that was previously deposited beneath the ice. The result, called ground moraine or till plain, is gently rolling land with low knolls and shallow depressions, possibly interspersed here and there with lakes and swamps.

When the rate of glacial advance about equals the amount of marginal melting, the ice continues to move forward, yet the terminus remains in the same place. As a result, large amounts of drift are deposited within a relatively narrow zone, with the volume of debris dependent upon the amount of sediment available in the melting ice, the rate of glacial movement, and the duration of this near-equilibrium condition. These rather special relationships happened frequently during the Pleistocene Epoch and produced numerous linear belts of higher, hilly land called end moraines. End moraines have added significance in that each records a former position of the glacier's edge. Studied collectively in terms of successive positions and geographic relationships, they reveal the evolving shape of the ice margin during the last deglaciation of the Pleistocene Epoch.

As the end moraines and till plains formed, numerous streams flowed outward from the glacier and deposited sand and gravel upward to a common level. The outwash plains thus formed slope gently away from the ice margin. Glaciolacustrine muds also slowly settled from ponded meltwater to flatten lake bottoms. Deglaciation drains many of these lakes to reveal flat, relatively low, swampy areas that differ markedly from either the linear belts of rolling hills or the well-drained, gently sloping outwash plains.[7]

Europe is a good place to study glacial effects because during the Pleistocene it was host to both glaciers in the Alps (and other high mountains) and a Scandinavian-centered continental ice-cap that reached well into the British Isles, Germany, and Russia (see Map 8.2 above). It may also be of added interest to those concerned with relationships between terrain and war because so many battles have been fought on Europe's glaciated landscapes.

Military Consequences

THE ALPS, WORLD WAR I

The theory that the earth's crust consists of moving, yet rigid plates that collide with one another may be applied to the formation of the European Alps (see Map 6.1).[8] Earthquake and volcanic zones, along with contorted rock strata, indicate that ongoing crustal compression is forming Europe's highest and most rugged mountains. At some places such formidable mountain ranges have served as effective buffers for competing nations, such as the Pyrenees between France and Spain and the Andes separating Argentina and Chile. Elsewhere, however, mountainous terrain may seem vital enough to fight over. Examples include military encounters between the former Soviet Union and both China and Afghanistan, and boundary disagreements among China, India, and Pakistan. Wherever this happens, control of key corridors and mountain passes is of great tactical importance. This, in turn, makes it necessary to secure adjacent uplands and mountains. The recurring principle is that rugged, mountainous terrain often provides good advantage to the adequately trained and well-positioned defensive force. Nowhere is the relationship between an alpine landscape and battle better illustrated than in the World War I fighting along the mountainous frontier between Austria and Italy.

The Alps. The exact origin of the word *alp* is obscure, but the term now refers to the mountain meadows above steep-walled valleys and below the bare rock and ice of higher summits. It is also the proper name for the "Alps" as we know them today, the mountain group that extends in a giant arc northward along the French-Italian border, then northeastward through Switzerland and northern Italy and eastward into Austria. To understand the nature of these mountains, we must look back about 150 million years. Then southern Europe and northern Africa not only had different outlines but were also separated by the Tethys Sea, a water body much wider and longer than the present Mediterranean Sea.

The Tethys Sea served as a collecting basin for large amounts of sediment that eventually solidified to rock.[9] According to plate tectonic theory, Europe and Africa are now advancing slowly toward one another. This movement compresses thick intervening marine sediments into enormous folds so complex that, in some places, their crests are overturned and the rocks have broken to slide along giant thrust faults. This deformation has, however, been sporadic rather than constant, resulting in erosional episodes between times of renewed compression. Nearby volcanoes and numerous earthquakes indicate that these mountain-building forces are still at work today.

Geologists suggest that this compressional movement has reduced the width of the area between Africa and Europe by about 250 km (150 mi); such great compression is accommodated not only by folding and faulting but also by the uplifting of material from deep within the earth's crust. Erosion has been so great that more than three-fourths of the uplifted rock has been carried away by diverging streams such as the Danube, Po, Rhine, and Rhone, only to be redeposited as younger sediment elsewhere. As a result, igneous and metamorphic rocks, originally formed far beneath the crust, are now at the surface in a central alpine zone (Map 8.3).

As such, the Alps may be divided into three major linear geologic realms: an interior of uplifted crystalline rocks, bordered on both sides by severely deformed sedimentary strata. Because the rocks differ in susceptibility to weathering, the eroded structures are reflected in the terrain as linear valleys and high ranges, with most paralleling the overall arcuate trend of the crystalline Alps. Thus, for the most part, this massive mountain system presents not one, but several, topographic barriers to north-south movement in this part of Europe.

Although only midway between the equator and the North Pole, the high altitude of the Alps made them ideal sites for Pleistocene glaciers. Even today they contain more than 1200 separate ice fields. During colder times, ice

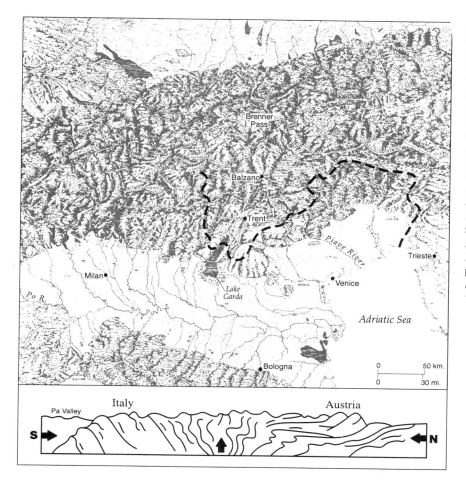

MAP 8.3. Map and north-south geologic cross-section of the central Alps. The dashed line extending from near Trieste to the Swiss border represents the average position of the Austrian-Italian front during most of the war. (Map modified from diagram by A. K. Lobeck in Army Map Service, *Continental Europe—Strategic Maps and Tables*, pt. 2, *South Central Europe*, U.S. Army, 1943.)

covered all except the highest peaks and filled most valleys to near capacity. The glaciers formed because the mountains forced moist air from the nearby Mediterranean Sea and Atlantic Ocean upward, thus lowering the temperature enough to produce abundant snow. Because 50 percent of global water vapor is in the lower 2000 m (6500 ft) of the atmosphere, more snow fell on the mountain slopes and valleys than on the highest summits.

The Pleistocene's alternating warm and cold climates repeatedly formed and melted glaciers in pockets and basins high on the sides of mountains and ridges.[10] Periodic glacial erosion enlarged these features into steep-walled, amphitheater-like forms, or cirques. Because of their flanking and lower position, the growth of cirques also sharpened mountain peaks into spires, or horns (for example, the Matterhorn), and high serrated ridges (aretes) with slightly lower notches, or passes, known as cols (Figure 8.2).

As these glaciers moved downward, they coalesced with others eventually to form flows of ice in valleys originally carved by the work of streams. Unlike the rivers that eroded only the bottommost part of the valleys to create a V-shaped profile, glaciers scoured both the base and the walls to modify the original

FIG. 8.2. Diagrams showing the development of features produced by mountain glaciation. (Modified from drawings by W. M. Davis, "Glacial Erosion in France, Switzerland, and Norway," 1900; A. K. Lobeck, *Geomorphology: An Introduction to Landscapes* [New York: McGraw-Hill, 1939]; and Alan Strahler and Arthur Strahler, *Introducing Physical Geography* [New York: Wiley and Sons, 1994].)

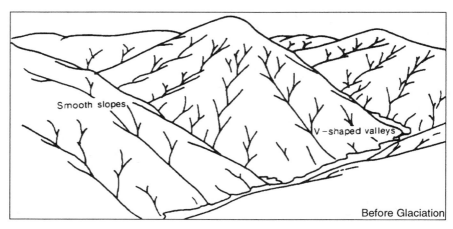

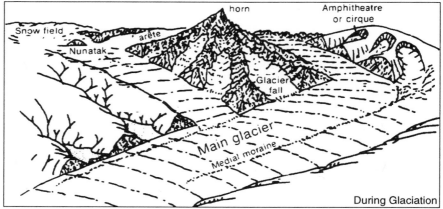

valley into the form of a U. At some places, the widening and deepening also produced elongated rock basins, the larger now occupied by picturesque lakes such as Iseo, Como, and Garda. Some valleys were so markedly eroded that their walls are now nearly vertical, leaving today's adjacent tributary streams left hanging high above.

In the Alps, most ice-eroded valleys parallel the dominant arc-shaped compressional structural trends. But some preglacial stream valleys also developed along less-dominant cross-cutting zones of crustal weakness. Glaciers modified these features as they flowed out from the mountains to deposit debris on adjacent plateaus and plains. Deep erosion along one of these cross-cutting zones of weakness converted several river gorges into one north-south U-shaped valley, unlike any other in the Alps, because it contains only one relatively low barrier, Brenner Pass (1500 m; 4500 ft). This valley offers the easiest and most direct passage between Italy and central Europe, and it has been of great military importance for centuries. Julius Caesar, Attila (King of the Huns), and Napoleon Bonaparte are among the warriors who used it. But the greatest battle for control of this corridor and the nearby alpine terrain with its numerous ice fields lasted three years and was fought between Austrian and Italian soldiers during World War I.

The War. Italy entered World War I during May 1915 by declaring war on Austria; geography quickly determined the lines of battle. Inhabiting a peninsula and fighting a land-locked enemy, the Italians were threatened nowhere along their extensive coastline. Furthermore, Italy's western and northwestern boundaries with ally France and neutral Switzerland were also secure. Thus the conflict was restricted to the northeast frontier and consisted of two major salients: a large, Italian-occupied, eastern bulge anchored on the south by the north shore of the Adriatic Sea; and an extensive, southward protuberance centered on the Trentino Alps, which placed Austrian troops in the high, southern fringes of the mountains overlooking the densely populated and fertile Po Basin. The west end of the battle line was anchored on Stelvio Pass (2715 m; 9050 ft) near the then-common boundary of Austria, Italy, and Switzerland and dominated by the Adamello and Ortler Alps. Between these two salients, the battle line extended along the topographic and structural trends of the Dolomite and Carnic Alps (see Map 8.3 above). Thus, with the exception of the far-eastern front north of Trieste, the entire front was within the deeply glaciated Alps.

As if diabolically contrived to equalize matters, the two fronts offered attractive offensive opportunities to the Austrians in the west and the Italians in the east. Terrain in the eastern front was somewhat more suitable for maneuver, and successful Italian advances here could lead to the fertile and vital Hungarian Plain. In contrast, the center of the western salient rested on the glacially eroded and tactically vital north-south valley containing Brenner Pass.

The Italian plan was to push the western alpine front northward with hopes of securing more depth and better defensive positions in the mountains, while at the same time mounting a major offensive in the east designed to threaten the enemy's heartland. In contrast, Austria's favorable position in the southern alpine borderlands offered an invasion route leading to the Po Basin; an advance there would threaten lines of communication and the rear of Italian forces to the east. Other more advantageous, large-scale opportunities for military operations were not available to either side. Thus, the geography that defined the theaters of war for Austria and Italy resulted in long-term battle in some of Europe's highest mountains.

During the first three weeks of the war Italian forces assumed the offensive in both salients. Northward movement in the western Alps was especially important because forward positions of the Austrians permitted observation of some Italian operations in the mountain valleys and foothills, and even the northern edge of the Po Valley. But aggression here quickly slowed as each advance brought higher, more rugged mountains and increasingly difficult logistical problems. Tortuous winding roads with steep grades, narrow passages, and numerous switchbacks were built into the glacially scoured valley walls to reach the higher battlegrounds on the uplands and alpine meadows. Vehicle size was extremely limited and where the roads ended, further movement toward the higher serrated ridges, cirques, and alpine peaks was by trail and small aerial trams. Special alpine ski forces dressed in white fought on both snow and ice. In the Ortler Alps, nearest Switzerland, Austrian troops occupied the Matterhorn-like summits of Cima di Campo (3480 m; 11,600 ft) and Konigspitz (3800 m; 12,660 ft).[11]

In these highest areas, most movement and resupply was by foot, in air with a third less oxygen than at sea level. With less atmospheric filtration and absorption, the effects of solar radiation could be debilitating. Storms and high winds with blowing snow were common, while pleasant days were rare. As the seasons progressed, temperature varied greatly between sun and shade, between night and day, and with daily weather changes.

Even so, the glacially scarred terrain presented the greatest challenge. The location of some military sites foiled even the most skillfully directed artillery fire. Furthermore, artillery shells shattered bedrock and multiplied their shrapnel effect; ricochet was common. Movement across a snowfield or bare rock surface offered little or no concealment from the enemy. Such conditions encouraged tunneling. Some excavations went through rock; others even extended through glaciers, requiring bridging where deep crevasses were encountered.

The open, steep-walled valleys, rugged, barren upper mountain slopes, exceedingly high ridgelines, and sharp, crested peaks provided favorable positions and visibility for defensive forces. It was in this setting that the Italian advance was halted within a few weeks, not to be effectively renewed for more than three years. Although the ground gained was modest in area, it later became increasingly valuable because it provided needed maneuvering depth in the mountains, permitting the Italians to establish several successive defensive lines during the following year.

Although the Italians made minor advances, the Austrians retained their position of advantage in the Trentino. Both flanks of their western salient lay along some of the highest and most defensible mountain terrain in Europe. Equally important, a mere 50-km (30-mi) advance southward across the Asiago Plateau would take the Austrians out of the mountains and onto the Po

Plain to threaten both Venice and lines of communication for Italian troops to the east.

Austria's mountain offensive began in mid-May 1916. It required a large preliminary buildup of men and materiel that moved into the area, some from other fronts, mainly by way of Brenner Pass and the north-south U-shaped Adige River valley. More than 300,000 men and 2000 artillery pieces (including three 420mm guns) were involved in the operation. Anchored on the right by Lake Garda (made militarily famous by Napoleon more than 100 years before) and on the left by the Dolomite Alps, the Austrians pushed southward at the center to recover territory previously lost. The advance overran several of the enemy's defensive lines and pushed into Italy, ever closer to the southern edge of the Alps. But the offensive also brought increasing attrition, distance, and resistance. Gradually the drive slowed and then finally halted in distant view of the Austrian objective—the Po Basin—at the Italians' last major mountain-defensive line.

Within a month this Austrian offensive was suspended. Eventually, as a result of counterattacks and the removal of Austrian troops to fight on the Russian front, the Italians recovered about half the ground they had lost. The front then gradually stabilized with the widespread establishment of even stronger defensive positions. The question had become, "How can armies move forward in these rugged mountains?"

By September 1917, the Austrians and the Italians had fought 11 major battles in the other salient along the Isonzo River, but those battles had resulted in only modest changes in the front. Then, in late October 1917, the Battle of Caporetto (Twelfth Battle of the Isonzo) to the east changed everything. With German military assistance and utilization of Oskar von Hutier's new deep-penetration storm-troop tactics focusing on key defensive positions and lines of communication, the Austrians moved across the Isonzo River and the rugged landscape near Caporetto to follow valleys that opened onto Italy's northeast lowlands. Success was dramatic. In just a few days the battle line was pushed 80 km (50 mi) westward across the plain to within 30 km (20 mi) of Venice.

This advance immediately threatened the flank and rear of Italian troops to the north in the Carnic Alps, forcing their withdrawal to the west and south. The Austrians also renewed their offensive in the west, forcing an Italian retreat from much of the Dolomite Alps. By the time the Austrians and Germans were finally stopped at the Piave River near Venice on 12 November, the Italians had lost more than half their Alpine territory. Thus, the front changed more in 20 days than it had in the previous two years. Even so, the Italians still controlled the vital southern outlet of the Adige Valley that extends north to Brenner Pass.

For the next six months, battle lines remained essentially static while other factors changed dramatically. German troops were removed to support their final 1918 offensives on the Western Front. Russia was then out of the war, releasing Austrian troops from that battleground; the Italian commander had been replaced; and the Austrians were experiencing increasing internal political problems at home.

In June 1918, during this time of internal change and crisis, the Austrians launched their final offensive. Instead of concentrating on one effort, they attacked on both fronts: southward in the Alps toward the Po Plain and westward across the Piave River toward Venice. Each failed with significant losses, further weakening an overextended army and a disintegrating government.

As conditions on the Western Front became more favorable for the Allies during the summer and early fall of 1918, the Italian troops were reorganized and resupplied. In October they mounted a two-pronged attack, one thrust northeastward on the coastal plain, the other north toward the Dolomite Alps. At the same time, smaller offensive efforts took place here and there along the front. During the first few days of action, little progress was made. But the Austro-Hungarian Empire was in the process of collapse. The effect was to create insurmountable problems for an exhausted and now desperate army. As the Italians, along with some newly arrived English and French troops, advanced, the Austrian army began a withdrawal that even in the defensible mountainous terrain became a rout. Surrender followed two weeks later.

During World War I, battle lines in the mountains changed markedly only twice after the initial and modest Italian advance: first during the Battle of Caporetto in 1917 and then with the final Austrian retreat in 1918. One was, at least partially, the result of new tactics or weapons, while the other resulted largely from external events or conditions. Neither was an innovative design for Alpine warfare.

There is no doubt that mountain battlegrounds present unusual difficulties for the soldier, especially when on the offense. In addition to the demands of the rugged topography, maneuver and supply are major problems. But of all extensive mountain landscapes, the high glaciated terrain presents a most formidable challenge with its ice fields, rarefied atmosphere, changeable and often poor weather, bare rock, problems of exposure, and steep and dangerous slopes. As is frequently the case, properly trained, equipped, and positioned defensive forces had distinct advantages in the Alps during World War I. One important reason for this was the distinct nature of a mountainous landscape modified long ago by effects of moving ice.

THE ROAD TO MOSCOW

Many roads lead from Western Europe to Moscow, but the one that passes northeastward through Berlin, Warsaw, Minsk, and Smolensk is the most direct and best situated.[12] Following slightly higher ground much of the way, it crosses few rivers and approaches the Russian capital through the River Gate (or Land Bridge), a natural gap between streams (Map 8.4). Over the centuries many have followed this route, including three invaders from Europe seeking to conquer Russia by defeating its army and occupying Moscow.

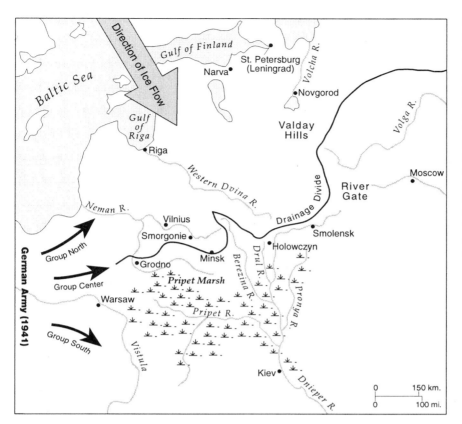

MAP 8.4. The continental drainage divide of northeastern Europe. The large arrow shows the direction of flow of glacial ice from the Scandinavian ice-cap. The three smaller arrows show the central thrust of three German army groups in 1941. Note that by following the drainage divide east from Grodno one can proceed to Moscow without crossing a major river.

But what accounts for this road on higher ground with fewer streams and a River Gate that guided different armies over three centuries? Are not other roads equally advantageous and usable? The answer involves a continental-sized glacier, the Scandinavian Ice Sheet, that modified the northern European landscape more than 10,000 years ago (see Map 8.2 above).

This last great European ice sheet had several predecessors. Parts of these earlier glaciers flowed southeastward over northwestern Russia, terminating at one time in two adjacent lobes that extended southward into the Dnieper and Don River lowlands (Map 8.4).[13] This ice not only changed the land, it also disrupted long-established and well-integrated drainage lines. In addition, glacial meltwater deposited large amounts of clay, silt, sand, and gravel outward from the glacier's margin to form widespread areas of nearly flat, poorly drained land. These low, wooded marshlands with water-saturated substrata, called polesyes by the Russians, vary greatly in size and are incredibly numerous and widespread in the glaciated areas of European Russia. The largest of these, the Pripet Marshes, are so extensive that they have repeatedly affected military tactics because of dense vegetation, poor drainage, and generally unstable land.[14] Only when frozen are polesyes firm enough to support heavy off-road traffic. As a result, several invading armies have purposely avoided the Pripet Marshes. But the larger marshes also offered a base of refuge and resistance for Russian defenders.

The last of the Scandinavian ice sheets, like its North American counterpart, reached its greatest extent about 20,000 years ago by merging with smaller glaciers in the northern Ural Mountains, Siberia, and the British Isles. Even the highest land was covered, making the glacier in some places at least 2400 m (8000 ft) thick and possibly exceeding 3000 m (10,000 ft).

Apparently the early center for maximum accumulation of snow and ice was in the Kjolen Mountains along the common border of Norway and Sweden. Here the land presents a formidable barrier that forces moisture-laden, prevailing westerly winds with their moist cyclonic storms upward over crest areas ranging from 1000 m (3200 ft) to more than 2000 m (6500 ft). With the colder temperatures of the last ice age, large amounts of snow fell to feed glaciers largely in the windward areas of these mountains. Then, as the extent and height of the glacier increased and the mountains were buried by an ice sheet, the center of maximum snow accumulation gradually migrated eastward to a position over what is now the Gulf of Bothnia.

The preexisting topography had an important effect on both the movement of the glacier and its marginal characteristics. Ice flowing west modified preexisting valleys in resistant rock into the spectacular fjords of Norway. In World War II, these fjords were militarily important in the German invasion and later efforts to protect elements of their navy.

The steep westward gradient from the mountains of Scandinavia favored relatively rapid ice movement toward a terminus effected by ice flotation and iceberg formation somewhere in the unfrozen Norwegian Sea. In contrast, ice flowing southward, after traversing the narrow Baltic depression and depositing debris to form much of what is now the Danish Peninsula, was somewhat obstructed by the higher ground inland in northwest Europe. But southeast of the ice-cap there was neither a higher land barrier nor a warmer ocean, a condition that favored glacial movement well into northern Europe and adjacent Russia.

The ancient igneous and metamorphic bedrock of the Fennoscandian Shield beneath the center of the ice sheet was especially resistant to glacial erosion. Here the glacier polished and grooved the underlying surface, leaving extensive areas without soil after deglaciation. Farther to the southeast, however, the glacier moved across much weaker sedimentary formations. The weakest of these were removed to form lower areas (similar to the Great Lakes basins) now occupied by the Baltic Sea and the related gulfs of Finland and Riga along with the White Sea. Most of the material eroded from these areas was transported farther south to be eventually deposited at the glacier's margin.

As the last Scandinavian ice sheet reached its maximum extent in northeastern Eurasia the rate of advance was approximately equaled by marginal melting. At that time the glacier's terminus fluctuated within a narrow zone now marked by the width of massive end moraines that can be traced from Denmark to Berlin and then eastward to Warsaw, Minsk, and Smolensk. In

Russia these drift hills separate the Pripet Marshes to the south from a northern lake district dominated by ponds, bogs, and polesyes.[15]

Much of the glacial debris within this hilly belt was deposited upon older Pleistocene sediments, which in turn rest upon somewhat higher bedrock—a combination that gives the landscape greater prominence and a fresh, or newly formed, appearance. The terrain within this zone is, however, by no means homogeneous. Instead, it is a variable complex of landforms and soils consisting of rolling morainal hills, scattered, gravelly outwash plains, and numerous poorly drained depressions and lowlands. And between Smolensk and Moscow, the higher tract of land veers northward to form the prominent Valdai Hills. From here a more subdued but rolling landscape on older glacial deposits extends onward to Moscow.

Though not particularly high in altitude, this great, linear, complex mass of glacial drift forms a drainage divide of continental proportions. It separates the headwaters of the eventually south-flowing Dnieper and Volga Rivers from the northern drainage of the Volchya and Western Dvina Rivers. Thus, following the belt of hills eastward, it is possible, by traveling along the drainage divide, to avoid stream crossings and enter the famous River Gate or Land Bridge (between the Dnieper, Western Dvina, and Volga Rivers) west of Moscow.[16]

But the trace of this drainage divide is not straight. Instead, because of variations in the altitude of the underlying bedrock surface, lobate patterns in the glacial margin, and differences in the amounts of deposition from the ice, it is sinuous, shifting closer to the southern margin of the terminal glacial deposits at some places but being more northern elsewhere. This variability is especially well shown by the far northerly source for headwaters of the south-flowing Berezina River. Thus, when traveling along the hilly tract between Smolensk and Minsk, one must either cross the Berezina River or avoid it by detouring a considerable distance to the north, a condition of vital importance and great peril to Napoleon's army in November 1812.

Over the centuries this great line of low hills, being flanked on the south by the lower Pripet Marshes and on the north by the poorly drained lake district, presented two especially advantageous opportunities in terms of movement and transportation. Those going north or south could travel by water to the headwaters of certain streams and then portage across the hills to continue their journey by boat. Minsk and Smolensk owe part of their existence to their position along well-used portages. More important, the higher, better-drained ground, with few rivers to cross, was the most favored overland route for those traveling east or west. It also became the central line of march for three invading armies aimed at Moscow.

Charles XII. The first major attempt to invade and conquer Russia from the west was led by Sweden's Charles XII early in 1708. During the previous decade, Charles had been trying to maintain and expand Swedish dominance of the Baltic area. He was opposed by an alliance (1696) among Denmark, Poland, and Russia. In response, by 1707 he had defeated the first two members of this triumvirate. Although then only 26, Charles was an effective tactician, leading his relatively well-equipped and organized troops in many battles. But the Swedish leader was not a skilled strategist. Apparently he embarked on this campaign with no clear plan except to defeat the Russians in a decisive battle and force the empire in Moscow to surrender. Thus, rather than directing efforts against the Baltic coast, which was closer to home and more easily supplied by sea, he moved inland, intending to find, pursue, fix, and defeat the enemy, then occupy their capital.

Charles's eastward march began by crossing the Vistula River north of Warsaw on Christmas, 1707. Following the more direct roads on the upland of glacial deposits, he was in Grodno by late January and proceeded through Vilnius to rest his troops at Smorgonie. At this point two primary options were available. The troops could either move north into the district of numerous glacially formed lakes and swamps, toward the old city of Novgorod and on to Saint Petersburg (both cities important to Russian connections with Europe), or continue east toward Moscow. The former axis of advance was safest, but it brought no assurances that the Russian army could be engaged or that it would result in collapse of the government.

Considering Charles's penchant for battle, it is not surprising that, as the weather improved, he moved southeastward toward Minsk and then toward Smolensk on the Moscow Road. Along the way he shifted his line of march along a somewhat more southerly and less direct route that entailed two river crossings, each large enough to be used as a military barrier by the defenders. However, at the Berezina River he outflanked the Russians and then crossed the Drut River without significant opposition. From there Charles moved northeast to Holowczyn, where Russia's Peter the Great placed 20,000 troops in defensive positions.

Here, in the first major engagement of an invasion now more than six months old, attacking Swedish foot soldiers and mounted cavalry broke through the enemy center to win the battle. The Russians, however, were not routed; they were merely displaced and their 3000 casualties could be replaced. But Charles's loss of 1300 valuable soldiers and much materiel was of considerable concern because reinforcements and supplies were now far to his rear.

In early August, after resting the army at Mogilev, Charles continued to move east, crossing the Dnieper River and then the Pronya River, in his approach to Smolensk. As the Swedish forces advanced deeper into Russia, problems multiplied. The retreating Russians destroyed foodstuffs and continually threatened the Swedes' flanks and rear. The vast length of the march began to have a serious effect. Decreasing supplies and ammunition were of increasing concern. The summer was coming to an end. Although more than 10,000 Swedish reinforcements had been ordered forward, they were still far to the west, and Charles did not know when they would arrive. And Moscow was still more than 400 km (250 mi) to the east.

Charles now faced a crucial decision. A winter advance through forests and polesyes to the capital would certainly be prolonged and difficult. Linkage with the reinforcements who were carrying needed supplies was desirable, but it would involve withdrawal. Encampment to wait for the new troops would invite Russian buildup and harassment in a barren land. None of these choices were acceptable to Charles. Instead he decided to move south toward Ukraine with its slightly less severe winter and somewhat better forage. Here he also hoped to find sympathetic Cossacks opposed to Peter the Great.

Between Smolensk and Ukraine lies a wide and poorly drained floodplain of the Dnieper River and several of its major tributaries, as well as a glaciated terrain associated with the easternmost part of the Pripet Marshes. Charles's march through this area was accompanied by numerous setbacks, including misdirection and the loss of supplies and weapons in the polesyes and rivers. These same conditions confronted the following column of reinforcements who, after several disastrous battles, burned their materiel and sank their artillery. With only 6000 Swedish soldiers remaining, this force eventually joined Charles in Ukraine in poorer condition than the army they were to reinforce.

After marching on Russian soil for more than a year, Charles now faced another enemy; the severe Russian winter. During the winter of 1708–9 he lost more than 3000 men to the cold and disease. Furthermore, Charles had yet to engage the Russians in a decisive battle. And now he was deep in enemy territory with dwindling resources.

In an attempt to resolve these problems, during the following spring Charles maneuvered his 18,000 remaining troops into position for an attack on Poltava with its Russian fortress and supplies. With the town under siege, a battle developed nearby in June 1709. At first the fighting went fairly well for the Swedish soldiers, even though they were greatly outnumbered. But as the small size of Charles's army and their lack of artillery became apparent, the Russians moved to victory. They, not Charles, won the decisive battle far, far away from the enemy's intended path, the road to Moscow.

Initially the slightly higher glacially formed hills of western Russia presented a favorable line of march for the Swedish invaders. However, with the change in direction, increasing distance, more rivers to cross, and progressing seasons came ever-increasing friction. Then the decision to turn south toward Poltava took the army off the better-drained roads and trails into the morass of polesyes and river floodplains. Charles's decisions and the nature of the terrain combined to produce a debacle that, about 100 years later, Napoleon studied in detail. Somehow he concluded that he could do what Charles XII could not.

Napoleon. Charles XII and Napoleon Bonaparte were very different types of military leaders. The Swedish king was essentially a warrior concerned with leading the fight. In contrast, the French emperor directed from the rear and placed great emphasis on strategy and logistics as well as tactical movements. On this basis we might expect their Russian campaigns to be quite dissimilar. Yet they both started on the same path and Napoleon ended up making some of the same choices and miscalculations as his predecessor.

The French invasion of Russia began on 23 June 1812 with 600,000 men.[17] By following the route north of the Pripet Marshes that led to Vilnius, Napoleon could threaten both Moscow and Saint Petersburg. From Vilnius he sent a smaller force north to Riga and moved most of his troops eastward toward the River Gate (see Map 8.4 above). The advancing force, now numbering 420,000, was too large to follow a single road. Instead, Napoleon's troops were organized into corps that, although separate, were coordinated. They all moved eastward along different routes within the slightly higher glacial deposits, thus avoiding the more difficult terrain and river crossings to the north and south.

By the second week of July, the emperor was past Minsk; in the middle of that month he had crossed the Berezina River. Napoleon, like Charles more than a century earlier, then had to make an important decision. Should he establish a camp here, secure this part of Russia, and prepare for the coming winter and a renewed 1813 offensive? Or would it be better to enter the River Gate and move on to Moscow about 500 km (300 mi) to the east?

There were many factors to consider. By now the advancing French army had been reduced to about 185,000 soldiers, largely because substantial forces were required to protect the flanks, rear, and numerous supply depots. With each step eastward lines of communication grew longer for the French and shorter for the Russians. The summer was slipping away. And because of Russian avoidance, a major battle had yet to occur.

Withdrawal was unacceptable to Napoleon, partly because it might encourage adversaries at home and abroad who threatened his government. To remain in this harsh land until the following year would require large amounts of supplies not readily available; it would also give the Russians time to better organize and increase their forces. Like Charles, after resting the troops, Napoleon decided to march on. But he would not turn south toward Ukraine. Instead, and in accordance with the original mission, the French would continue eastward, enter the River Gate, and follow the higher ground to Moscow. Few, if any, recognized that they were also, with time, advancing deeper into the realm of the fierce Russian winter.

Their path led first to Smolensk, where on 17 and 18 August a confused, sprawling battle (much influenced by terrain and streams) occurred that resulted in further Russian withdrawal. Napoleon pursued on an eastward course that eventually took his army off the deposits of the last Scandinavian ice sheet and onto the lower and less hilly terrain underlain by older glacial sediments lying between the Valdai Hills and Moscow. Within this landscape, about 120 km (75 mi) west of Moscow is the village of Borodino.

Until this point the Russian troop movements involved occasional harassment and retreat but avoided a major battle. Near Borodino a small stream valley and rolling, glacially formed hills presented some reasonably good defensive positions that the Russians had time to improve before the French arrived. Furthermore, Napoleon's pursuing army now numbered about 140,000 men, compared with 120,000 Russians. The odds were getting even and a new Russian commander, Mikhail Kutusov, decided to make a stand.

While approaching Borodino, Napoleon must have hoped that the time had finally come for the decisive engagement. In surveying the battlefield he planned to advance and then hold at the enemy's center while mounting a strong flanking maneuver on his right. The battle, which began at dawn on 7 September, proved to be the largest of this campaign. Through several attacks, the Russians lost some ground, especially on their left. But the French were never able to rout their enemy. By late afternoon more than 40,000 Russian and nearly 30,000 French soldiers were dead or wounded. By sunset little terrain had been won or lost. That night the Russians once more withdrew, and the next day Napoleon occupied an abandoned battleground.[18] Although the Russian forces were again displaced, they were by no means defeated. Instead, they remained organized and effective under Kutusov, somewhere to the east along that long road to Moscow.

Again the French pursued eastward, meeting no major resistance as they advanced on the enemy's capital. But the capture of Moscow must have been a major disappointment to the French. As they moved into a city torched and abandoned by the Russians, no enemy army, decisive battle, Tsar, or victory was in sight. Gradually it must have become clear to all French soldiers that the terribly long and difficult road had led them to a cul-de-sac rather than a triumph (see Map 4.7).

After several weeks in the city, Napoleon recognized that his major objectives in Russia were unattainable. He ordered his army to return to France, eventually following the same roads used during their advance. But the emperor had waited too long. A new and more formidable enemy, one that he had never seen before and could neither master nor escape, was steadily growing in strength; it was, of course, the Russian winter.

The withdrawal from Moscow of about 100,000 soldiers began in mid-October. A large force first moved south, rather than west, either as a diversion or with hopes of capturing a Russian supply depot at Kaluga. After a brief battle with Russian forces at Malo-Yaroslavetz, Napoleon turned north and then west to reach the original invasion route. Once again on terminal deposits of the Scandinavian ice sheet, the return march passed the rolling terrain at Borodino, where thousands who died nearly two months before remained unburied. As the withdrawal continued, troops from outposts and flanking positions rejoined Napoleon; simultaneously he also lost more men to battle, disease, straggling, desertion, and the elements.

By the time he passed through Vyazma on 31 October 40,000 were gone, and another 18,000 were lost on the way to Smolensk, which the army reached on

9 November. As the march moved on, the weather rapidly deteriorated. Night-time temperatures plummeted, snow arrived, enemy harassment increased, and casualties mounted. The withdrawal became a rout. One heroic but tragic part of the saga involves the source, size, and trend of the largest Russian river that the retreating French had to cross: the Berezina.

With about 50,000 remaining troops, Napoleon moved along the most direct line from Smolensk to Minsk. He was followed by an equal number of stragglers. On 25 November, near Borizov, he reached the south-flowing Berezina River, with headwaters along the drainage divide a considerable distance to the north. Here the river, made large by numerous upstream tributaries, is much too deep to ford. Not yet completely frozen, it was choked with masses of floating ice, and the bridge he intended to use had been destroyed. Of at least equal concern, elements of the Russian army were at Napoleon's rear, on his flanks, and along the planned line of march on the far side of the river.

Napoleon maneuvered a short distance northward for a more favorable crossing point. Now elements of the terrain, the weather, and enemy tactics all came together to present one of the greatest challenges that can face military engineers in combat. Two bridges, one capable of supporting heavy artillery, had to be constructed quickly. This required soldiers to enter the river and build underwater supports. By working continuously through the night the bridges, such as they were, were ready for traffic on 26 November. Most who labored in that icy stream likely died of exposure or drowning.

Rapid movement was essential if 50,000 soldiers were to cross the river in a short period. Even at an average rate of 500 men per hour for each bridge, it would take two days to evacuate all of the troops. But the bridges were narrow, makeshift structures that regularly broke down, stopping traffic and requiring repeated repairs (Figure 8.3). Movement of artillery, wagons, supplies, and baggage further slowed the process and congested the bridges. Russian military pressure increased. Even so, the structures supported traffic through 27 November and part of the next day, when after coming under direct enemy fire, the French intentionally burned the bridges to the water line to limit Russian pursuit. In the end, many soldiers and practically all of the 50,000 stragglers remained behind on the river's east bank at the mercy of both the winter and the pursuing Russians. And those who tried to swim the river probably perished like the many engineers who entered the frigid water to build the structures two days before.

The crossing of the Berezina River north of Borizov cost Napoleon about 20,000 soldiers. Many more of the stragglers died. Stunning as these numbers are, the most impressive fact is that the crossing was achieved at all; a lesser army would have failed here. The ability of the French to salvage something from this disaster appears to be the result of effective leadership, loyal, dedicated, and skilled troops, and lack of Russian military initiative. But the problem itself developed because of the line of march Napoleon selected, in com-

FIG. 8.3. Passage de la Béresina (26–29 November 1812), composition and lithography by Victor Adam (1842), depicting westward crossing of the Berezina River by Napoleon's troops. (Paris: Bibliothèque Nationale. Photograph by Hachette.)

bination with the effect of glacial deposition on the establishment of river courses more than 10,000 years ago.

As the retreat continued through December 1812, French troops were continually pursued by the Russians and severely punished by the increasingly harsh weather. Only a small fraction of 600,000 men who left France to march toward Russia ever returned to their homeland. By the time the remains of Napoleon's army reached home, more than 1000 pieces of artillery, 200,000 horses, and 570,000 men had been left behind in the glaciated terrain of northwest Russia.

Hitler. Nearly 130 years later, officers of Germany's invading 4th Panzer Army at a Berezina River crossing near Borizov pondered the original purpose of several submerged structures built into the river bed. They were, of course, looking at some last remains from one of Napoleon's two bridges. But could the terrible experiences of the French and Swedes in Russia be of any significance to the Germans? Great increases in military firepower and mobility, along with the airplane, had done much to change warfare. Many thought that Germany's political leadership, fervent nationalism, advanced industrialization, professed racial superiority, powerful army, and revolutionary blitzkrieg tactics would make things completely different this time.

Yet the Pripet Marshes remained a major military barrier, diverting armies to the north and south. The polesyes, lakes, and bogs to the north were still formidable obstacles. The great distances, compounded by obstructing rivers, persisted as major military problems. Moscow was once again a primary objective, and the best way to get there was the same: over the slightly higher but rolling terrain of glacial deposits. But in early July 1941 at the Berezina River, the advancing German army, like Charles the XII and Napoleon, had yet to appreciate the full challenge of the long road to Moscow.

In planning for the invasion, German forces were arranged in three groups. Army Group South would move south of the Pripet Marshes, capturing Ukraine and its great resources. Army Group North would advance northeast to secure Riga, Narva, and Leningrad and control all of the Baltic coast with its numerous ports. Most important, Army Group Center would attack across the slightly higher terrain north of the Pripet Marshes (previously followed by Charles and Napoleon) and advance eastward, using flanking tactics by one or both wings, taking Minsk and Smolensk, passing through the River Gate, and finally occupying Moscow (see Map 8.4 above).[19]

The German attack began on 22 June 1941; Napoleon had waited until the twenty-third. The glacial sediments were dry and supported all varieties of traffic. Army Group Center reached Minsk in only five days, whereas Napoleon took 15. The Germans approached Smolensk on 16 July and, like Napoleon who arrived on 17 August, engaged the enemy in a fierce battle. But unlike the two-day encounter in 1812, the 1941 battle lasted until 5 August. Here, in a wide-sweeping double-envelopment, more than 300,000 Russian troops were captured and uncounted were wounded or killed. But in doing so the German movement toward Moscow was delayed by three weeks.

Problems rapidly increased for the Germans. As the front progressed eastward, it also widened. More distant flanks needed to be protected, the shape of salients had to be carefully managed, encircled forces eliminated, and lengthening supply and communication lines kept intact and open. In addition, the Pripet Marshes presented a special dilemma. Impossible to traverse using blitzkrieg tactics, they remained unsecured, a largely unoccupied sanctuary for thousands of Russian partisans who, in organized bands, would later wreak havoc on nearby German transportation lines and supply depots.

The need to reorganize and resupply, plus increasing Russian resistance and the diversion of armored forces to the northern and southern frontal sectors, slowed the German advance in the center. With the fall season came cold cyclonic rains that turned roads on clay-rich, unconsolidated glacial deposits into quagmires, slowing and sometimes stopping the army's advance altogether. In the campaign's first two months the Germans moved eastward more than 650 km (400 mi). They advanced less than half that distance in all of September, October, and November combined.

Napoleon reached Moscow by mid-September. Not until 5 December did a few German units get within sight of the city. Here, as Army Group Center came to a halt, the Russians launched their first large-scale winter counterattack. Although this war would continue for more than three years, the Army Group Center would never capture Moscow. In the end, after fighting forward for more than 1000 km (620 mi), they remained 15 km (9 mi) short of their main objective.

Charles, Napoleon, and Hitler started out with the same mission: defeat the Russian army and impose their will on the country. Their initial plans to achieve these objectives were identical: engage and destroy the enemy in a decisive battle and occupy Moscow. Although the three men were dissimilar and much had changed from 1707 through 1941, the fate of each was strongly influenced and adversely affected by the same basic geographic conditions. Their lines of march were inextricably related to specific characteristics of the terrain. All three armies suffered miserably from the weather and climate. Their failure to bring a reluctant, elusive, and withdrawing Russian army to decisive battle while still possessing superior forces, and their inability to deal with lengthening lines of supply and communication, were the direct result of available space and increasing distance. Although the setting for battle in European Russia is vast and diverse, it is clear that geographical factors narrowed the strategies and focused the tactics of all three invaders in a similar fashion. Some of these factors also contributed greatly to their demise.

Conclusions: Assessment and Modern Impact

Although glaciated mountains and plains are quite different in detail, both landscapes present formidable tactical and logistical problems. All of the world's highest mountains have in some way been made more rugged and formidable by glacial erosion. Furthermore, it is not uncommon that some of these ranges serve as sites for political boundaries. Where friendly nations are involved, such borders serve as useful boundaries. But when these borders become zones of dispute leading to conflict, as has often happened, the mountains form a battleground where maneuver is always difficult. Even a cursory look at maps showing political borders in the high mountains of Asia, Europe, and South America clearly illustrates the potential importance of such terrain, considering the sensitive and changing relationships that exist between countries. Who would doubt that battles will again be fought along some of these borders in high glaciated mountains?

Mountain combat has shown that maneuver, battle, and logistics can be strongly influenced by both large- and small-scale glacial effects on the land. Major corridors take on added importance but may be difficult to secure and defend. A few suitable niches high in ice-carved mountains may effectively control movement of much larger forces in nearby valleys and narrow passes. The rugged terrain with its variable weather, snow and ice, and thin atmosphere may be as big a problem as the enemy. Experience shows that to counter these elements, mountain troops require special training and equipment, a period of acclimation to high altitude, and a clear understanding of this challenging terrain.

DRUMLINS AND BUNKER HILL

Drumlins are elongated, streamlined, elliptical hills with trends parallel to the flow direction of the glacier that formed them. With long axes often measuring more than a kilometer and commonly more than 15 m (50 ft) high, the features formed beneath an active glacier either through molding of preexisting basal material or from plastered deposition of sediments by the ice. Whether in Eurasia or North America, they always appear in swarms rather than isolated features. One of these drumlins is the most famous topographic feature of the American Revolutionary War. It is, of course, Bunker Hill (Map 8.5).

There are many drumlins in the Boston area, most with shapes indicating that the glacier that formed them flowed toward the east-southeast. Locally well-known hills with names like Spring, Winter, Prospect, Telegraph, Tower, Beacon, and Copp's are all drumlins. Two more are immediately north and west of the place where the tidal estuaries of the Charles and Mystic Rivers join to form Boston's modern Inner Harbor. The crest of the southernmost and lower drumlin, Breed's Hill, is about 20 m (65 ft) higher than the nearby rivers. The somewhat larger Bunker Hill, which is over 30 m (100 ft) high and more than a kilometer in length, coalesces with the north part of Breed's Hill to together form a streamlined, two-humped ridge, the most prominent in the area. Thus, both hills directly overlook the original site of old Boston along with its harbor facilities.

During early June of 1775, about 7000 British troops were posted in Boston. On 16 June Col. William Prescott, with more than 1000 Massachusetts rebels, occupied the two drumlins on the north side of the Charles River estuary. After some debate they excavated a large redoubt in the unconsolidated glacial sediments on the higher part of the more southerly, but lower, Breed's Hill. They also dug trenches and built obstructions that extended northeastward down the flank of the drumlin. Such overlooking positions threatened both the harbor and city to the south, but they were also easily visible to British observers on another drumlin, Copp's Hill, in Boston.

In response, Gen. William Howe, with about 2200 British troops and 12 artillery pieces, moved to engage the revolutionaries. Rather than attempting a flanking maneuver that might have been highly effective considering the position of the Americans on a narrow-necked peninsula, Howe (delayed somewhat by the tide) eventually crossed the estuary in full view of his enemy to land a short distance east of Breed's Hill. Here the British commander prepared to make a frontal attack while also exerting pressure on the rebels' left. Such a tactic, however, required uphill movement toward entrenched troops whose visibility from higher parts of the drumlin was good. Observers on Copp's Hill in Boston saw the first two British attacks fail and sent 400 additional troops. With these reinforcements the third attack captured the redoubt and trenches, largely because many of the Americans

---

In contrast, plains and hills modified by continental glaciation are important because of their vast extent, subtle yet significant topographic differences, variety of soils and sediments, and overall effect on trafficability which can change dramatically with the progression of the seasons. Though Napoleon's troops and Hitler's Army Group Center marched many hundreds of kilometers, they never left the glaciated landscape, either in their advances toward

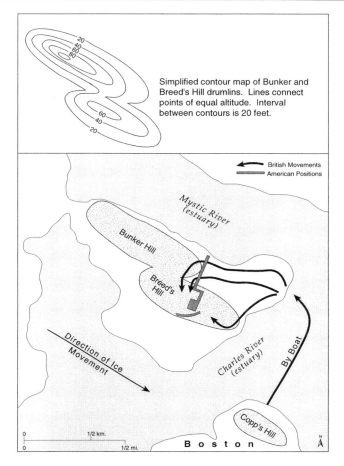

Simplified contour map of Bunker and Breed's Hill drumlins. Lines connect points of equal altitude. Interval between contours is 20 feet.

British Movements
American Positions

Mystic River (estuary)

Bunker Hill

Breed's Hill

Direction of Ice Movement

Charles River (estuary)

By Boat

Copp's Hill

Boston

0   1/2 km.
0   1/2 mi.

N

were now out of ammunition. Most of the rebels, however, escaped by withdrawing northwestward from Breed's Hill and along Bunker Hill to leave the peninsula.

By the time the battle ended, British casualties (226 dead, 828 wounded) more than doubled that for the Americans (140 dead, 301 wounded). Although the colonists had been tactically defeated, they clearly demonstrated that, though outnumbered, they could fight courageously and inflict severe losses against well-trained professional soldiers. But they also had a great tactical advantage on that drumlin. Most of the Massachusetts troops had unobstructed views from well-protected and higher positions of a continually exposed enemy making an uphill frontal attack. Clearly the location, shape, and composition of the drumlins had much to do with the Battle of Bunker Hill.

Moscow or in subsequent retreats. And Charles XII escaped it only by moving south toward his army's dissolution at Poltava.

Militarily, one difference between the present and the past is that landforms, sediments, and soils of many glaciated plains and mountains have now been mapped, some in great detail. Furthermore, large-scale topographic charts are much more abundant, and incredibly detailed remotely sensed terrain data has become commonplace. Although relationships may be complex, the glacially eroded bedrock surfaces, various types of Pleistocene deposits, and landforms associated with both now can—with appropriate training—be accurately interpreted and scientifically assessed for military purposes. This potential capability will continue to improve in the future. The commander who has accurate information regarding the regional and localized constraints and opportunities that a glaciated landscape presents to adversaries in battle may also have a definite, yet possibly unsuspected, advantage over the enemy.

CHAPTER 9

Peninsulas and Sea Coasts

Anzio and Inchon

We drew up a list of every conceivable natural and geographic handicap—
and Inchon had 'em all.

LT. COMDR. ARLIE CAPPS, TASK FORCE 90

During the late summer of 1950, while United Nations forces faced destruction along the Pusan Perimeter in southernmost Korea, Gen. Douglas Mac-
Arthur was launching one of the most spectacularly successful amphibious
operations in military history. Within weeks after his troops landed at Inchon,
the North Korean army, so close to a resounding victory far to the south at
Pusan, was routed north of the 38th Parallel. The invasion was an extraordinary blend of audacity and skill, demonstrating expert exploitation of apparently forbidding hydrography. A deep turning movement to the rear of the
enemy quickly severed lines of communication that passed through Seoul,
trapped enemy forces in a giant pincer, and, for a short time, completely
changed the balance of power in the Korean conflict.

The astonishing success at Inchon, South Korea, contrasts starkly with the
near-disaster at Anzio, Italy, during World War II. Of all the major Allied
invasions in the European and Pacific Theaters, Anzio was the one beachhead
that was most nearly "thrown back into the sea." For three months British and
American troops clung to a small enclave on Italy's Tyhrrenian coast, barely
withstanding the punishing attacks of Field Marshall Albert von Kesselring's
forces. Directed by Adolf Hitler to remove the "abscess south of Rome," six
German divisions repeatedly attacked an invasion force that had landed essentially unopposed and registered great initial success, but soon found itself
desperately clinging to a small and exposed beachhead.

These two contrasting amphibious assaults highlight the potential risks involved in conducting one of the most complex military operations. The nature
of this type of battle is unique, where movement-to-contact involves equipment, tactics, and maneuver skills of both sea and land combat. It is also a
technically demanding arena where thorough knowledge of the coastal processes and terrain is essential and may spell the difference between victory and
defeat.

Amphibious warfare was most fully developed during World War II. The
Pacific Theater was essentially a long series of coastal assaults as U.S. Army and
Marine forces swept west and north toward Japan, hopping from island to

island, bypassing and isolating, where possible, enemy strongpoints. From Guadalcanal to Okinawa, the planning, timing, and execution of these exceptionally difficult and complicated military maneuvers were perfected as never before. Though fewer in number, even larger-scale amphibious operations were conducted in the European Theater at Sicily, Salerno, and Anzio, leading to the "longest day" at Normandy, an amphibious operation of such magnitude that it is likely that its equivalent will never be seen again.

There are few examples of major amphibious assaults before World War II. In World War I, the British–Australian–New Zealand fiasco at Gallipoli (now Gelibolu, Turkey) was one of the Allies' most costly and tragic defeats. This assault on the Dardanelles failed primarily because of faulty doctrine, ineffective logistics, and poor interservice coordination, leading B. H. Liddell-Hart to comment that

> a landing on a foreign coast in the face of hostile troops has always been one of the most difficult operations of war. It has now become much more difficult, indeed almost impossible, because of the vulnerable target which a convoy of transports offers to a defenders' air force as it approaches the shore.[1]

In the American Civil War, Gen. George B. McClellan conducted an initially successful amphibious movement of troops when he launched his Peninsular Campaign of 1862 by moving 75,000 men and accompanying equipment with a fleet of approximately 400 ships southward on Chesapeake Bay to Fort Monroe in Virginia without serious incident or accident (see chapter 6). His debarkation, however, was unopposed and presented vastly different circumstances from the amphibious assault: a seaborne attack onto a defended hostile shore. McClellan's movement was characteristic of amphibious operations before World War I. Rarely were shorelines major battlegrounds, although small-unit operations and raids are commonplace in military history.

Contested amphibious assaults involve the projection of military power from the sea onto a hostile shore. Today the planning, rehearsal, and execution of this type of maneuver generally involves close coordination and timing among air, sea, and landing components. Physical features of the landing site must be carefully considered both in planning and execution. The most important hydrographic variables are wave energy, tidal range and height, offshore currents, beach characteristics, and near-shore bathymetry. Today these factors, though complex, can generally be measured, assessed, and predicted. By careful analysis of appropriate intelligence information, an accurate hydrographic and topographic representation of the landing site can be accurately drawn. Collectively these variables can be crucial in determining the time and place of the attack, and knowledge of related characteristics is essential in the planning and execution of the successful amphibious assault. Ignorance or neglect of these physical factors can lead to disaster.

Facility of exit from the beach is an essential characteristic of a suitable

An early, well-recorded amphibious operation was Caesar's "invasion" of Kent. The landing site, a stretch of beach near Dover, was selected because the absence of cliffs facilitated movement inland. The large galleys and transports, carrying foot soldiers, grounded near the shore and, after overcoming formidable resistance, Roman troops secured the beachhead. The galleys were then drawn up onto the beach while transports were anchored offshore. An evening storm literally blew follow-on cavalry forces back to the French coast, while high waves, coinciding with a September spring tide, destroyed the beached galleys and drove the anchored transports onto the shore, battering the wooden ships to pieces. Twelve ships were destroyed and Caesar, having made repairs to the others but now without supplies, limped back to France.

From Edward B. Clancy, *The Tides* (Garden City, N.Y.: Doubleday, 1968).

landing site. Furthermore, in larger operations the inland landscape should allow space for logistical buildup and maneuvering of the assaulting force. Port facilities and a road network capable of moving troops and material forward are also vital factors.

Beach material and the presence or absence of dunes are important when selecting a site for an amphibious assault.[2] The trafficability (ability to support vehicle movement) over dunes and beaches is a function of beach slope and support capacity of the shore-zone clastics. These, in turn, are related to the supply of sediments and nature of wave energy. Steep beaches composed of coarse sand and gravel generally have poor trafficability. In contrast flat, moist, fine-sand beaches are more likely to support equipment movement. Good landing-site selections must take beach and shore-zone trafficability into consideration. Poor on-shore vehicle mobility can seriously impair an operation; severe examples include the steep gravel beach in the 1942 Allied raid on Dieppe and the volcanic sand at Iwo Jima (see chapter 10), greatly hampering both track- and wheeled-vehicle mobility in the assault.

Tides, Waves, and Surf

A shoreline is essentially one of nature's battlegrounds where, ever since the seas originated, waves constantly attack the land with storm surges that often exceed 10 m (33 ft). The assault is incessant, with the wave energy of sea and swell at an exposed coastal location varying continuously, seemingly without pattern, over the years. However, one type of wave is always predictable. This water-level oscillation is the tide, and the rhythmic rising and falling of the sea is the heartbeat of the oceans.

Every coastline of the world experiences tides of varying degree, dependent primarily on the relative positions of the Sun and Moon and the neighboring hydrography and coastline geometry. Tidal ranges may vary from the 15-m

(50-ft) monsters at the Bay of Fundy, Canada, to nearly imperceptible sea-level fluctuations. They may approach the shore as a barely noticeable long-period wave, or surge up the mouth of a river in the form of a bore, such as the 8-m (25-ft) near-vertical tidal wave of China's Chien-Tang River that rushes upstream at 13 knots or the Amazon's 5-m (15-ft), 12-knot tidal bore called the pororoca. Regardless of the circumstances or the approach, the tide is an important coastal process that sculpts the shoreline and alternately uncovers and submerges the features of the near-shore zone.

In fact, the periodic rising and falling of sea level is caused mainly by the gravitational attraction of the Sun and the Moon. These bodies provide the energy, or pull, that produce the tidal waves. Newton's law of gravity, showing that the gravitational attraction between two objects is directly proportional to their masses and inversely proportional to the square of the distance between them, helps explain the phenomenon. However, this gravitational attraction does not alone produce the tidal activity on Earth. Individual particles at Earth's surface are attracted to the Moon and Sun with slightly different magnitudes due to their varying distances from these celestial bodies. Newton's law, in fact, indicates that, due to its enormous mass, the gravitational pull of the Sun on Earth exceeds that of the Moon by over 177 times. Yet it is the Moon that creates the dominant tidal force, with the Sun being only about 46 percent as effective in generating tides. Why is this? It can be shown that Earth's tidal force, although derived from Newton's law of gravitation, is inversely proportional not to the square of the distance between two attracting bodies but to the cube of this distance. Since the Sun is 390 times farther from Earth than the Moon, the Moon becomes our primary tide-producing body.

Each particle on Earth experiences both a gravitational attraction toward the Moon (as expressed by Newton's law) and an acceleration (force) due to its circular motion about the center of mass for the Earth-Moon system. This is called a centripetal acceleration, and it is equal in magnitude and direction for all particles on Earth. The gravitational attraction of the Moon, however, is greater for particles closer than those more distant from it. Thus the difference between the gravitational and centripetal forces at any point provides the resultant force that is responsible for tidal activity on Earth. Furthermore, this force is inversely proportional to the cube of distance between the two attracting bodies.

As a result, two tidal bulges appear on Earth at all times. Considering only the gravitational attraction of the Moon, one bulge is directly toward the Moon while the other, on the exact opposite side of Earth, extends away from the Moon. These two distensions, or waves, maintain their relationship with the Moon as Earth rotates, giving the appearance that the tidal wave is progressing east to west, counter to Earth's rotation, at an equatorial speed of approximately 1600 km/h (1000 MPH). Theoretically they collide with Earth's shorelines twice a day with two high tides, or wave crests, and two low tides, or wave troughs.

Because the Moon revolves in the same direction that Earth rotates, any point on the planet must travel slightly more than one complete turn to again arrive at its previous position relative to the Moon; this makes the tidal (lunar) day 24 hours and 50 minutes long. The lunar high tide at a particular section of beach, therefore, will occur approximately 50 minutes later each day.

The Sun's tidal effects, although smaller, are also important because they may complement or diminish lunar effects. Twice a month Earth, Moon, and Sun are aligned (syzygy); then the Sun's gravitational pull enhances that of the Moon. This occurs when the Moon is in its full and new stages and is called spring tide, which are bimonthly occurrences of the greatest tidal ranges. As days pass, the solar and lunar forces gradually progress out of phase until Earth, Moon, and Sun are 90 degrees in opposition (quadrature), reducing the tidal range by approximately 20 percent below the average. The time between the greatest and least tidal variation is approximately 14 days, or fortnightly. Additionally, about once a month, the Moon, because of its elliptical orbit, comes closest to Earth (perigee), increasing the tidal range to about 20 percent above its mean. From this it follows that, in combination, perigee and syzygy may add approximately 40 percent to tidal variations.

Over the longer term, the greatest tide-producing forces occur roughly every 1600 years when both Sun and Moon are simultaneously in perigee, and spring tides are the highest when Sun, Moon, and Earth all have the same orbital plane. This condition will recur about the year 3300, the last occurrence being in 1700, during the Little Ice Age.[3]

Tides are generally imperceptible at sea but become clearly apparent as they reach shallow water. It is the configuration of the shoreline, water depth, and geometry of the ocean basin and continental shelf that determines tidal range and whether tides will be semidiurnal (two high tides and two low tides of approximately the same height), diurnal (one high and one low tide each day with a period[4] of 12 hours and 25 minutes), or mixed tides (two high tides and two low tides per day, with strong inequalities in heights on successive tides). As examples, the Atlantic Coast of the United States experiences semidiurnal tides, the Gulf of Mexico diurnal tides, and the West Coast of the United States has mixed tides, the most common type throughout the world. Tides occur at every shore with predictable regularity and amplitude, making them an essential planning ingredient for amphibious operations.

Surf condition is another important consideration for coastal assaults. Formed by wind blowing over the water, waves travel across the open ocean with little energy loss until they approach the shore. When moving into shallows, the wave slows because of friction with the bottom, is compressed, becomes unstable, and then breaks when wave height becomes about three-quarters of water depth. Consequently, the near-shore bathymetry is an extremely important factor in explaining areas of high and low wave energy and the offshore location where the breaking occurs.

As waves move into shallow water and their speed is reduced, their length

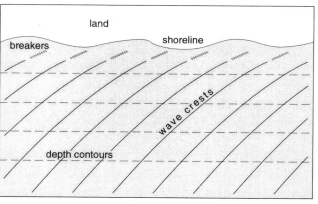

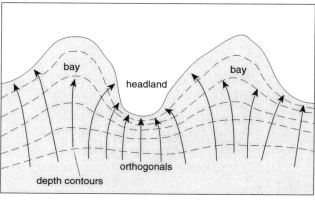

FIG. 9.1. Wave refraction along smooth and irregular coasts. Note that in (a) waves approaching a smooth coast from an angle curve shoreward as the water shallows. Along irregular shorelines (b) waves converge, as shown by the arrows, on the headlands, thus concentrating most of their energy there rather than on the adjacent bays.

between other incoming waves is shortened. Assuming offshore depth contours are approximately parallel to the shoreline and the waves are approaching the coast at an oblique angle, that part of the wave closest to the beach will be moving more slowly than the part in deeper water. The seaward portion, now traveling faster, swings landward, while the landward crest bends and slows. Consequently, along smooth shorelines the wave will tend to approach the shore more directly, a process called refraction (Figure 9.1a).

Wave crests therefore respond to the configuration of the near-shore bathymetry and the slope of the seafloor near the shore. For example, as waves approach a headland on an irregular shoreline they are refracted toward the promontory and diverge from adjacent embayments. Highest energy is thus concentrated on these protrusions, while lower energy favors sediment deposition in the zone of divergence within the embayments (Figure 9.1b).

These differing situations may require important consideration when planning an amphibious assault. For example, conditions within embayments can be expected to be more favorable for landing craft than the high-energy environments near headlands while enemy gun positions on the promontory can fire directly at landing craft as they pass shoreward into the bay.

SALERNO TO ANZIO, ITALY, 1944

For four months in 1944, a semicircle strip of Italian seashore fifteen miles long and seven miles deep kept much of the world on tenterhooks. For the Germans it was an abscess to be lanced. To the Romans it was a long-pending hope of deliverance from unwanted occupation. To the British it was inconceivable that what was happening there should be happening. To the Americans it was agony, more proof that we should never have allowed ourselves to become so thoroughly enmeshed in the Mediterranean toils.[5]

On 9 September 1943, Allied forces landed at Salerno, Italy, in Operation Avalanche. The primary objective for this amphibious landing was Naples, the largest port and best harbor on the Italian west coast, 80 km (50 mi) north of Salerno (Map 9.1). Salerno was chosen as a landing site because of its favorable sea approach, its proximity to Naples, and its being within range of fighter air cover from Allied-controlled Sicily. The chosen location for this amphibious assault was a 32-km (20-mi) stretch of sandy beach bisected by the River Sele just south of Salerno. The Plain of Salerno, a nearly flat and widely cultivated lowland, borders this large crescent-shaped beach. Here the coastal plain is narrow, dominated by adjacent mountains. Only two mountain passes to the

MAP 9.1. South-central Italy. (Map modified from diagram by A. K. Lobeck in Army Map Service, *Continental Europe—Strategic Maps and Tables,* pt. 2, *South Central Europe,* U.S. Army, 1943.)

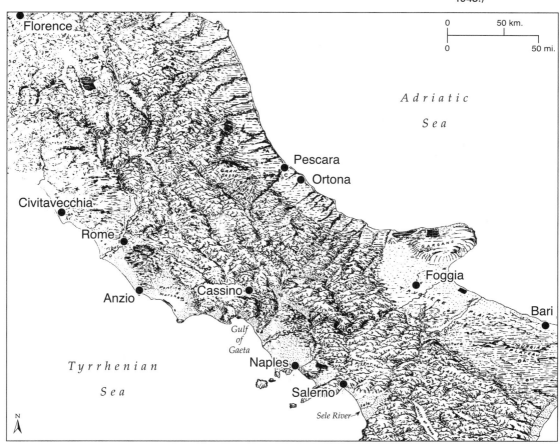

north, notched into the Sorento Peninsula ridge, provide adequate tactical exits from the beachhead.

Lt. Gen. Mark W. Clark, commander of the Fifth Army, would have preferred to land at the Gulf of Gaeta, north of Naples, where the terrain was more suitable for maneuvering. The German Commander-in-Chief South, Gen. Albert von Kesselring, also expected the landing to be at Gaeta and had positioned a strong opposing force there. This potential landing site was discarded, however, because of a large offshore bar that paralleled the beach and because Gaeta, being farther north, stressed the range of Sicily-based fighter aircraft.[6]

Nevertheless, Salerno was a favorable amphibious site. The embayment (Gulf of Salerno) offered protection from high-energy waves due to wave refraction. Additionally, the gulf was deep, allowing landing craft to run aground at or very near the beach, and avoiding the necessity for troops to wade ashore against expected heavy resistance. Only one shoal (shallow) area, where the Sele River deposited sediment transported from the mountain regions to the east, was present along this gulf shore. The major drawback of Salerno was the limited routes of egress from the beachhead. The Germans tenaciously defended these narrow defiles leading from the triangular-shaped coastal plain, even as the bulk of Kesselring's forces prepared to pull back toward Rome.

While the U.S. landing at Salerno was vigorously opposed, Kesselring knew that his position there was eventually untenable. The British 8th Army under Gen. Sir Bernard L. Montgomery was advancing northward on the east side of the Italian peninsula with the objective to link up with Lt. Gen. Mark W. Clark's Fifth Army. Within days, however, the Germans had concentrated six divisions around Salerno and inflicted heavy casualties on the two-corps landing force.[7] Nevertheless, the U.S. Fifth and British 8th Armies joined on 16 September, and captured Naples on 1 October.

While withdrawing, the Germans had taken great advantage of the exceptionally rugged river-dissected terrain which, with the onset of the annual winter rainy season, were torrentially flooded in some places. The rains began on 29 September, quickly turning the flat coastal plain into marsh and mud incapable of supporting off-road vehicle movement.[8] "Rain, rain, rain," wrote Maj. Gen. John P. Lucas in his diary on 8 October, "the roads all so deep in mud that moving troops and supplies forward is a terrific job. Enemy resistance is not nearly as great as that of Mother Nature."[9] The weather, terrain, and an exceptionally skillful enemy resulted in some of the most difficult winter fighting of World War II.

Lt. Gen. Heinrich von Vietinghoff, Kesselring's 10th Army commander, was directed to fight the Allies as cheaply and as long as possible to buy time for the construction of defensive positions south of Rome. Defending and withdrawing in a series of lines that stretched from the Adriatic to the Tyhrrenian Sea, the Germans fortified the terrain by constructing strongpoints and fighting positions to block the limited passages through the mountain ridges. To meet

Hitler's commitment to defend south of Rome, von Vietinghoff brilliantly executed a delaying action through the Barbara and Bernhardt Lines to the Gustav Line. Here the German army meant to hold.

The Gustav Line, defended by eight German divisions, extended across the Italian peninsula from the Gulf of Gaeta to Ortona and was the strongest of the German prepared positions. Three rivers—the Rapido, Garigliano, and Gari—and the extremely steep and rugged terrain of the Matese and Apennine Mountains, were heavily reinforced by Todt units, German labor battalions. The Germans were very familiar with the natural defensive barriers of this part of Italy and had reinforced this terrain during the preceding three months while the 10th Army delayed. Concrete and steel emplacements protected German guns that covered the mountain passes. Fire directed from uplands and mountain tops, such as Monte Cassino, brought the Allied advance to a standstill.

In early January Gen. Sir Harold Alexander's 15th Army Group, with the U.S. Fifth Army attacking on the west and the British 8th Army attacking on the east, battered their way through the Winter Line to face the Germans' most formidable defenses in Italy at the Gustav Line. Frontal assaults on these exceptionally well-prepared positions would unquestionably be costly and time-consuming. The U.S. Fifth Army, consisting of eight divisions, advanced only 11 km (7 mi) in six weeks from the Volturno River to the Rapido River, at a cost of 16,000 casualties, before being stopped.[10] In addition to the German defenses and rugged terrain, the worst winter in many years proved another formidable adversary. "A biting wind, cold, clammy fog, virtually incessant rain, rocky ground, no shelter, insufficient blankets, cold food, and accurate German mortar and artillery fire made life miserable."[11] The stalemated front called for tactics that would allow Allied armies to regain the initiative and break through enemy defenses. Here, and in Korea seven years later, an amphibious envelopment was the solution decided upon.

The most vocal proponent for an amphibious assault around the Gustav Line was Prime Minister Winston Churchill. Insistent that Rome be taken, he was also convinced that a major landing operation leading to the capture of the capital city would "astonish the world and certainly frighten Kesselring." Churchill believed a landing behind the Gustav Line, thereby threatening German lines of communication, would cause Kesselring to withdraw from his prepared positions to avoid encirclement and allow the 15th Army Group to break through the Gustav Line and drive on to Rome. Whether the landing at Anzio would have taken place without Churchill's insistence is questionable; he was certainly a strong and powerful advocate for this venture.

An amphibious assault at Anzio was first considered at an Allied conference in Bari on 8 November 1943, after it became apparent that the German army was going to strongly defend southern Italy. The United States, however, had reluctantly committed forces to the Italian campaign despite the warning of

Gen. George C. Marshall that "the Mediterranean is a vacuum into which America's great military might could be drawn off until there is nothing left with which to deal the decisive blow against the Continent."[12]

Originally, the invasion plan called for one unit, the U.S. 3rd Infantry Division, to land at Anzio on 20 December to disrupt German supply lines and eventually capture Rome. The 3rd Infantry Division was to link up with the attacking forces from the south within seven days. As the weather worsened and the Allied offensive was stopped at the Gustav Line, the Anzio operation was canceled. At the Teheran conference, however, Churchill resurrected the plan, this time with a landing force of two divisions plus separate units and supporting elements. On Christmas Day 1943, the decision was made to land at Anzio, about 100 km (60 mi) behind the German defenses and 55 km (35 mi) south of Rome. Operation Shingle was underway.

The amphibious assault on Anzio had several objectives. The first was to weaken the Gustav Line by forcing Kesselring's response to the threat behind the line. A second was to disrupt German supply lines from Rome. The third was to capture the capital city eventually. A single, well-defined objective for the assault itself, however, was absent.

Unlike many World War II amphibious landings on small islands in the Pacific, an attacker can exercise considerable choice when selecting an assault site on a large land mass. In Italy this option involved numerous locations covering many kilometers on both the Adriatic and Tyhrrenian coasts. As desirable landing site criteria were applied, however, favorable locations dwindled until decisionmakers were left with only a few choices.

The Italian peninsula is dominated by a central spine of rugged mountains, consisting of eroded volcanic rock, flanked by intermittent tracts of coastal lowlands. At the latitude of Rome, the Adriatic (east) shore zone consists of steep ridges that dip toward the sea, leaving an irregular and narrow coastal plain only several kilometers wide. This land is dissected by rivers, generally about 30–50 km (20–30 mi) apart that flow eastward from the mountains to form a series of barriers for north-south movement. Furthermore, once ashore, a large amphibious force would have difficulty moving inland from the coastal plain across the mountains to Rome (see Map 9.1 above).

The western (Tyhrrenian) shore near Rome is topographically more diverse. At many places the volcanic mountains slope abruptly to meet the sea, while elsewhere a coastal plain extends over 40 km (25 mi) inland, making maneuvering of a multidivisional force feasible. These areas, however, are few. Two of them—Anzio and Civitavecchia—met most landing-site criteria, including suitable beaches, deep water in the shore zone, and a location within the range of fighter support bases in southern Italy. Civitavecchia was considered the more favorable of the two, except for the fact that it was 65 km (40 mi) northwest of Rome. Allied planners considered this to be too far behind the Gustav Line for a rapid link-up with forces attacking from the south.

It is worth noting that this same objection was voiced to MacArthur regarding his selection of Inchon as a Korean conflict invasion site. The Joint Chiefs of Staff had recommended a landing closer to Pusan, once again to facilitate rapid link-up. MacArthur rejected this notion on the grounds that the envelopment must be deep to preclude North Korean reaction from Pusan to oppose the landing. It would be interesting to know if the Anzio experience influenced MacArthur's Inchon decision.

Although Anzio was eventually selected largely because of its proximity to the then-static front lines, other considerations may have helped to account for the final decision. First, the coastal plain here is relatively broad, extending inland for about 40 km (25 mi) with a breadth of roughly 25–30 km (15–20 mi). This was sufficient space for the landing, beachhead, buildup, and breakout maneuver planned for the two-division invasion force. Additionally, the road network inland and toward Rome was adequate to allow for exploitation of initial success of the invasion.

The amphibious invasion at Anzio was coordinated with a frontal assault on the Gustav Line. The U.S. Fifth Army and two divisions from the British 8th Army were to attack across the Rapido and Garigliano Rivers. In addition, a corps of the French Expeditionary Force would assault the German left (east) flank on 2 January to secure that portion of the Gustav Line north of Cassino. On 17 and 18 January, the British 10th Corps crossed the Garigliano River and established a bridgehead but was unable to break through the German lines as Kesselring rushed reserves from Rome to his threatened flank. The main attack was then conducted by the U.S. II Corps on 20 January to establish a bridgehead across the Rapido River and break through into the Liri Valley. The attempted crossing met with disaster.

The intent of these attacks was, first, to force the Germans to move troops north of Rome southward to reinforce the Gustav Line and, second, to break through German defenses and link up with the amphibious force. The landing at Anzio was to take place at 0200 hours on 22 January to support the anticipated breakthrough of the Gustav Line and cause Kesselring to abandon his prepared positions because of a threat to his rear. What went wrong?

Crossing of the Rapido River, a most difficult assignment, was the responsibility of the 36th Infantry Division, a unit badly mauled at Salerno. The river was in flood stage after three months of periodic heavy rain. Approximately 20 m (60 ft) wide and 3 m (9 ft) at its deepest, it had a velocity of about 13 km (8 mi) per hour. Both banks dropped vertically 1–2 m (3–6 ft) to the water's surface, further complicating the launching of the assault boats. Additionally, the German 15th Panzer Division was well dug in on the far shore.

The assault was conducted during darkness in a thick ground fog, making command and control difficult. The line of departure for the attack was about 1.5 km (1 mi) from the river across an open, marshy floodplain guarded by enemy minefields, artillery, and machine-gun fire. In the attack some troops became lost and units intermingled in the darkness and fog. Disorder was rampant. Even so, portions of several companies did manage to gain a foothold on

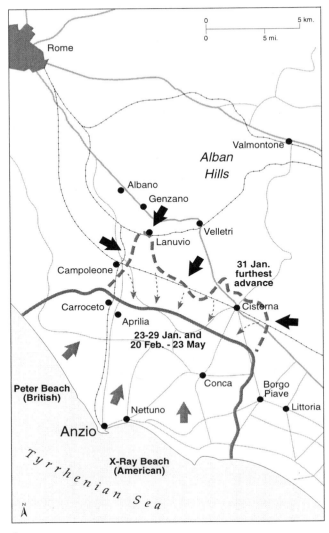

a.

the German side of the river before withdrawing in the face of counterattacks. The next day they tried again and this attack also failed because of poor execution and intense resistance. In these two attacks alone the 36th Division sustained 2128 casualties, which prompted a congressional investigation. Meanwhile, the offensive of the Fifth Army on the Gustav Line was completely bogged down by 22 January, the day of the Anzio landing by VI Corps.

Effective penetration of the Gustav Line was an essential part of the amphibious assault plan. When it failed, VI Corps was forced to fend for itself at Anzio. The VI Corps, a force of approximately 40,000 British and American soldiers, landed on two beaches. The British 1st Infantry Division and three commando battalions went ashore on Peter Beach at the far left of the assault site, which was by far the most difficult of the Anzio beaches (Map 9.2a).[13] Here, steep, sandy cliffs along the shore forced the British to extend their land-

b.

ing area northward. The beach and offshore gradient was very low, approximately 1:120.[14] Waves approaching this type of bathymetry break a considerable distance offshore, only to re-form and break a second and perhaps a third time, significantly complicating landing operations. In addition, large offshore sandbars grounded some landing craft, forcing the assault troops to disembark and wade the final 100 m (330 ft) to the beach. Finally, the sand formation at Peter Beach was too soft to support wheeled vehicles, and the inland dunes further restricted movement. Rear Adm. Thomas Troubridge, commander of Task Force Peter, cited these beaches as "the worst in his experience."[15]

Fortunately for the British and American forces, the landing at Anzio was unopposed. American beach X-ray was to the south and east of Peter Beach. Three battalions of Col. William Darby's 6615th Ranger (U.S.) force landed at 0645 hours and secured the port of Anzio by 0815 hours. The 3rd Infantry

Division went ashore about 6.5 km (4 mi) to the east of the city. The landing was achieved with fewer complications than anyone had dared hope. The weather was clear and the seas calm. The beaches at Anzio were virtually free of German defenders. General Lucas's VI Corps had apparently achieved total surprise.

Kesselring knew that an amphibious attack was being planned by the Allies. The concentration of landing craft and supplies at Salerno left little doubt that there would be an attempt to outflank and possibly envelop the Gustav Line. What he did not know was where and when the landing would come. This is the one major tactical advantage of the amphibious assault—to be able to project a force ashore at a time and place of your choosing. Success, however, requires careful and thorough planning along with effective execution to prevent a disaster such as the one that occurred at Gallipoli during World War I.

Initially the landing force is always vulnerable. Supply and reinforcement can be interdicted by enemy air and sea action or by severe weather, leaving the assaulting troops stranded on the shoreline. The landing force is generally "light," short of armor and artillery augmentation. The first few days of a major amphibious operation, therefore, are generally the most critical.

The coastal plain inland from the beaches at Anzio is gently rolling and cultivated, rising slowly to meet the western slopes of the Alban Hills. These hills provided observation and fields of fire over the two main supply roads from Rome to the battlefields of the south. Unless the Allies controlled the Alban Hills, approximately 30 km (20 mi) inland from Anzio, they could not control the communication routes leading to Rome. The hills are crowned by the dormant volcano, Colli Laziali, with a summit of about 1000 m (3300 ft). There is some question whether General Lucas was directed to secure these hills or merely told to "advance" on them.[16] Whichever the case, despite the lack of resistance on D-day, VI Corps proceeded inland cautiously, limited their advances to preset objectives, and awaited an anticipated counterattack.

The main battleground inland from Anzio was divided into two distinct topographic tracts. The east and southeast sectors facing the Americans consisted of farmland reclaimed from swamps by an elaborate system of drainage ditches built largely during the Fascist regime. The right (south) boundary of this region was the Mussolini Canal, the main drainage channel for a network of smaller canals and ditches excavated to drain the northern edge of the Pontine Marshes for agriculture. These marshes had defied reclamation for centuries and are frequently referred to in Italian history as breeding grounds for pestilence and mosquito-borne disease. The Mussolini Canal was a substantial military obstacle and helped secure the 3rd Infantry Division's right flank. Approximately 55 m (180 ft) wide and 5 m (16 ft) deep, with steeply sloping sides, it also formed a near-perfect tank barrier.

The northern and western portion of the coastal plain, in the British sector, was militarily much more formidable. The area was thickly wooded and cut by deep ravines and narrow gullies formed by rivers and streams incised into the soft volcanic soils. Some of the drainage features were more than 10 m (33 ft) deep with steep to near-vertical sides. During the rainy season, they generally contained rapidly flowing streams a meter or more in depth.[17] The area was referred to as "wadi country" by the soldiers, a carryover terminology from the North African campaign.[18] Consequently, mobility was quite limited in this region, especially for armored units. Across much of the beachhead area, "digging in" to any depth was futile because of the high water table that at many places reached to within a meter (3 ft) of the surface.

Allied intelligence had expected the landing to be strongly opposed. Initial D-day resistance was anticipated to be from one to two enemy divisions. In fact, on 18 January the Germans had stripped two divisions from the Rome and Anzio region in response to the Fifth Army assault on the Gustav Line. Elated by their good fortune, the task force quickly landed and consolidated their first day's objective, a perimeter of approximately 11 km (7 mi) in radius centered on Anzio. Expecting a counterattack, they then took up defensive positions.

German reaction to the landing was swift. Activating a carefully devised plan called Case Richard, Kesselring had 20,000 troops en route to Anzio before darkness fell on the Anzio D-day. Part or all of 13 German divisions began converging on the area. They came from southern France, the Balkans, northern Italy, and Germany. Perhaps most significantly, elements of six divisions and a corps headquarters (I Parachute Corps) were withdrawn from the Gustav Line to meet this threat to the rear.[19] The I Parachute Corps reached Anzio by 1700 hours on D-day and established command.[20]

While the amphibious envelopment at Anzio was considered too deep by some of the planners, it was not nearly deep enough to prevent Kesselring from exploiting an interior line advantage and rapidly shifting forces from the Gustav Line to Anzio. Unlike MacArthur's very deep turning movement at Inchon, which totally excluded the possibility of North Korean reinforcement from Pusan, the Anzio landing was not far enough behind the static front to prevent the rapid redeployment of German forces to the beachhead.

Once ashore, Major General Lucas chose a conservative course of action, consolidating the beachhead instead of boldly dashing forward to secure the Alban Hills on D-day before the Germans could converge on Anzio. There is little question that these strategic heights, only 30–40 km (20–25 mi) from Rome, could have been occupied on that first day. There is considerable doubt, however, that General Lucas's forces could have held these positions in the face of the strong, rapid, and effective German responses. Lt. Gen. Mark W. Clark commented:

There was criticism of our failure to advance deeply inland in the beginning, but in my opinion we most certainly would have suffered far more heavily, if not fatally, had our lines further extended against reinforcements the enemy was able to move in rapidly.[21]

Winston Churchill, the primary proponent for the Anzio operation, was not so kind. He commented, "I had hoped that we were hurling a wildcat on the shore, but all we had was a stranded whale."[22] Whether through lost opportunity or well-founded caution, militarily the Allies' beachhead stabilized and then stagnated. By 29 January, VI Corps had 60,000 troops ashore, but these were matched by an approximately equal number of Germans in more favorable positions.

Fortunately, the Allies controlled the air and sea at Anzio and suffered relatively minor losses to the Luftwaffe. Allied air power, however, was not able to seal off the beachhead from enemy troop units moving to counter the landing. Operation Strangle, the code-name for the air operation aimed at interdicting road and rail movement of German troops toward Anzio, failed. The problem was partly due to the weather. Heavily overcast skies badly limited Allied air operations during these crucial first few days. During the winter of 1943–44, Allied air operations were limited to an average of only two days a week of flying over the Italian battlefields because of rains and overcast conditions.[23]

This foul weather also brought periodic rough seas at the landing site. On 26 January a storm largely disrupted beachhead resupply as portions of landing piers were destroyed, and 13 LCIs and LSTs were beached by the waves. Even so, by late January Lucas thought his force and logistical base were strong enough for a breakout from the beachhead.

The attack reached its farthest advance on 31 January (see Map 9.2a above). But the 3rd Infantry Division spearheading the assault bogged down. Features that had been interpreted as hedgerows on aerial photos were actually drainage ditches overgrown with brush and proved impassable for armor—a very costly mistake. Meanwhile, the U.S. 1st Armored Division encountered marshy ground that gave way beneath the heavy tanks, and steep-walled ditches impossible to cross by combat vehicles. The division, therefore, could not follow and exploit sizable advances by the British 1st Division in this area. The attack lost its spearheading armor force not to the enemy but to the terrain.

Possibly the greatest tragedy was the annihilation of Darby's Rangers. In the 3rd Infantry Division, two battalions of Colonel Darby's three-battalion force preceded the assault by attempting to infiltrate the German lines. Making their way 6.5 km (4 mi) along the Pantano ditch to Cisterna under cover of darkness, they intended a surprise attack against German defensive positions. Instead they were greeted at dawn by an unexpectedly strong force in the form of the Hermann Goering Division. With planned support from the bogged-down 3rd Infantry Division unavailable, the Ranger force was virtually eliminated. Of the 767 men of these two battalions, only six escaped. The unit that had served so well in Sicily and Salerno ceased to exist.

During February the Germans launched three furious counterattacks in an all-out effort to drive VI Corps from the beachhead. Driven back to the last major defensive line, the Allied forces held, later prompting General Kesselring to say, "Anzio was the enemy's 'Epic of Bravery.' Our enemy was of the highest quality."[24]

Both at Anzio and on the Gustav Line stalemate set in and lasted until mid-May. Finally, Monte Cassino fell to the Allies on 18 May, and VI Corps, now under Gen. Lucius Truscott, broke out of Anzio on 23 May (Map 9.2b). After months of fighting, Allied troops at last entered Rome on 4 June, just two days before Normandy. During the period from 22 January to 22 May, U.S. forces sustained 23,800 casualties, the British counted 9203.[25] The strategy, championed by Churchill, had failed.

The Anzio invasion was unsuccessful for a number of reasons. Clearly Kesselring's ability to deploy a substantial force against the beachhead in a very short period was a major factor. In fact, his execution of Case Richard is one of the most impressive displays of generalship of World War II. But it is important to note that the relative shallowness of the Anzio insertion allowed Kesselring to move troops from a static front to meet the Allies quickly. Perhaps General Lucas could have been more aggressive and occupied the Alban Hills on D-day, but in doing so he ran the real risk of losing a vital part of his force to the German response.

Probably the operation's biggest deficiency, or error, was the inadequate size of the amphibious force. VI Corps had to operate independently because of the failure to break through the Gustav Line. The force was too small to threaten seriously Kesselring's lines of communication, primarily because he rapidly and effectively countered the landing. Essentially all Anzio did was open a second front in Italy. This, of course, drew German forces from Western Europe prior to Normandy. Ironically, it was to some degree the preparations for the Normandy invasion which limited the number of landing craft, and consequently the size of the assaulting force, available for Anzio.

Ocean hydrography and shore-zone characteristics favored a landing at Anzio, factors that Allied planners recognized and analyzed correctly. But failure to assess the inland terrain adequately, especially the natural and human-made drainage features, immediately added an unexpected obstacle to an undersized force. Thus an initially strikingly successful amphibious assault quickly became a dismal and costly failure. Poor application of the principles of amphibious warfare, the shallowness of the envelopment, the inadequate size of the assaulting force, effective generalship by the opposition, failure to make a coordinated breakthrough on the Gustav Line, and VI Corps' lack of a clear objective all combined to add struggle and suffering to the longest campaign (Italian) in American military history. Not one of these characteristics reappeared several years later at Inchon, a place where most experts thought a large amphibious operation was not feasible because of bathymetric and coastal conditions.

INCHON, SOUTH KOREA, 1950

The situation for South Korea during the summer of 1950 was desperate. On Sunday, 25 June, at 0400 hours, North Korean forces invaded the South. Trumpets blaring and Russian T-34 tanks rumbling, approximately 100,000 troops swept across the 38th Parallel. The main attack followed the Uijongbu Corridor, an ancient invasion route leading straight to Seoul.[26] The Republic of Korea Army was no match for the well-equipped, highly trained Communist forces. Achieving total surprise, the North Korean People's Army smashed the Southern Korean forces and quickly secured routes southward. In retrospect, all indications for the impending invasion were available but, as in the Battle of the Bulge, were either ignored or misinterpreted. As a result, the South Korean army and Gen. Douglas MacArthur's Far East Command were totally unprepared.

U.S. military disposition in 1950 was not particularly healthy. Demobilization after World War II was almost frantic. Army strength was at 591,000, with approximately 108,000 in MacArthur's Far East Command. Four understrength divisions, the 24th, 25th, 7th, and 1st Cavalry, were in Japan, with parts of other divisions in Hawaii and Okinawa. Marine Corps units were even more depleted, with the Corps fighting for its very survival.

On 1 July, Task Force Smith, two battalions of the 24th Division, landed in Korea from Japan with orders to "head for Taejon. . . . Block the main road as far north as possible" (Map 9.3).[27] The remainder of the 24th Division and the 25th and 1st Cavalry Divisions were rapidly deployed as they became ready, but did little to stop the advancing North Koreans. The necessarily piecemeal commitment of U.S. forces was marked by a curious mix of ignominy and extraordinary valor, and by the end of July, U.S. and South Korean forces had been battered and pushed into a small enclave whose perimeter was not far from Pusan, a city on the southeastern shore of Korea.

Gen. Walton H. Walker, commander of the U.S. Eighth Army at Pusan, then issued a statement that his forces would "stand or die" along the then-existing line.[28] Bolstered by an influx of UN troops and the arrival of the 1st Marine Brigade and the 2nd Infantry Division, the line stabilized but was by no means secure. UN forces were surrounded, three sides facing the enemy and their backs to the sea. Pressure continued through August as 10 North Korean divisions pounded the 230-km (145-mi) arc enclosing only 10,360 km² (4000 sq mi). Combatant strength was relatively equal by this time, but UN forces were still in a decidedly defensive posture.

For 82 days UN ground forces waged a defensive battle, as they retreated to the Pusan Perimeter. The North Korean advance was at last halted in late August, but pressure against the enclave continued without pause. UN reinforcements arrived steadily, but only so much combat power could be brought to bear against the compact perimeter. Additionally, Walker's forces were now conditioned to the defense and had not been able to gain the initiative and break out of the cordon. The situation was set for one of the boldest military operations in history, one that would instantly turn the tide of the war.

MacArthur saw the situation clearly. Recognizing the enemy's extended and vulnerable supply line, he conceived a deep turning movement that would

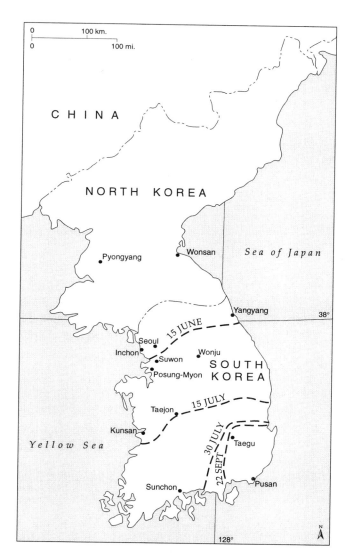

MAP 9.3. Selected battle lines in the Korean conflict, June to September 1950.

Peninsulas and Sea Coasts

209

strike at the heart of the enemy's logistical trains at Seoul. He concluded that an amphibious landing at Inchon would save countless lives and greatly shorten, if not end, the conflict.

The selection of Inchon as the landing site was certainly not the obvious choice. "Make up a list of amphibious don'ts and you have an exact description of the Inchon invasion," commented Rear Adm. James Doyle's communications officer, Comdr. Monroe Kelly.[29] MacArthur selected the site over the objection of the Navy, the Joint Chiefs, and Gen. Omar Bradley, chairman of the Joint Chiefs. Without question MacArthur was the only active duty U.S. commander who could have challenged such formidable opposition and prevailed.

"Perhaps the principal and most sobering hazard which every naval and marine planner who examined the charts of the Inchon area found was the miserable geography."[30] This "miserable geography" was due largely to hydrographic factors at Inchon, South Korea's second largest port and a city of 250,000.

On 23 August 1950, MacArthur hosted a conference at his Tokyo headquarters. In attendance were Adm. Forrest Sherman, Gen. J. Lawton Collins of the Joint Chiefs of Staff, and most Navy flag officers in the Far East. Their attitudes regarding the proposed Inchon landing ranged from skepticism to vocal disapproval. For over an hour, members of Admiral Doyle's Amphibious Group 1 staff briefed the assemblage on the specifics: beaches, tides, intelligence, currents, channels, landing craft, gunfire and air support, and boat waves. At the conclusion of these briefings, Admiral Doyle rose and said, "General, I have not been asked nor have I volunteered my opinion about this landing. If I were asked, however, the best I can say is that Inchon is not impossible, but I do not recommend it."[31] MacArthur's reply lasted 45 minutes, his eloquence swaying even the most skeptical.

The tides were the Navy's gravest concern. Their 10-m (33-ft) range at Inchon is among the largest in the world and they are semidiurnal (two highs and two lows of approximately equal height each day). As the water level rises and falls, currents up to eight knots are generated, equal to the maximum speed of some of the landing craft. Low tide at Inchon exposes vast oozing mud flats that extend seaward 5500 m (18,000 ft) from shore. Landing craft would need to approach Inchon's 4-m (14-ft) seawall at high tide to accommodate landing ships with tanks that have 9-m (29-ft) drafts.

Both the date and hour of the invasion were therefore fixed by the tides. Only on 15 September, 11 October, and 3 November would the water be deep enough to support the landing at its desired time. The critical situation at the Pusan Perimeter dictated the 15 September date. One further hydrographic condition compounded the problem. Access to Inchon was along Flying Fish Channel, a 80-km (50-mi), narrow, tortuous, shallow (6–10 fathoms or 11–18 m) incision into the mud flats.

Thus the hydrographic factors of tide, current, mud flats, and constricted approach combined to make Inchon one of the worst possible places for an invasion. That is precisely why it may have been the best. MacArthur saw these major physical handicaps as advantages, reasoned that the North Koreans would never expect a major assault at Inchon, and had, from their experiences in World War II, perhaps the finest group of amphibious experts ever assembled. It was Inchon where MacArthur wanted to land, but the assurance of surprise was not his primary motive.

An amphibious assault should have a precise objective. MacArthur's was Seoul, South Korea's capital of two million people, approximately 40 km (25 mi) east of Inchon. The occupied city had become the transportation and communication center for the invading North Korean army. Severing the main north-south supply line here would wither the North Korean forces on the Pusan Perimeter. Seoul was the prize, and that is principally why Inchon was the landing site.

MacArthur understood, perhaps like no other commander in history, the value of amphibious operations. He insisted that to be decisive, the envelopment must be far behind the lines at Pusan.

Both the Navy and Army had proposed other assault sites. Admiral Doyle preferred Posung-Myon, 48 km (30 mi) south of Inchon on the west coast, objecting to Inchon's natural obstacles. General Collins and Admiral Struble also favored a more southerly landing site, Kunsan, 160 km by air (100 mi) south of Inchon, because they believed the depth of an Inchon turning movement was too great (see Map 9.3 above). MacArthur, almost haughtily, dismissed all challenges, so convinced was he of the need for a deep envelopment to prevent the North Koreans at Pusan from rapidly shifting forces to meet the amphibious threat. The Joint Chiefs of Staff approved the Inchon plan on 28 August; the day and hour of attack were set by the tide. The execution and spectacular success of this operation involved incredible daring, a great deal of skill, and some luck.

On 11 September, Admiral Struble's 7th Fleet, carrying Gen. Edward M. Almond's X Corps, left Japan headed for Korea. The immediate concern was not the landing, however; it was the weather. Ten days earlier Typhoon Jane had ripped across Okinawa with 100-knot winds and 12m (40-ft) waves. The typhoon's storm surge flooded the docks of the invasion fleet at Kobe and ripped seven ships from their moorings. Fortunately, however, damage was relatively minor. On 7 September, a second typhoon, Kezia, was detected in the Marianas, heading for Japan with 125-knot winds, and forecasted to cross the Korea Strait on 12 or 13 September, just as the amphibious task force was scheduled to leave Japan (Map 9.4). The would-be invaders had as examples of potential destruction the famous thirteenth-century Kamikaze (see chapter 1) and Typhoon Cobra, which had devastated Adm. William F. Halsey's World War II fleet in the Philippine Sea in late 1944, a single storm that capsized three destroyers, swept 150 planes from the decks of aircraft carriers, and claimed the lives of 790 men. And the 200-km/h (130-MPH) winds and 20-m (70-ft) seas inflicted so much damage that the fleet could not carry out its mission in the attack on Luzon. Kezia could scuttle the entire Inchon operation.

Delaying the invasion was out of the question because of necessary timing and the tides; it was either go or no-go. Admiral Struble, therefore, working on the prediction of his staff meteorologist that the approaching typhoon was likely to curve to the northeast, ordered the fleet to sea a day early on 11 September to escape its full impact. This weather prediction proved to be correct, and despite being battered by rough seas, the fleet arrived at Inchon on schedule.

The invasion had two phases. Just after dawn on 15 September, during the day's first high tide, the 3rd Battalion of the 5th Marine Regiment captured Wolmi-do Island, a nub of land connected to the mainland by a causeway, clearly the tactical key to Inchon.[32] Assisted by close and effective naval gunfire and air support, the Marines secured the island in less than two hours with only 17 wounded. Phase II of the attack was at Inchon itself and commenced

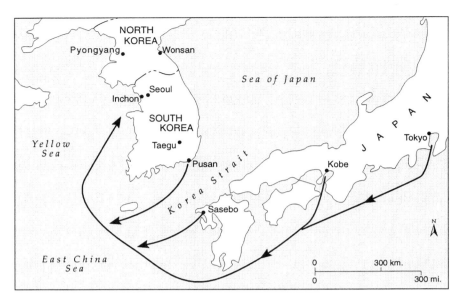

MAP 9.4. Sea routes of invasion forces involved in the Inchon amphibious operation, September 1950.

Map labels: NORTH KOREA, Pyongyang, Wonsan, Sea of Japan, Seoul, Inchon, SOUTH KOREA, Yellow Sea, Taegu, JAPAN, Tokyo, Pusan, Kobe, Korea Strait, Sasebo, East China Sea, 0 300 km., 0 300 mi., N

on the day's second high tide at 1720 hours. After beginning the assault, the Marines (the remainder of the 1st Marine Division) had only two hours to secure positions before darkness fell on this large Oriental city.

The main landing site at Inchon was split by the causeway connecting Wolmi-do Island to the mainland. Red Beach to the north, approximately 300 m (1000 ft) wide, was in reality not a beach at all. Instead, mud flats led to a 4.5-m (15-ft) protective seawall. Marines in tank-carrying landing vehicles used ladders to scale the wall and enter the city.

Approximately 6.5 km (4 mi) to the south was Blue Beach. Situated just south of the causeway and adjacent to a less urbanized area, it was also flanked by mud flats that at low tide extended 4 km (2.5 mi) offshore toward the channel. From their line of departure it took amphibious tractors about 45 minutes to cross the submerged flats at Blue Beach where another rock seawall had to be scaled.

Both beaches were assaulted at 1730 hours. Smoke blowing from fires in the city provided an effective screen for the Marines at Red Beach but obscured Blue Beach. Hampered by the poor visibility, lack of guide boats, inexperienced coxswains, and a strong longshore current, waves of assault boats crossed each other's paths en route to Blue Beach, causing units to land at the wrong location. Meanwhile, at Red Beach the landing proceeded generally as planned. D-day casualties were 20 men killed, 1 missing, and 174 wounded.[33]

Once ashore, the Marine regiments moved toward Seoul and its Kimpo Airport. Joined by the 7th Infantry Division on 18 September, X Corps captured Kimpo on D+3 (three days after the landing began), and the Han River was crossed on D+6. On 28 September Seoul was recaptured and Operation Chromite was completed. The cost was relatively low. In the 13 days 536 soldiers and Marines were killed.

The Inchon invasion took place after the most difficult period of fighting in Korea. The first half of September resulted in more casualties than any other 15-day period, before or afterward. The North Korean Great Naktong Offensive had just barely been repulsed, leaving UN forces at the Pusan Perimeter exhausted and critically short of supplies. All the U.S. and South Korean soldiers spread along the shrinking perimeter were in a defensive posture. Nevertheless, the operations directive of 11 September ordered the Eighth Army to break out from the cordon at 0900, 16 September, one day after the Inchon landing 300 km by air (185 mi) to the north.

Walker's Eighth Army in the south had trouble gaining the initiative. They advanced slowly through 22 September. But on 23 September, elements of the 1st Cavalry Division broke through the perimeter and raced north. North Korean forces began to crumble and the rout began. The "Gary Owen Battalion" linked up with elements of the 7th Infantry Division at 2000 hours on 26 September—D+11.

The pincer had closed. Victory was complete. The brilliantly planned and executed amphibious invasion, far behind the front lines at Pusan, completely reversed the complexion of the war. The North Koreans, fearing encirclement, retreated northward in disorder. UN forces rapidly advanced northward and by late October had moved to the 38th Parallel, poised to continue their attack into North Korea. By careful selection of the amphibious landing site and astute understanding of the hydrographic processes, MacArthur achieved one of the most decisive maneuvers in military history.

Conclusions

The resounding success at Inchon was a striking opposite to events at Anzio in 1943. MacArthur's audacity and expert execution in 1950 epitomizes the proper use of the amphibious assault. This deep, hard-hitting envelopment relieved General Walker's beleaguered command at Pusan, cut North Korean lines of communication, totally routed Communist forces on the verge of apparent victory, and changed the complexion of the Korean conflict. By contrast, Major General Lucas's amphibious force at Anzio on the Italian Tyhrrenian coast failed to achieve its principal objectives. The parallels of Inchon and Anzio are clear. Both attempted an envelopment designed to dislodge a powerful enemy in a favorable position by cutting their lines of communication. Inchon was bold and audacious; Anzio was cautious and predictable. MacArthur had a sufficiently strong two-division force to achieve his objectives; Lucas did not.

But the major difference lies in the geographic scale of the two envelopments. One permitted the enemy the option of rapid response and counterattack. The other was too deep to allow this. Another major difference was the nature of the enemy. Kesselring was prepared to react to the invasion. The North Koreans were not. The German general possessed far greater mobility and availability of forces than did the North Korean commander at Pusan. Misjudgment by Allied intelligence of Kesselring's ability to react, and the inability of air operations to isolate the beachhead, placed the Allied troops in severe jeopardy at the beachhead and contributed to the near-disaster at Anzio.

The physical nature of the landing sites could not have been more dissimilar. MacArthur used the forbidding hydrography at Inchon to great advantage. His selection of this site ensured surprise, precluded quick North Korean reinforcement, and allowed the rapid capture of the primary transportation hub of the North Korean army, Seoul. The tides, mud flats, and limited-access channel to the port all but excluded Inchon as an invasion site. It was the understanding of these seemingly overwhelming negative physical features that fostered MacArthur's victory.

The selection of Anzio was far more predictable. The short-lived surprise resulted largely from the timing of the invasion rather than the location. Generally, the hydrography at Anzio was favorable for amphibious operations. Once ashore, however, the landing force and follow-on echelons were confined to a relatively small area that limited the power the Allies could bring to bear on the defenders. Additionally, the terrain at the Gustav Line contributed to the failure at Anzio. The inability of the Fifth Army to break through this well-defended and superbly reinforced mountainous region doomed the VI Corps to fight alone at Anzio.

In both cases the site selection and physical features of the amphibious landing area made a significant and possibly determinant difference in the success or failure of the assault. Clearly, understanding the total battlefield environment is a vital part of fighting the battle.

Antoine Henri Jomini, the nineteenth-century military strategist, described the basis of amphibious operations. Two key precepts he included were to deceive the enemy as to the point of debarkation and to select a site of favorable hydrographic and terrain conditions. Often these two criteria are in opposition when assaulting a hostile shore. Such was the situation at Anzio and Inchon.

CHAPTER 10

Island Battles

Tarawa and Iwo Jima

My military education and experience in the First World War has been based on roads, rivers and railroads. During the last two years, however, I have been acquiring an education based on oceans and I've had to learn all over again. Prior to the present war I never heard of any landing-craft except a rubber boat. Now I think about little else.

GEN. GEORGE C. MARSHALL, 1943

The Marine invasion of Tarawa in November 1943 marked a first in modern martial history: A major seaborne assault was launched upon a wholly occupied and heavily defended atoll located thousands of kilometers from the homelands of the adversaries. Moreover, this battle was to serve as a prototype for an array of future amphibious operations by U.S. forces that moved from the Gilbert Islands to the Mariana and Marshall Islands and then to Iwo Jima and Okinawa. Tactically, it established the pattern of warfare that would take the conflict to Japan. For the United States Tarawa was an invaluable experience in the amphibious assault.[1] But it also demonstrated that Japanese defenders were going to fight to the death and that in all such operations casualties were going to be considerable (Map 10.1).

For most Americans World War II introduced the word "atoll" to their vocabulary. These mid-ocean islands are most commonly formed of an exposed reef complex composed of coral and algae. Together, these organisms secrete a stable but highly porous (20–50 percent) calcareous mass. As the reef grows, large cavities with connecting chambers are formed while openings provide channels between the surrounding sea and a largely enclosed lagoon. As coral dies, new coral builds atop to enlarge a reef that may eventually be hundreds of meters thick.[2]

The architect of the reef is the hermatypic coral polyp. This living creature secretes for itself a calcium carbonate, cocoon-like, outer structure. From this home, it feeds on zooplankton, the tiny, often microscopic animals of the sea. The coral grows upward, commonly about 1 cm (0.4 in) per year.[3] Physical circumstances that allow these coral colonies to flourish are fairly restrictive, now limiting their geographic distribution to latitudes between approximately 30 degrees North and 25 degrees South. Their extent is first governed by acceptable water temperature, the minimum for the coral being approximately 18–21°C (65–70°F) while temperatures above 35°C (95°F) will kill most coral. Coral also requires certain salinity conditions, living within a range of 27 to 40

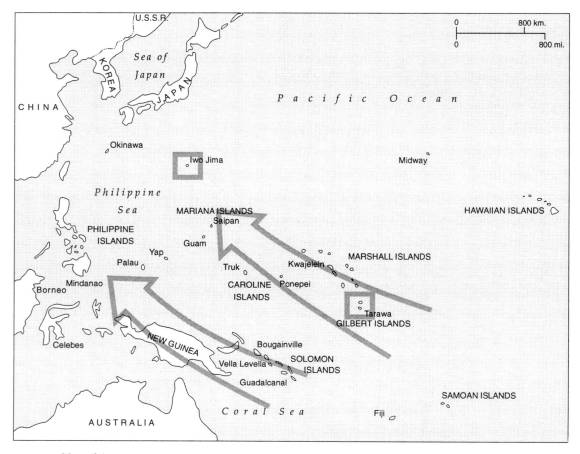

MAP 10.1. Map of the southwest Pacific Ocean with names as they were during World War II. Tarawa and Iwo Jima are within the squares. Arrows represent the two major U.S. military thrusts against Japan.

parts per thousand. Consequently, it is not likely to grow where rivers discharge large quantities of fresh water into coastal regions. Furthermore, sediment discharge from streams and rivers smothers growth, depriving the animals of necessary light and oxygen. Finally, because coral needs light it seldom survives at depths greater than 100 m (330 ft).

In 1842 Charles Darwin developed a hypothesis regarding the development of coral reefs. He proposed that the three types of reefs—the fringing reef, barrier reef, and atoll—represented different stages of coral growth associated with a volcanic feature. Initially, coral grows along the shores of an island volcano (where temperature, salinity, and light conditions permit), forming a fringing reef (Figure 10.1). These reefs experience the most vigorous growth, and generally become widest where oxygen content and nutrient levels are high, such as at a headland.

Although the volcano may disappear because of dormancy, erosion, and tectonic subsidence, or an increase in sea level, the living coral may continue to grow upward, with its top remaining at or near the surface of the sea. With time this process results in a volcanic island surrounded by a barrier reef that encloses a lagoon. Later, if the central island disappears completely, the continually growing coral remaining near the sea surface forms an atoll. The atoll is the most common form of coral reef, its circular, elliptical, or irregular form commonly enclosing a lagoon. Small islands may form on the atoll rim by wave-induced accretion of sediments derived from the reef itself. Such islands are characteristic of low elevation and are interconnected by partially submerged sections of the reef.

Most vegetation will not flourish on atolls because of wind and wave destruction and the infertile, lime-rich soil formed from the decomposed calcium carbonate reef. The adaptable coconut palm is one of the few exceptions; it is the dominant plant and often the economic basis for island inhabitants. While some atolls evolve under somewhat different circumstances, geologic evidence commonly supports Darwin's conceptual model developed more than 150 years ago.[4]

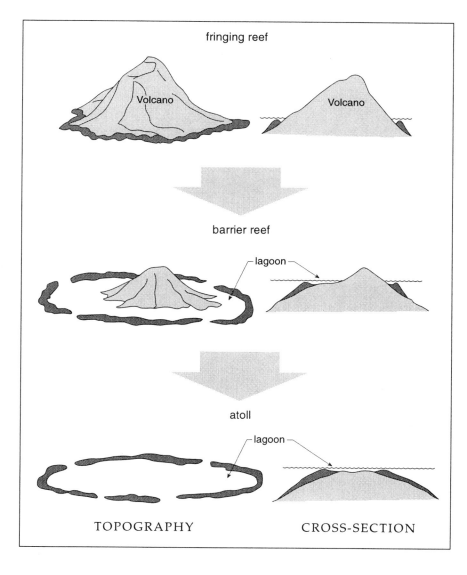

fringing reef

Volcano

Volcano

barrier reef

lagoon

atoll

lagoon

TOPOGRAPHY   CROSS-SECTION

FIG. 10.1. Three stages in the formation of reefs and atolls associated with oceanic volcanos according to Charles Darwin.

Most small islands far at sea owe their origin to volcanism. Aside from volcanoes that protrude above sea level, there are numerous submarine volcanic peaks called sea mounts and guyots.[5] In addition, hundreds of atolls are built atop submerged volcanoes that dot the tropical oceans. Most of the world's atolls are in the Indo-Pacific region, near deep sea trenches formed in tectonically active subduction zones at plate boundaries (see chapter 6). However, not all volcanoes are at the edges of plates. Other volcanic island chains are formed where one of the earth's plates moves over a stationary rising plume of molten rock (magma).[6] Like the linear, fragmented trend of Hawaii, the north-south trending Gilbert Islands are atolls associated with a volcanic plume, with Tarawa the largest of the 16 major islands (Map 10.1).

Tarawa, Gilbert, and Ellice Islands, 1943

Tarawa atoll is a roughly continuous triangular reef that supports numerous coral islands 2.5–3 m (8–10 ft) in elevation (Map 10.2). The atoll is about 30 km (18 mi) long on two sides with a somewhat shorter base. Although islands are numerous, five larger ones exist on both the eastern and southern part of the atoll. These two strings of islands are connected by a submerged barrier reef on the west (Map 10.2). Enclosed within the triangular reef is a large lagoon of varying depth.

The island of Betio is at the extreme southwest corner of the atoll. It is here that the battle for Tarawa took place. This island, former seat of the British headquarters in the Gilberts prior to Japanese occupation in 1941, is 4 km (2.5 mi) long and only 600 m (2000 ft) across at its widest point. Nowhere is the island more than 3 m (10 ft) above sea level, and its total area is roughly 2.5 km² (1 sq mi).[7]

In cross-section Betio resembles the typical atoll island. The submerged outer (ocean-facing) slope of the atoll is very steep, 35 to 45 degrees, and is subjected to high-energy waves that travel from the storm-tossed seas of the South Pacific to crash against the reef front. Atop this outer slope is the low coral (algal) ridge backed by a submerged reef flat that merges with the island's beaches. The beach consists largely of calcareous sand and gravel derived from erosion of the reef itself. The slope is gentle from the island beach into the inner lagoon and across the leeward reef-flat and then steepens to meet the lagoon floor.

Typical maximum lagoonal depths reach approximately 15–45 m (50–150 ft). Lagoon floors are not flat but consist of coral knolls and depressions with relief often of 9–18 m (30–60 ft).[8] Finally, and especially important during this military operation, a leeward reef flat, submerged only at high tide, often lies several hundred yards off the lagoon-side beach.

A major planning factor at Tarawa, and for every amphibious assault, is wave energy, or wave height, at the landing site. Waves are developed in stormy areas at sea. As winds blow across the water, energy is transferred to the sea

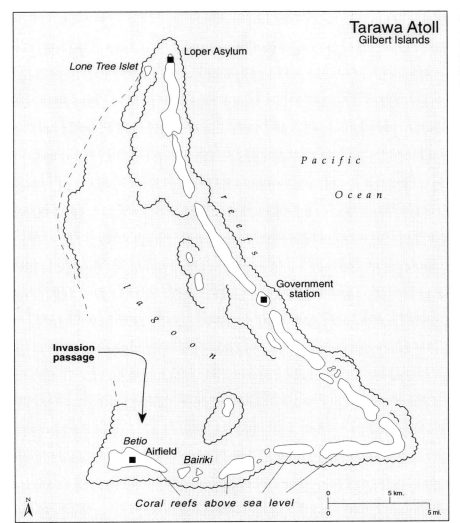

MAP 10.2. Tarawa and the amphibious invasion route followed by the U.S. Marines who attacked Betio.

surface to form the waves, their height, spacing, and frequency dependent principally on three variables: velocity, duration, and fetch. Velocity refers to the wind speed: the faster the wind, the higher the wave. There is also a direct relationship between wave height and duration, or the length of time the wind blows across the water. The longer the wind blows, the higher the wave. Finally, the greater distance the wind blows across the water in the storm-generating area—the fetch—the higher the wave. The three variables are interrelated; any one limits the effects of the others, but together they determine the height and spacing of waves.[9]

As the waves leave the tumultuous storm area, they separate into groups having similar characteristics. These are called wave trains, and they travel at an approximately constant speed and are relatively evenly spaced. The waves may retain these characteristics across thousands of kilometers of open sea, but their form and spacing change markedly as they enter shallow water. Ap-

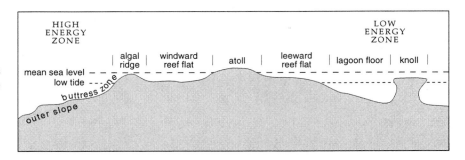

FIG. 10.2. Cross-section diagram of a barrier reef and atoll with an interior lagoon. The ocean is to the far left and the lagoon to the far right.

proaching the shore they begin to slow due to the interference and friction of the shallowing sea floor. Eventually, when the wave height is approximately equal to three-quarters the water depth, the wave becomes unstable and breaks.[10] The more gradual the offshore slope, the more opportunity the wave has to slow down. Conversely, the steeper the offshore slope, the faster the wave approach and consequently the higher the incident wave energy.

On a coral atoll, the ocean side of the reef has an extremely steep slope, giving waves little chance to slow before breaking on or very near the shore. These are characteristically high-energy shores, making them hazardous for amphibious assaults. All shorelines have characteristic wave energies and predominant direction of wave attack. These properties often vary seasonally, being a function of prevailing wind direction and velocity.

During the summer of 1943, the Allies fought the Japanese in four regions. In the South Pacific, Guadalcanal had been won, and U.S. forces were moving onto New Georgia and Vella Lavella toward Bougainville in the Solomon Islands. In the southwest Pacific, Army forces were pushing the Japanese from the jungles of New Guinea (see chapter 11). The war in China was ongoing, and U.S. forces had defeated the Japanese on Attu Island, driving them from the Aleutians. The decision was then made to initiate an offensive in the central Pacific. Tarawa was to be the first objective.

Tarawa was the only heavily defended atoll of the Gilbert Islands. Adm. Ernest King, Chief of Naval Operations, considered them of great significance because they were north and west of other islands in U.S. possession and immediately south and east of the important Japanese bases in the Caroline and Marshall Islands (see Map 10.1 above). He considered the capture of the Gilberts a necessary part of any serious thrust at the Japanese empire.[11] They were also the enemy's nearest base to American supply routes from San Francisco to Hawaii to Australia.[12] In addition, planes from Japanese airstrips in the Gilberts could cover the Samoan area and deter forces from striking the Marshalls.

It was, however, Adm. Chester W. Nimitz, commander of the Pacific Fleet, and Adm. Raymond A. Spruance, his chief of staff, who were largely responsible for selecting the Gilberts as the objective to initiate the offensive in the central Pacific that would move through the Marshalls, Carolines, and Marianas to the Japanese homeland islands of Iwo Jima and Okinawa. The Joint

Chiefs of Staff concurred, citing the geographic character of the central Pacific to be strategically decisive and that success in this region would sever the Japanese homeland from the overseas empire to the south.

Of immediate importance to American planners was that only larger land-based planes, and not carrier-based aircraft, could provide the stable platform needed for quality air-photo reconnaissance. Air photos were considered essential for the assault on the Marshalls, a lesson learned from earlier Pacific and North African amphibious operations. Admiral Spruance later commented that the United States could not attempt to capture any defended island without adequate aerial photographs, and those of Kwajalein (in the Marshall Islands) became available only after U.S. troops had taken the Gilberts and reconstructed the airfields on them.

U.S. forces had no more recent combat experience in amphibious operations than those in the earliest phase of World War II. In World War I American troops disembarked on friendly shores and were transported toward the front by train or truck. In fact there were few amphibious assaults in that war, the notable exception being the disastrous British–Australian–New Zealand effort at Gallipoli (now Gelibolu, Turkey).

Doctrine for amphibious operations, to be severely tested in World War II, was developed largely by the Marine Corps between 1922 and 1935.[13] A significant part of the planning, both then and now, involves detailed consideration of hydrographic factors. Antoine Henri Jomini, the nineteenth-century military strategist, commented that amphibious operations must be designed to deceive the enemy as to the site of debarkation and a beach must be selected that has hydrologic and terrain conditions favorable to the attacker.[14] Hydrographic and shore-zone variables that must be considered include wave energy, tides, beach slope, beach material, nature and composition of the near-shore currents, topography at the landing site, vegetation and material shoreward of the beach, and the weather on D-day. Several of these features were of extreme importance at Tarawa.

The exact date (D-day) for the Tarawa invasion was determined largely by the tide. The original target date, set by the Joint Chiefs at 15 November 1943 to coordinate with attacks in the south, central, and southwest Pacific, was delayed five days for what was anticipated to be more favorable, but certainly not optimal, tidal conditions.[15] A neap tide occurred at this time, producing the lowest tidal range of the month. More favorable conditions (spring tides) would not exist at a suitable time of day until 27 December, too late, it was decided, to coordinate with other operations.

Tidal information at Tarawa was provided to the fleet by British shipmasters familiar with the waters in and about the atoll. Charts for this area were about 100 years old, being drawn by Lt. Charles Wilkes, U.S. Navy, in 1841. Thus the largest U.S. invasion fleet ever assembled in the Pacific to date was basing its operation on very old charts and the memory of a few seamen. The consensus was that about 1.5 m (5 ft) of water would cover the coral in the lagoon north of Betio at flood tide, 0830 hours, 20 November. The entire operation was planned around that moment.

One dissenter, a civilian named Major Holland who had lived on Tarawa for 15 years and measured its tides for the British, predicted a "dodging" tide for 20 November, a term not familiar to Navy hydrographers or the U.S. Geological Survey. Dodging tides seem to represent an erratic episode of low, slack water. Holland's prediction, then, was that there would be no high tide on 20 November. Unfortunately, this forecast was not corroborated by scientists or by other seamen supposedly familiar with these waters and was largely discounted. The landing craft (LCVPs) had a draught of about 120 cm (4 ft), hopefully giving a 30-cm (1-ft) safety margin. But if Holland's prediction was correct and the assault boats failed to clear the reef, Marines would have to disembark and traverse 450–900 m (1500–3000 ft) of a rough, uneven, coral-strewn sea floor under withering Japanese automatic-weapons fire.

To offset the possible problem of insufficient water depth over the protecting reef, amphibious tractors would also be utilized, a tactically innovative use of this logistical craft. But no more than 125 of the amphibious tractors (LVTs) could be gathered before the invasion, enough to lift only the first three waves of Marines over encountered reefs. The remaining assault forces would have to hope that the tides were high enough to float their LCVP landing assault craft (Higgins boats) over the coral.

Landing beaches (Red 1, Red 2, and Red 3) were chosen on the lagoon (northern) side of Betio for two reasons (Map 10.3). First, because of the steep offshore slope, the southern ocean shore of the island was subject to high wave energy from deep-water waves. Landing here would present an extra danger of swamping in the consistently high surf. Also, a coral reef about 460 m (500 yd) off the beach presented special navigational problems. The second reason for not selecting this beach was that most Japanese obstacles and coastal guns were oriented in this direction, in readiness for an attack on the southern shore.

The defenses on Tarawa were formidable. Intelligence reports indicated that the island fairly bristled with weapons, including two 8-in Vickers guns captured at Singapore, two 75mm mountain guns, six 70mm cannon, and nine 37mm field pieces. In addition, the Japanese defenders had nine light tanks and hundreds of automatic weapons. Betio was garrisoned by approximately 5000 men, mostly from the elite 3rd Special Base Force and 7th Naval Special Landing Forces or Imperial Marines. Antipersonnel mines were placed along the beach and antiboat obstacles were constructed on the coral and in the shallow water shoreward of the reef. Wire and concrete obstacles were covered with well-coordinated fires, the result of 15 months of preparation. Concrete pillboxes with walls 2 m (7 ft) thick lined the island.

Preassault naval bombardment was not particularly effective despite the promise by the commander of the gunfire group, Rear Adm. Howard F. Kingman, to "obliterate" the island.[16] Marine Corps Gen. Holland M. Smith later "entered every pillbox and block house on the west end of the island and found

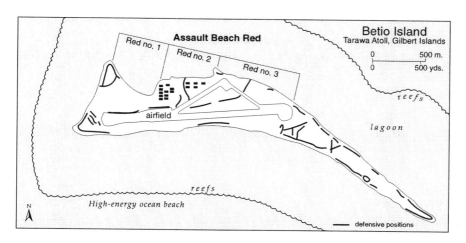

MAP 10.3. The assault beaches, major Japanese defensive positions, and airfield on Betio.

only one had been hit by naval gunfire." This was due in part to the very low profile of these structures, the flat trajectory of the naval guns, and the insufficient preparatory bombardment. Additionally, underwater obstacles remained essentially intact because demolition teams that proved so effective in later operations had not yet been organized.

Such was the situation as the 2nd Marine Division and elements of the 8th Marine Division prepared to debark the transports and load the assault craft at 0320 hours, 20 November 1943. A number of factors caused H-hour to be delayed from 0830 to 0900, ranging from mechanical problems and inexperienced crews to an unanticipated strong longshore (parallel to the shore) current that caused landing craft to drift far to the south after leaving the parent ships in the early morning darkness.

At dawn the minesweeper *Pursuit* led the first three waves of the invasion force through the lagoon inlet just north of Betio. At approximately 0900 hours, they crossed the line of departure (LD) and headed toward the shore. Their tracked vehicles crossed the reef with little difficulty even though it was partially exposed in some locations. Water depth over most of the reef was less than 1 m (3 ft).

The fourth and succeeding waves were not nearly so fortunate. The trackless Higgins boats had no way to negotiate the lagoon's reefs and many ground to a halt on the jagged coral. The only options were to remain in the boats and abandon the troops already on the beach or wade and swim ashore, a distance of 450–900 m (1500–3000 ft) across the jagged coral. Water depth was generally from 1 to 3 m (3–9 ft) but numerous depressions were 4–5 m (12–15 ft) deep. Some Marines, carrying over 30 kg (65 lb) of equipment, drowned after leaving the boats, while the others began their agonizing struggle toward the beach.

Those amphibious tractors that had survived the first assault began shuttling wounded back to the hospital ships. But by then, unit integrity had disappeared. Troops were mixed and officers separated from their commands.

Both in the water and on the beach, personal survival was the order of the day. Control was lost as the carefully designed landing plan collapsed. At 1007 hours, Maj. John F. Schoettel, commander of the 3rd Battalion, 2nd Marines, one of the three assault battalions, radioed the stunning words, "Unable to land. Issue in doubt. Boats held up on reef. Troops receiving heavy fire in water." When told by Col. David M. Shoup, commander of the assault regiment, to land on the neighboring beach, Schoettel replied, "We have nothing left to land."[17]

Throughout the day Marines made their way ashore and attempted to advance onto the island. Many found refuge behind a log revetment 1 m (3 ft) thick, some 6 m (20 ft) shoreward of the high-water line. Eventually the situation began to stabilize. Small units formed spontaneously, organized by noncommissioned leaders and junior officers, to become an effective fighting force.

As darkness fell at the end of the first day, a beachhead only 300 m (1000 ft) wide and 30–50 m (100–160 ft) deep was established, providing the setting for three more days of fierce and unrelenting combat. The battle for Betio ended the afternoon of 23 November. It had been a continuous assault for the entire time, with the combatants literally within meters of one another as the Marines slowly pushed across the tiny island.

The price of victory was high: 1009 Marine and Navy personnel died on Tarawa while 2101 others were wounded. Of the 125 amphibious tractors involved in the invasion, 90 were destroyed. More than 4700 Japanese defenders died—all this carnage on an island with an area of roughly 2.5 km² (1 sq mi).

America was stunned by the magnitude of the losses. A naval board of inquiry asked Adm. Kelly Turner, commander of V Amphibious Force, to justify his decisions regarding the tidal issue, a major factor for the high casualty toll. It was estimated that approximately 300 Marines died while trying to get ashore from landing craft grounded on the coral. Unfortunately, good hydrologic information was not available to Admiral Turner, and he elected to invade based on conflicting tidal forecasts. Because of the timing, he had little choice. In retrospect, however, it appears that the landing should have been delayed until more favorable tidal conditions existed, assuring clear passage of landing craft from ship to shore.

Despite the terrible losses, Tarawa was an important victory for the United States. American forces now operated from a base 1100 km (700 mi) closer to Tokyo. The airfield on Tarawa was quickly repaired, and soon reconnaissance and tactical missions were being flown over the next objective—the Marshall Islands. Tactically, Betio formed the basic textbook for future coastal assaults. Lessons learned from this coral atoll battle certainly saved many lives as the doctrine for amphibious warfare was being perfected as never before. The technical aspects of amphibious warfare, including landing-site selections and wave and tide forecasting, became routine as U.S. forces worked their way west and north toward Japan.

Iwo Jima and Volcano Islands, 1945

While amphibious doctrine was developed at Tarawa, it was perfected at Iwo Jima (now Sulfur Island) (see Map 10.1 above). This was the epic struggle of Adm. Chester Nimitz's island campaign in the Pacific. Here the growing power of the U.S. Pacific Fleet fully met the dedicated but increasingly desperate forces of Japan. The importance of Iwo Jima was clear to both commands and both knew that it must be assaulted. No deception or surprise was involved. Once engaged, for more than 30 days these combatants literally fought to the death, often in hand-to-hand combat with battle lines that frequently were indistinguishable. It was one continuous assault from D-day, 19 February 1945, until the 21 km^2 (8 sq mi) of barren volcanic sand and rock was finally declared secured on 16 March.

The island is part of the Nanpu Shoto Group, an archipelago extending in a north-south direction. Iwo Jima lies on the southern tip of this group, approximately 1000 km (620 mi) south of Tokyo, directly on the long-range bomber route between Tokyo and the southern Marianas. The Japanese used airbases on Iwo Jima to intercept U.S. aircraft traveling, because of the extended range, without fighter escort (Map 10.4). Additionally, the radar installations on Iwo Jima allowed the home islands time to prepare for the increasingly frequent bombing raids on Japan.

Four basic reasons were given for Operation Detachment, code-name for the Iwo Jima operation:

1. to attack the homeland of the Japanese empire;
2. to protect our bases in the Marianas;
3. to cover our naval forces searching for approaches to the Japanese islands; and
4. to provide fighter escort for very long-range air operations.[18]

Long-range U.S. B-29 bombers, called Superfortresses, had been striking Japan from bases in the Marianas since November 1944. Their missions were complicated by a number of factors. Japanese planes from Iwo Jima were raiding U.S. airstrips, damaging the runways and planes, as well as intercepting the B-29s en route to Japan. Aircraft damaged in the target area had no place to go except into the sea with little hope of rescue, should they not be able to make the 1100-km (700-mi) return trip to Tinian or Guam in the Mariana Islands. As a measure of its importance, in the first three months after Iwo Jima was secured, 850 B-29s made emergency landings with their crews, totaling over 9000 airmen. It is estimated that more than 20,000 crewmen in crippled aircraft landed safely on Iwo Jima before the war's end.

The Joint Chiefs directed Nimitz to occupy positions in the Nanpu Shoto, which consists of three major island groups. From north to south, these are

MAP 10.4. The landing beaches, Mount Suribachi, several U.S. offensive positions between 21 February and 16 March 1945, and the three airfields on Iwo Jima.

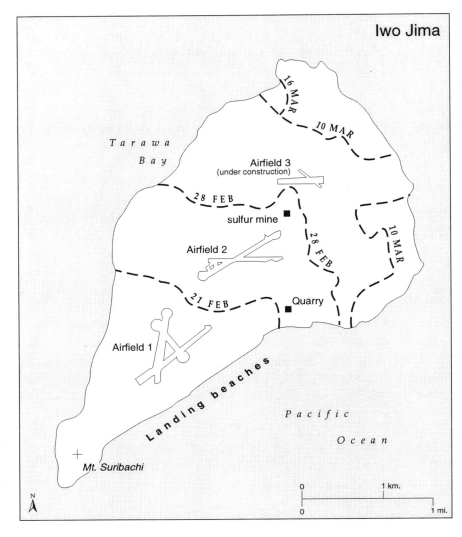

the Izo Shoto, the Ogasawara Gunto (Bonin Islands), and Kazan Retto (Volcano Islands). All are part of the island arc system of the western Pacific, one of the most tectonically active regions in the world. This is also the location of the ocean's greatest depth, where the ocean trench systems plunge more than 11,000 m (36,000 ft) below the sea surface. The deepest of these waters is the Challenger Deep in the Mariana Trench, of record depth at 11,034 m (36,200 ft).

Recall that island arcs are caused by plate subduction—one plate diving, or subducting, beneath the other. In the Pacific Basin, it is the heavier Pacific plate that is sliding beneath the lighter, continental Eurasian plate. This activity, involving the enormous mass of these giant bodies, results in incredible energy transfer, often manifesting itself in earthquakes and volcanism (see chapter 6).

Typically, in this system, the deep trench bends convexly toward the ocean. While some of these trenches are V-shaped, most have a flat, relatively narrow floor several kilometers wide. It is here that the oceanic plate is actively moving beneath the continental plate at a rate of several centimeters each year. Landward of the deep trenches lie the island arcs formed from material scraped from the subducting plate and subsequent volcanic activity.

Molten material, formed at great depth in the subduction zone, eventually rises to the surface through mechanisms still not fully understood. Predominantly andesitic in nature, some of the magma extrudes through the crust to form volcanoes similar to those of the Cascade Range in Oregon and Washington. Behind the volcanic islands, toward the mainland, lies the back-arc, or marginal, basin, an area of relatively shallow water overlying continental crust. Iwo Jima is part of such an island arc complex with the bottom adjacent to Bonin Trench 9810 m (10,780 ft) below sea level. In contrast, the Shikoku Basin, landward of the Bonin Islands, has the relatively shallow depths characteristic of back-arc basins.

Nimitz needed to select an island where beaches were suitable for an amphibious assault and space and terrain were adequate for several airfields. Only Iwo Jima fully met these requirements. Natural vegetation on the island was sparse, consisting only of coarse grasses and small shrubs. Located at 24° 44′ North latitude and West 141° 22′ East longitude, its climate is subtropical with a cool season from December through April; February, the invasion month, is the driest time of year. Sparsely populated in 1938 by slightly more than 1000 civilians of Japanese ancestry, the island derived its economy from a sugar refinery and sulfur mining. It had been a Japanese possession since 1861.[19]

The island is approximately 8 km (5 mi) long on the southwest-northeast axis, with a width that varies from about 4 km (2.5 mi) to less than 1 km (0.6 mi) and an area slightly less than 20 km² (8 sq mi). Dominating the southwestern part of the island is Mount Suribachi, a volcano that rises to 165 m (550 ft) and overlooks Iwo's only two beaches considered suitable for amphibious assault. These beaches are on opposite sides of the narrowest portion of the island and vary in width from 45 to 150 m (150–500 ft), space barely sufficient for a major landing. In between is the lowest land on the island, flanked by Mount Suribachi immediately to the southwest. Toward the northeast is higher ground that slopes upward to a low plateau. The plateau, which is about 1.5 km (1 mi) in diameter and bordered by a steep, rugged sea-cliff to the north, is dissected into a system of gorges and ridges that radiate from the central high point of 116 m (382 ft), making the terrain rough and rocky. Any force attacking from the lowland, either northeast or southwest, fought an uphill battle all the way.

The choice between the two possible landing sites was not difficult. The southeast beach was better because February's prevailing winds from the northwest produce much higher waves on the northern shore. Even so, wave energy on the other beach could be hazardous for amphibious operations during severe weather. Additionally, because of cooler ocean temperatures no surrounding coral reef was present to break incoming waves. While simplifying access to the shore, this also permitted high-energy waves to break directly onto the beach.

Although judged adequate, the landing site was far from ideal. Beach material is a coarse, unconsolidated volcanic ash, and the shore-zone slope steep. Assault troops, upon disembarking from landing craft, sank to their calves in the soft sediment. Once ashore, traction was poor in the gritty volcanic ash and sand. Moving inland, the amphibious tractors encountered a steep-faced scarp 0.5–6 m (2–20 ft) high. Scarps of this nature form at varying levels from storm waves. At high tide wave energy is concentrated higher on the shore and at the maximum extent of the tidal advance a terrace may form, its height dependent upon the beach material, beach slope, and wave energy. The slope of the beach between the high- and low-tide level, part of the shore platform, is variable. As a generalization, these platforms are gently sloping, in the range of 1:100.[20] At Iwo they were much steeper than average.

Waves, ash, and terraces severely limited mobility and took their toll. Landing craft, pounded by heavy surf, faced the danger of broaching and swamping. Trucks and jeeps sank axle-deep in the ash and were often abandoned in the soft volcanic material the size of buckshot.[21] Even tracked vehicles had difficulty getting ashore and could be stopped completely by the difficult terrain.

The defense of Iwo Jima was the final responsibility of Lt. Gen. Tadamichi Kuribayashi, commander of the 109th Infantry Division. With 21,000 soldiers, he had transformed the island into a veritable fortress, one of the most elaborate and effective examples of terrain reinforcement ever accomplished. The defense was dictated principally by the topography. Natural features, predominantly caves, along with artificial defenses, were woven into a deadly net of mutually reinforcing fields of fire.

Kuribayashi's plan was simple but effective. The beaches were to be lightly defended. Although pillboxes were present on both shores and in the vicinity of airfield no. 1, Kuribayashi considered these low-lying areas indefensible because of U.S. air operations and naval gunfire superiority. Mount Suribachi and the low plateau to the northeast, however, were heavily fortified with commanding observation of the two obvious beach-landing sites. Departing from the traditional Japanese defensive doctrine of all-out counterattacks against the beachhead, he defended the high ground in the north and south with a series of in-depth positions. These were occupied before D-day with orders to defend to the death.[22] The defenses of the southern part of the island, Mount Suribachi, and the northeast section of the island were conducted independently.

Mount Suribachi was the primary objective of the 28th Marines of the 5th Division. The volcano, being newly formed, had a crater at the top and was essentially devoid of vegetation. The entire mass bristled with enemy emplacements, both natural and constructed. The lower slopes of Suribachi were armored with well-prepared defensive positions, all mutually supporting. With little cover, the Marines needed to take each position individually, often by frontal assault using flame-throwers and demolition. Caves, tunnels, and chambers honeycombed the mountain, one being large enough to hold 90 soldiers.

After the first day of the Suribachi assault (D+1), the Marines had advanced only 185 m (200 yd). And so it continued slowly up the volcano's steep slopes. A cold rain, combined with the sand, repeatedly jammed weapons and often reduced effectiveness to single shots. Caves were closed by bulldozers and explosives, sealing many of the defenders alive in their graves. On D+4, the American flag was raised over the mountain, taken at the cost of 510 U.S. casualties and more than 2000 Japanese dead, many entombed in more than 1000 caves.

The relatively quick capture of Suribachi not only relieved some pressure from the beachhead but also boosted morale for the Americans and dealt a psychological blow to the Japanese. Attention then turned fully to the other, larger part of the island. The first airfield, lying immediately inland from the beach, was taken on D+1, and the Marines swung northeast into the deep defensive belt of Kuribayashi's stronghold. Japanese positions stretched completely across the island. No flank was accessible. Gains were measured in meters, and casualties were many.

For three days the struggle here was a virtual stalemate. The beaches were still subjected to the murderous Japanese fire. A south-tracking middle-latitude cyclone brought foul weather and curtailed air support because of the limited visibility. Especially high surf, with wave heights exceeding 3 m (10 ft), broke across the landing beach and extended all the way to the first terrace. The strong backwash pulled materiel and smaller landing craft to sea, reducing the rate of reinforcement and resupply. During the first four days of the campaign, 200 landing craft were lost to the breakers and offshore currents; those attempting to disembark troops and equipment filled with water and sand as the ramps were dropped. Adequate purchase for anchors was impossible because of the soft sand. Landing craft pounded by the heavy surf capsized, and many Higgins boats were broached by the incoming waves. Only amphibious tractors were fully functional during this period of high seas.[23]

At last, on D+5 (the fifth day after the landing), the Americans effected a breakout toward the northeast and once again established momentum. In the area fronting the 21st Marines, a sector 900 m (3000 ft) wide by 65 m (200 ft) deep, mopping-up parties counted 800 separate Japanese fortifications in the soft volcanic soil and rock, all mutually supporting and most directly connected by wire communications. Kuribayashi's best had been almost good enough.[24] With further advances the Marines took airfield no. 2 and continued the slow, bloody, constantly uphill battle.

The terrain north of airfield no. 2 was dominated by Hill 382, a barren knob that radiated a series of ridges and deep ravines cut in the soft volcanic soil. It had been the site for Japanese radar installations and was the highest elevation aside from Suribachi. This hill and adjacent ridges became known as the "Meat

Grinder" to the men of the 23rd Marines. The direction of assault was cross-compartment, requiring the Marines to traverse ridges and ravines running at right angles to their direction of advance. This is unquestionably the least desirable path, preferred movement being along the trend of ridges or neighboring valleys or compartments. The assault at first met suspiciously little resistance, with the Marines rapidly moving to the summit with relative ease. Once there, however, they were subjected to murderous fire from every angle. The Japanese had anticipated that this feature would be a primary objective and had registered their weapons on its summit. In addition, while hurriedly moving to the hilltop, the Marines had unknowingly passed expertly camouflaged pillboxes, bunkers, and hidden caves. The Japanese trapped four platoons atop the hill, and the Marines desperately fought their way back to their lines. This same scene was repeated elsewhere many times as the Japanese mastery of this terrain became increasingly apparent. Repeatedly, Marines would overrun enemy positions only to have the enemy at their back.

For 26 days the battle continued. At last, on 16 March, the island was considered secured, but mopping-up was to continue for months. In all, 6766 Americans would not leave the island alive and another 19,189 left wounded. One in every three Marines to land on Iwo Jima was a casualty. Approximately 20,000 Japanese soldiers also died in this—the bloodiest island battle in military history.

Conclusion

An opposed amphibious landing on a small island has several unique considerations when compared to an assault on a large land mass. Prior to Tarawa, U.S. forces had landed only on friendly, undefended, or lightly protected shores. Such was the case at Fort Monroe, Virginia, in 1862, France in World War I, and North Africa in 1942. Major battles took place only after American troops had secured and consolidated a beachhead and establishment of a strong logistical base. On the smaller Japanese-occupied Pacific islands, beginning with Tarawa, this could not be done. Landing sites were limited and all beaches were heavily fortified. The islands were virtual fortresses, and once the campaign was initiated, it was an assault from beginning to end. It was the closest, most violent form of combat.

The similarities and contrasts of Tarawa and Iwo Jima should now be apparent. Although both islands owe their origin to volcanism, they represent variations on Darwin's evolutionary model of the atoll. Both illustrate the importance of considerations involving tides, wave energy, and offshore bathymetry. While complex, these factors are either interpretable or predictable. By thorough analysis, an accurate hydrographic evaluation of potential landing sites can be rendered. Failure to do so accurately could easily lead to a repeat of the U.S. misjudgment at Tarawa.

The evolution of amphibious techniques in the Pacific war was extraordinary. Planning and timing of movements involved months. Staging was done across thousands of kilometers of ocean. Conflicting priorities of a two-front war made logistics a constant challenge. Closer coordination among air, naval, and assault forces was enhanced. The technical aspects of selecting a landing site were rapidly improved with some scientists trained in wave-forecasting procedures at Scripps Institute of Oceanography to take some of the guesswork out of predicting surf conditions. By war's end U.S. forces were more experienced and skilled in amphibious operations than any other in military history.

The next major amphibious assault after Okinawa was the 1950 Inchon operation in Korea (see chapter 9) and that success depended heavily on lessons learned during World War II. Since then there have been, among others, shore-zone operations in Vietnam, Falkland Islands (Islas Malvinas), the Middle East, Central America, and Somalia. All the while, however, the feats of World War II and Korea may have become blurred with time. While the Marine Corps still maintains proficiency in sea-to-land operations, the U.S. Army has, in struggling with its responsibilities and priorities, largely dismantled its amphibious forces. Although some may question this policy, all can be sure that there will be other military amphibious operations in the future.

CHAPTER 11

Hot, Wet, and Sick

New Guinea and Dien Bien Phu

Bob, I want you to take Buna or not come back alive.

DOUGLAS MACARTHUR TO ROBERT EICHELBERGER

In addition to the enemy, forces fighting in the humid tropics must deal with oppressive temperature, overabundant moisture, and a myriad of related diseases. Sapped by the heat and constantly threatened by body-attacking organisms, the soldier soon finds that coping with the elements is every bit as challenging as meeting the opponent on the battlefield. With a year-long growing season, plus abundant sunlight and rainfall, vegetation thrives, creating densely tangled jungles and multicanopied forests that often impede movement, deployment, and communications. Careful selection of battle positions becomes critical. Poorly placed foxholes and bunkers can become cesspools that may jeopardize, rather than protect, their occupants.

These conditions also severely test medical-support systems. During 1944 in the China-Burma-India operations, 90 percent of the U.S. casualties were from disease, whereas only 2 percent resulted from combat. In the southwest Pacific, 83 percent of the casualties during World War II stemmed from ailments. During 1969, in Vietnam, more than two decades later, the figure was 67 percent for U.S. troops.

Tropical conditions not only endanger the healthy soldier, they also make it difficult for the disabled to recover. Above all, the pervasive and enervating heat and moisture drain soldiers' energy, test their ability to pursue the enemy, and may even lower their determination to win.

Following the example set by Gen. Ord Wingate of the British army in the early stages of the war in China and Burma, Brig. Gen. Frank D. Merrill prepared the 5307th Infantry Regiment (Merrill's Marauders) in 1943 for operations in the same theater. Operating to the rear of the Japanese, Wingate's foes were both enemy soldiers and the tropical environment. Although his forces were tactically aggressive, most of the casualties were nonbattle, which entailed the added requirement of assigning combat troops to stay behind to protect and transport the sick.

At the same time some Philippine and U.S. troops remained operational and effective in the Philippines by avoiding direct combat with the Japanese. In fact, 95 percent of Mindanao was in the hands of these "guerrillas." Yet for these isolated fighters the environment was as threatening as the enemy. Because medicine and facilities were essentially nonexistent, even a slight injury might soon result in infection and possible death. About 20 years later U.S. troops in Vietnam found that modern medicine and training had yet to overcome the diseases, parasites, and continuous oppression of the tropics.

All ground operations in swamps, jungles, and rainforests are regularly threatened by such a variety of pestilence that cannot be controlled. Combat in lowlands and bogs is often made doubly worse by heavy rains and runoff that slow movement, flood positions, exacerbate sanitary problems, threaten equipment, and lower the soldiers' fighting effectiveness. The challenge is to minimize the debilitating effects of the humid tropics while at the same time making maximum use of the cover and terrain in fighting the enemy.

Ramifications of combat in the humid tropics are made vivid by two related battlegrounds. The first is in New Guinea where, during World War II, Australian, U.S., and Japanese forces maneuvered widely in one of the most rigorous and devastating tropical environments imaginable. The second is Dien Bien Phu in 1954, where a single isolated and exposed French outpost in the center of a lowland was surrounded by jungle, rainforests, mountains, and an enemy committed to its destruction.

The Humid Tropics

For many the word *tropics* brings to mind a dense, lush jungle created by heavy rains that beat mercilessly on a sweltering landscape. Actually, vegetation varies greatly in the humid tropics, depending on differences in insolation, seasonal distribution of precipitation, topography, drainage, and soil. Sunlight, being relatively abundant year-round, brings constant warmth and perpetual summer. Because the average temperature for all months exceeds 18°C (64.4°F), a result of the always high angle of the noon sun, the annual range may be only a few degrees at many places.[1] Therefore the largest temperature variation takes place between day and night rather than from month to month.

The second defining factor of the humid tropics is that total yearly precipitation exceeds the annual evaporation potential, the latter determined largely by the year's temperature regime. Where this occurs, the surplus moisture is readily available to plants and may accumulate to form lakes and swamps or drain away in rivers. All of the humid tropics are not the same; some areas have abundant year-round precipitation while elsewhere the rainfall is distinctly seasonal.

The flora, fauna, soils, and drainage of the tropics are adjusted, in an interconnected and apparent way, to the yearly distribution of precipitation. Because of this, three distinct tropical rainy climates can be identified: tropical rainforest, savanna, and monsoon. Tropical rainforests exist in equatorial areas that are always under the influence of rain-producing mechanisms, most often the Inter Tropical Convergence Zone (ITCZ) along with intermittent convection cells (see Map 1.1). The year-round abundance of moisture and the perpetual growing season combine to favor abundant, yet intensely competitive, plants that form multistoried canopies of leaves that produce such shade that it limits growth of low plants and ground cover. However, in areas where local environmental factors limit the growth of large trees, dense, ground-hugging jungle or thick, high tropical grasses may thrive from the sunlight that reaches the ground.

Tracts of savanna tend to exist north and south of the tropical rainforests. In these areas, rain produced by the ITCZ alternates seasonally with dry conditions related to the tradewinds and high-pressure Hadley cells, the result being distinct wet and dry phases with each lasting several months (Map 1.1). More extensive than any other tropical humid climate, the vegetation generally consists of widespread grassland mixed with scattered trees and shrubs.

The third type of wet tropical climate, less extensive than the other two, is most often referred to as the monsoon (derived from an Arabic word for "season"). It, too, has an alternating dry and wet season but the amount of rainfall far exceeds that of the savanna. Although present elsewhere on a smaller scale, the tropical monsoon is most extensive in the equatorial coastal areas of Southeast Asia and on some of the offshore islands, such as the Philippines. Its meteorological origin is traditionally explained on the basis of alternating high- and low-pressure centers that form over Asia during the winter and summer respectively. With the changes in pressure, drought-producing winds blow from land to sea (October to March). These are followed by opposite winds (May through September) that bring moisture-laden, rain-producing, maritime tropical air over the land. The result is an extremely moist, wet season followed by several months of drought.[2] Although the dry season is prolonged, its effect is lessened by copious rainfall that falls during the rainy period. As a result, in many areas of monsoon climate the vegetation is similar to that of the tropical rainforest.

TROPICAL HUMID CLIMATES AND THE MILITARY

Although unfamiliar to most U.S. citizens (Hawaii and southernmost Florida being the only states with tropical climates), the humid tropics have been important in several American military operations, most notably in World War II island campaigns in the southwest Pacific, and later in Southeast Asia. Others less obvious, but historically significant, include the Spanish-American War in Cuba, and projects in Central America, including the construction of the Panama Canal. The major implication here is that U.S. forces have conducted a variety of operations in tropical environments so unfamiliar that each required an initial period of adjustment and all were accompanied by health-threatening conditions. The same was true for troops from many other middle-latitude countries, such as the English in equatorial Africa and India, the French in Southeast Asia, and the Japanese as they extended their empire early in World War II.

The abundant moisture and high temperatures of the humid tropics favor rapid chemical decomposition, or weathering, of bedrock. The resulting residuum, perhaps many meters thick, provides the parent material for clay-rich, lateritic soils. When wet this deeply weathered material becomes unstable, complicating construction, and the related mud can drastically slow over-riding traffic. In addition, low-lying tropical areas underlain by alluvial soils often become impassable when saturated during rainy periods. Meanwhile, in rugged areas the surplus moisture, which is often great, runs off the land to erode steep-sided gullies, deep canyons, and large valleys.

Where moisture is continually abundant, deciduous broad-leafed evergreen forests may completely cover the land, impeding movement but favoring cover and concealment. In the savanna climate, where vegetation must cope with an annual drought lasting several months and grasslands mixed with trees and shrubs are widespread, conditions are more favorable for cross-country travel and air operations, especially during the dry season.[3]

Heat and moisture also provide breeding grounds for disease-transmitting organisms, so newcomers to these areas are continually threatened by tuber-culosis, typhoid, cholera, malaria, parasitic hepatitis, schistosomiasis, and a variety of diarrhea diseases. Intense heat exacerbates the situation, stressing the body and reducing its natural defenses. Scratches and wounds quickly fester and often cause unexpectedly severe medical problems. Chemoprophylaxis and mosquito repellents can reduce the incidence of malaria but preoccupied soldiers often fail to adhere carefully to the preventive regimen.

The same conditions favor abundant small noxious creatures such as rats, fire ants, scorpions, centipedes, and poisonous snakes, which add to the threats to health and life. Although statistics indicate that relatively few casualties result from snakebites, there are enough venomous creatures in the tropics to cause concern among individuals operating in these areas.

Heat, moisture, parasites, disease, and pests make the humid tropics a con-tinuously hazardous environment. Add combat to the list of threats and the combination becomes ominously challenging to any soldier who fights here, regardless of rank or position. This is illustrated by the following two examples of warfare in humid equatorial realms; the first is the tropical rainforest of New Guinea and the second is the monsoon climate of northwestern Vietnam.

The Campaign in New Guinea

The psychological factors resulting from the terrain were also tremendous. After a man had lain for days in a wet slit trench or in the swamp, his physical stamina was reduced materially. This reduction served to make him extremely nervous and to attribute to the unfamiliar noises of the jungle specters of Japanese activity. These reactions preyed on his mind until he was reduced to a pitiably abject state, incapable of aggressive action.
REPORT OF THE COMMANDING GENERAL, BUNA CAMPAIGN

By March 1942, Japanese forces had invaded Papua New Guinea, New Britain, and the Solomon Islands to threaten Australia and its most direct line of communication with the United States (Map 10.1).[4] The primary peril to Australia was centered on a Japanese effort to control New Guinea and occupy its south coast city of Port Moresby (Map 11.1). In late July Japanese troops landed at the north shore ports of Gona and Buna to begin a southward drive across the Owen Stanley Range toward Port Moresby.

Meanwhile, defending Australian army elements on the north side of the island, outnumbered and operating in extremely rugged terrain, withdrew southward across the mountains to within 40 km (25 mi) of Port Moresby. Here, with the advantage of good defensive positions and short supply lines, the Australians held, stopping the enemy offensive. Meanwhile many Japanese soldiers remained strung out along mountain trails through some of the most rugged terrain and dense rainforest on the planet.

Six months later, U.S. and Australian units were in control of the eastern end of the world's largest tropical island. In the interim, thousands of lives were lost. Although the combat was fierce, most of the casualties resulted from disease.

When the fighting began, the Japanese commander in New Guinea, Gen. Himosho Horii, believed that because his troops had more experience in the tropics they could overcome all challenges. On 22 July he began his move from Buna toward Port Moresby, 225 trail km (140 mi) to the south, sure that his highly motivated soldiers could handle both the resisting Australians and the high mountains. When the battle was over it became apparent that Horii had not adequately assessed the capabilities of his army, the strength of his enemy, and the demands and friction of the environment.

The Japanese advance inland from the coastline first encountered a low, flat, poorly drained coastal plain covered by jungle and rainforest. Most of this area was slightly lower than the coastal beaches. As a result, water from heavy rains became trapped between the coast and the mountains, the result being widespread swamps and a water table that was everywhere close to the surface.

MAP 11.1. Allied-Japanese conflict area in eastern Papua New Guinea.

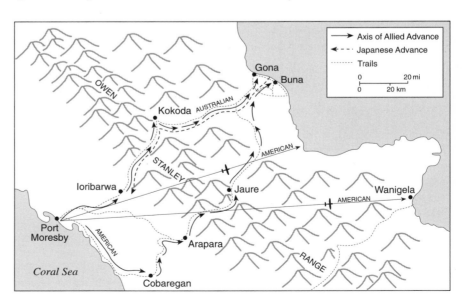

WHY IS THERE AN OWEN STANLEY RANGE?

The answer lies in plate tectonics. Although Papua New Guinea is second only to Greenland as the world's largest island, geologically it is a part of Australia, being separated only by the relatively shallow Arafura Sea. Together these landmasses consist of continental-type rocks that rest upon the northward-moving Indo-Australian plate. This huge section of the earth's crust is converging on the eastward-moving Pacific plate in the vicinity of New Guinea (Map 6.1). The result is a complex boundary involving great compression, intense crustal deformation, and the upward construction of high, east-west-trending mountains which, spine-like, extend along the full length of the island; the easternmost segment makes up the Owen Stanley Range. All passes through the range are craggy and high while nearby summits are commonly above 3500 m (11,500 ft), the highest being Mount Victoria and Mount Albert Edward at 4035 m (13,238 ft) and 3990 m (13,090 ft), respectively.

Deep and rapid weathering favored by the moist, warm climate, plus great relief, steep slopes, high stream gradients, and abundant runoff have eroded innumerable ravines and canyons into the mountains, making them especially rugged. Some of the material removed from the mountains has been deposited along the island's north and south shores to produce narrow, nearly flat, poorly drained coastal plains that are barely above sea level. It was this dynamic and complex geographical setting that set the stage for World War II fighting in New Guinea. Thus, from mid-1942 to mid-1944, a great human struggle on the surface paralleled enormous long-term subterranean geological turmoil that produced New Guinea in the first place.

Cross-country movement here created a quagmire that quickly slowed and tired foot traffic pushing through the muck.

Once across the coastal lowland the Japanese then had to climb 2000 m (6500 ft) up the steep, jungle-covered north slope of the Owen Stanley Range and slip through a pass in this 3000-m-high (10,000-ft) chain that forms the central part of the island. However, upon reaching the north slope of the Owen Stanleys to begin their climb, Horii's troops found only a few primitive jungle trails, hacked out by the native people who roamed the hills in small bands. Few of these tracks were suited to the movement of supplies and vehicles, meaning that practically all sections of usable trails had to be improved and widened.

When the Japanese troops ascended the Owen Stanley Range, they found mountain-climbing through the thick vegetation more difficult than expected. The attendant problems of heat, insects, disease, and increasing altitude multiplied their discomfort. The steeper slopes required the construction of more than 20,000 steps. Movement in and across mountainous streams washed the advancing troops in water polluted upstream by feces of natives, their own predecessors, and animals.

As the Japanese advance continued, the size of their vanguard shrank. Limited Allied air interdiction added to their problems. Evacuation of sick and wounded to the rear diverted additional manpower. Occasional heavy rains

turned the numerous mountain streams into raging torrents that threatened
the safety of the troops and required time-consuming, deliberate stream-
crossing operations. As the soldiers climbed higher, the thinner and much
cooler air added another dimension to their discomfort. Sickness increased so
that by 2 September when General Horii's forces reached Kokoda Pass, high in
the Owen Stanleys, their number had been reduced by one-third.

Now the Japanese troops had to descend the range's even steeper south
slope. Meanwhile, the small element of Australians fell back in the face of their
numerically superior enemy eventually to establish favorable defensive positions at Imita Ridge near Ioribarwa, about 40 km (25 mi) northeast of Port
Moresby. When the Japanese finally arrived at the ridge, their supply lines were
stretched to the limit. It was impossible to keep up a steady flow of food and
ammunition. In addition, many of the troops were malnourished, exhausted,
and suffering from all the problems associated with prolonged exposure to a
hostile environment.

Recognizing the enemy's weakness, Gen. Douglas MacArthur then ordered
an attack on General Horii's advanced positions. He also increased bombing
on their north slope supply lines. Fearing that his remaining troops would be
killed or wounded, and knowing that he could no longer depend on support
from Guadalcanal (where Allied forces were increasingly successful), Horii
withdrew northward while announcing to his troops:

> For more than three weeks . . . every unit forced its way through deep forests
> and ravines and climbed scores of peaks in pursuit of the enemy. Traversing
> knee-deep mud, climbing up steep precipices, bearing uncomplaining the
> heavy weight of artillery ammunition, our men overcame the shortage of
> our supplies and are succeeded in surmounting the Stanley Range. No pen
> or work can depict adequately the magnitude of the hardships suffered.[5]

With the 7th Australian Division and elements of the U.S. 32nd Division in
pursuit, Japanese withdrawing through the rainforest and jungle of the northern coastal plain were also plagued by the oppressive heat, continually high humidity, dense vegetation, and overabundant moisture. General Horii himself
drowned attempting to raft across a back-bay near Buna. When his forces finally
closed on the north shore of Papua New Guinea, fewer than 500 of the 4000
soldiers who had begun the operation in July remained in fighting condition.

The 3500 U.S. and Australian forces who pursued the Japanese were not in
much better shape. Their resupply efforts had been a nightmare and the farther they went the worse it got. As supplies left Port Moresby, they were first
carried a short distance by road in three-ton trucks. Where the road narrowed,
materiel was transferred to quarter-ton trucks for a 5-km (3-mi) inland haul.
At the next transfer point, supplies were loaded on pack animals and 1000
native carriers for another 5-km (3-mi) trek. Finally, porters carried the food,
water, and ammunition forward to the troops.

While the Australians followed the retreating Japanese via the Kokoda Trail, elements of the U.S. 32nd marched eastward along the coastal trail to Cobaregan and then north on the Kappa Kappa Trail to cross the Owen Stanleys through a 3000-m (10,000-ft) pass near Jaure (Map 11.1). The Allied units then converged on about 5500 Japanese who, by 15 November, occupied a defensive perimeter around Buna. As fighting over the Owen Stanley Range ended, MacArthur entered into one of the most interesting phases in his military career. It was, of course, the campaign to secure the whole north coast of Papua New Guinea, and it began with Buna.

As the bulk of the American and Australian troops followed the Japanese over the Owen Stanley Range, two regiments of the U.S. 32nd Division flew to plantation airstrips southeast of Buna in a 30-day air operation that began in mid-October 1942. On 19 November, two of these air-lifted battalions from the 128th Infantry began an attack on Buna from the east which quickly disintegrated.

Meanwhile, Australian and American troops moving cross-country and approaching Buna and Gona from the south encountered nothing but swamps and jungle. In contrast, the Japanese held the slightly higher ground along the coast, the better-drained land giving them great advantage in an area averaging 300 cm (120 in) of rainfall annually and where more than 20 cm (8 in) fell during a single day in January 1943 (Figure 11.1).

As fighting intensified, the front lines were increasingly indistinct. For the Allies, Japanese soldiers seemed to be everywhere, but their strongest positions were on the shore-zone terrain. Here troops could move from place to place quickly, and numerous bunkers constructed of coconut logs and sand provided added protection and a superb defensive perimeter. In looking at Japanese positions, MacArthur's staff reported that "every contour of the terrain was exploited and the driest stretches of land were carefully chosen to be occupied and fortified, making it impossible for the Allies to execute any lateral movements without becoming mired in swamp." Lt. Gen. Robert Eichelberger, U.S. Corps Commander, called the Japanese terrain utilization "perfect" and "brilliant."[6]

In contrast, the Allied forces were somewhat disoriented by the jungle and terrain. Their use of aircraft and artillery support was limited, in part, because of possible fratricide. The dense vegetation also prevented them from determining the exact position of the enemy's front lines and outposts. Aerial photography was next to useless in locating Japanese positions, fog and afternoon rain limited aircraft missions, and the land supply route was still tenuous. Furthermore, Eichelberger, whose mission as ordered by General MacArthur was to "take Buna or not come back alive," found that troop morale was extremely low.[7] His soldiers were continually wet (and once wet, often cold) and uncomfortable. They were unable to excavate suitable protective positions because of the marsh and high water table. The swamps were a jumble of twisted, slime-covered roots and nasty muck. The abundant snakes and other similarly threatening creatures actually were not a direct hazard, but practically all of the newly arrived soldiers feared them. Leeches up to 2.5 cm (1 in) were ubiquitous. The mosquito was considered as formidable an enemy as the Japanese. Jungle noises were strange and frightening. Hot food was rarely available. Most soldiers had torn and tattered clothing from the long march. "Jungle rot," covering arms, bellies, chests, crotches, and armpits, was widespread. Dysentery, dengue fever, and malaria were on the increase. All told it was not uncommon for 250-man companies to be operating with only 40 to 50 effective soldiers.

Determined to improve the discipline, health, and morale of his troops, General Eichelberger concentrated on logistics and sanitation. Through a Herculean effort, the Army Air Corps supplied badly needed ammunition, clothing, and food. He also ordered that all troops shave regularly and be dressed in clean uniforms. Slowly the internal strength and morale of his forces improved. Even so, staying healthy in the Papua New Guinea jungle remained difficult and required constant attention to personal hygiene.

Combat operations were also severely challenged by the environment. Communication among adjacent units was generally poor or nonexistent. The moisture of the swamps often shorted out splices and internal connections in

FIG. 11.1. Two soldiers examining a Japanese pillbox in New Guinea. (U.S. Army photo, photographer unknown, 21 December 1944.)

SURVIVAL IN THE JUNGLE

En route to the attack on Singapore, the Japanese Imperial Army Headquarters produced more than 40,000 copies of a standard operating procedure (SOP) which digested voluminous data on operations in the humid tropical environment (*Singapore: The Japanese Version,* trans. Tsuji Masanobo [New York: St. Martin's Press, 1961]). Flavored with anticolonial and anti-American propaganda, it focused on the environmental problems to be encountered. The following paragraphs are paraphrased from that SOP.

Always you may be killed in battle; yet there are other dangers peculiar to the present campaign which you must heed. A great variety of deadly diseases and the great enemy, the malarial mosquito, are lying in wait for you. It is historical fact that in all tropical campaigns since ancient times, far more have died through disease than have been killed in battle. To fall into a hail of bullets is to meet a hero's death, but there is no glory in dying of disease or accident through inattention to hygiene or through carelessness.

Water is your savior; one man will consume ten liters a day. You must sleep well and eat well. There is danger the troops will weaken through sheer lack of sleep. The sudden drop of temperature in the latter half of the night, when sleeping in damp clothes, will give you a chill and lead to diarrhea. Native settlements are nests of bedbugs, fleas, and infectious disease. If you discover a dangerous snake you must kill it: swallow its liver raw and cook the meat.

Westerners, believing themselves to be superior people, but very effeminate and very cowardly, have an intense dislike of fighting in the rain or in the mist or at night (although they believe night is excellent for dancing). We must seize upon night; upon it lies our great opportunity. Nature is the enemy of equipment as well as people. Iron rusts, leather mildews, the glass mists. You can get dirty water anywhere, but pure water is not so readily available since the natives defecate and urinate quite freely in all lakes and streams. Even the water which the natives use for drinking is full of germs. Malaria is to be avoided at all costs. The malarial mosquito is rarely found in the jungle away from the sea. It is most commonly found by clear mountain streams or places near the coast where the sea water and river water intermingle. Take your anti malarial medicine as directed. To avoid sunstroke, the best preventives are ample stocks of drinking water, sufficient sleep, and a well filled stomach. Do not exist on an unbalanced diet which can lead to beri-beri. Eat as many fresh vegetables and fruit as you can.

---

the audio equipment, disrupting contact by wire and radio. Shrapnel from Japanese artillery tore away many other wire communication lines not already ruined by the water. Most messages had to be hand-carried, sometimes by circuitous routes through the jungle and swamp. The two U.S. combat regiments, although separated by only a few kilometers, were a two-day march apart. Fields of fire were nonexistent, and all thoughts of mobility were challenged by the realities of close-quarter, infantry jungle warfare.

By sheer force of personality and an emphasis on troop leadership, General Eichelberger and his subordinate commanders rallied their soldiers for numerous attacks on the Japanese at Buna and Gona.[8] Allied interdiction of resupply vessels and aircraft further threatened the enemy forces. Six U.S. tanks were put ashore east of Buna for use on the higher land along the coast. Gradually the situation that began in near despair evolved into one of growing confidence. Between 30 December 1942 and 3 January 1943, Eichelberger's corps wiped out organized resistance in Buna, and by 22 January 1943 the area was secured. Now began a series of westward-moving amphibious operations on New Guinea's north coast that captured airfields while bypassing Japanese strongpoints. The result would be successive victories at (using World War II names) Salamauao, Lae, Finschmafen, Aitape, Hollandia, and Wakde, among others, all on the north coast of Papua New Guinea.

Looking back on the battle, General Eichelberger noted that nature as well as the enemy was pitted against his units and that only through the exceptional effort of his troops was the campaign a success. In summary he said, "We were prisoners of geography."[9]

The battle for Buna was important for several reasons. It eliminated the Japanese presence in eastern Papua New Guinea, greatly reduced their threat to Australia, and put the emperor's forces on the defensive. It also demonstrated to Australian and U.S. commanders that continued success would require conquering the tropics as well as the enemy. But the price had been high. During the Buna campaign, the U.S. 32nd Division suffered 10,960 casualties—8286 from disease. The Australians had equally heavy losses, with six soldiers reported sick for every three battle casualties. For the Japanese, malaria, beriberi, diarrhea, skin disease, and malnutrition, together with battle losses, gradually reduced their strength in New Guinea to one-seventh of its original size. The oppressive humid tropics had taken their toll and proved to be a master on the battlefield. To Eichelberger, the casualty count was a "high purchase price for the inhospitable jungle."[10]

Dien Bien Phu: Hell in a Very Small Place

Dien Bien Phu was a very strongly fortified entrenched camp. But on the other hand, it was set up in a mountainous region, on ground which was advantageous to us and decidedly disadvantageous to the enemy. Dien Bien Phu was, moreover, a completely isolated position, far away from all the enemy's bases.

VO NGUYEN GIAP[11]

The cessation of hostilities between the United States and Japan in 1945 brought only a brief respite from warfare in Southeast Asia. The French, anxious to regain their colonial rule over the resource-rich countries of Indochina, quickly moved into the region to reassert their influence. Guerrilla forces organized by the native leader, Ho Chi Minh, created the Viet Minh to challenge the French, and from 1946 until 1953, running gun battles took place between them for control of the land.

The French held the largest cities: Hanoi, Saigon, Da Nang. The Viet Minh controlled the countryside, squeezing ever more tightly the supply lines that linked the cities and pressuring their fellow Vietnamese, Lao, and Cambodians to join with them in overthrowing the French. By mid-1953 the French were anxious to counter the military successes of the Viet Minh, in the hope that they would be able to negotiate with their enemy for control of the country. Laos had also been threatened by the Viet Minh, and the French saw a need to establish a strong outpost near the Laotian border to confront an expected major Viet Minh advance toward the capital of Laos. On 20 November 1953, airborne forces of the French army parachuted into the village of Dien Bien Phu in northwest Vietnam to establish a fortified camp that would block large-scale Viet Minh movements and assert French control of the region (see Map 3.4).

The village is in northwest Vietnam's largest valley, which is about 25 km (16 mi) by 10 km (6 mi), the longest dimension extending north to south. This lowland is encircled by low mountains that rise from foothills to heights of 700–900 m (2400–3000 ft). In all directions from the village the landscape is similar: jungle and rainforest on alternating belts of high ground and deep, narrow, stream-eroded ravines. To the descending paratrooper, Dien Bien Phu must have resembled an open spot in the bottom of an elliptical bowl, with jungle-covered sides. The only road in northern Vietnam capable of supporting even moderately heavy traffic, Highway 41, ran through the lowland and Dien Bien Phu. The French reasoned that any large conventional force moving to or from Laos would have to pass through this valley.

Once landed, the French airborne forces were about 300 km (190 mi) from their base at Hanoi. The nearest city was 160 km (100 mi) away, and most sections of road extending to Dien Bien Phu were unsecured and subject to interdiction at will by the Viet Minh. The army could be resupplied only by air. A small airfield at Dien Bien Phu could be expanded, but all incoming and outgoing planes would be vulnerable to enemy fire as they passed low over the surrounding hills.

French reconnaissance concluded that the village would be safe from hostile artillery. They reasoned that in order to place rounds on Dien Bien Phu, the enemy would have to position its weapons on a slope inclined toward the village and would thus be exposed to direct fire from "superior" French artillery in Dien Bien Phu. They also believed that the rugged terrain and dense vegetation would preclude positioning heavy artillery on the surrounding hills. Both conclusions proved completely erroneous.

To worsen matters, the French also incorrectly concluded that weather would not be a major problem. The highlands of northwest Vietnam have a tropical monsoon climate. From October to March the winds are generally out of the north, making rainfall, at the most, only sporadic. Sometime in April, however, the winds reverse to bring the southwest monsoon with downpours generally yielding more than 150 cm (60 in) of rain within a five-month period. But the French did not intend to stay that long. They planned to parachute into Dien Bien Phu in the middle of the dry season and be out of the area before the heavy precipitation began. Unseasonable December rains in 1953 indicated that the weather pattern was not normal; they might also have been an ominous warning of what was ahead.

The initial French parachute assault was successful, although the surprised Viet Minh at the site did inflict some casualties on the descending soldiers. As planned, soon after the French secured the area, the airfield at Dien Bien Phu was operational and a smaller auxiliary airstrip was constructed at an outpost a short distance to the south. During December and January the French reinforced their outposts and sent patrols into the surrounding hills.

Soon it became apparent to the French that to complete the mission they would have to stay in the valley longer than had been planned. This also meant that defensive positions and fortification had to be improved and that resupply of materiel and food would rely totally on aircraft. Recognizing that preparing all their bases against artillery fire would require some 36,000 tons of materiel that could not be supplied by air, the French commander focused on strengthening the main airstrip and its nearby headquarters and hospital. Gradually the original offensive mission was changing into one of survival.

Anticipating the approaching wet monsoon, the French commander, with concern, pointed out to his superiors in Hanoi that much of the stronghold's center portion could be flooded during the coming rainy season and, if so, strongpoints within the camp would be separated by the Nam Youm River. Furthermore, communications within Dien Bien Phu could be severed; rainfall might saturate the sandbags in the fortifications (adding weight) and endanger the bunkers; visibility would be greatly reduced by the monsoon rains; and adequate aerial resupply would be highly questionable under such conditions. But the French command did little to meet these approaching threats.

While the troops at Dien Bien Phu were fortifying their positions as best they could, the Viet Minh were also at work. As they did again some 20 years later with the Ho Chi Minh Trail, they moved many people, abundant supplies, and much heavy equipment along the narrow paths in the jungle and rainforest surrounding Dien Bien Phu. Soon the hills were teeming with activity. Although French intelligence sources were cognizant of some movement, they could not determine the size and location of the enemy forces. They were also largely unaware of the numerous Viet Minh firing positions being built on the forward slopes of the hills facing Dien Bien Phu.

Nearly complete vegetative concealment made visual and photographic aerial reconnaissance, including infrared photography, just about useless to the French. In addition, French maps of this remote region were not sufficiently accurate for the precise placement of their artillery fire without direct observation of the enemy and careful registration. And the multicanopy foliage diminished the impact of French shelling that did occur. Later, when the rains began, napalm dropped on suspected Viet Minh sites was less effective since the waterlogged vegetation overhead limited the spread of fire and offered some protection to those on the ground.

On 13 March 1954, the Viet Minh began their siege of the fortress at Dien Bien Phu. Eleven days into the battle, the French headquarters in Hanoi notified the forces at Dien Bien Phu that the impending rainy season would significantly slow the Viet Minh attack and reduce their ability to carry on sustained operations. The French hoped the coming monsoons would be an advantage to their encampment. Instead the arriving rain made their situation even worse.

The 1954 monsoon began earlier than expected. Heavy rains of late March and early April quickly created a morass for the French, filling their trenches with water, creating severe disease and sanitary problems, and threatening the stability of the bunkers and shelters.

Viet Minh artillery fire striking soaked and stressed bunkers brought chaos as mud, sand, timber, and water poured in on the occupants. The thick monsoon clouds forced the French resupply aircraft to fly into the area at very low levels. As a result the surrounding and well-hidden Viet Minh antiaircraft batteries, which far exceeded the number the French had expected, were deadly accurate. Anything that landed was quickly fired upon by artillery batteries that had been individually sited, dug into the mountains, and covered with vegetation rejuvenated by the monsoon rains. The French had counted on being able to find and destroy those guns. They were completely mistaken. In looking back, the French command concluded:

The battle of Dien Bien Phu underscored the value of a good set of artillery positions. On the enemy side, the batteries had the benefit of downward views and, above all, their field of fire was reduced because they concentrated their fire in the shallow basin of Dien Bien Phu. In this way, the batteries could be set up in casements, could be completely camouflaged and thus take advantage of remaining unpunished.[12]

Because of this understanding of the battlefield environment, all of the 24 105mm Viet Minh guns installed around Dien Bien Phu remained undamaged throughout the siege. Week after week, poor flying weather in the area limited French air drops. The heavy rains overflowed the Nam Youm River, which ran alongside the airstrip, and flooded parts of the base. The French commander was particularly concerned for the wounded soldiers:

One must add to all this . . . the continuous rain which causes complete flooding of the trenches and dugouts. The situation of the wounded is particularly tragic. They are piled on top of each other in holes that are completely filled with mud and devoid of any hygiene.[13]

Some of the open sores of the hospitalized wounded were infested by white maggots that were rampant in the nearby cemetery. Reductions in aerial resupply drastically decreased the amount of medicine and food available to troops already suffering from disease and hunger. Bernard Fall, a French writer who detailed many of the tragedies of the operation, appropriately labeled Dien Bien Phu "Hell in a very small place."[14]

While the wet monsoon increased both the suffering of the French and the difficulty of movement and resupply by the Viet Minh, it also reduced French air force interdiction of enemy supply lines. As a result, while materiel for the French dwindled, Viet Minh soldiers and recruited coolies continued to deliver supplies and disassembled heavy equipment through the back-breaking work of thousands. Meanwhile the French garrison could be reinforced only by airborne and, as a measure of their courage and dedication, some of these arriving troopers had never jumped out of an airplane before.

The siege raged throughout April and into early May. Finally on 7 May, nearly 7000 exhausted, wet, and almost totally debilitated French soldiers surrendered to the Viet Minh. The capitulation ended France's attempt to control this remote region and presaged their eventual withdrawal from Indochina. Nearly 2000 French soldiers died in the valley and more than 6000 were wounded or sick.

Considering the overall situation, the French army unwisely assumed an untenable position within a terribly unfavorable geographical setting. Troops were placed in an open area at the bottom of a poorly drained valley surrounded by hills that offered excellent concealment. Expectations that the Viet Minh would be incapable of maneuvering effectively within and around the lowland were totally incorrect. Only six months after the French parachuted into the area, all that remained left as prisoners of the Viet Minh. Isolated by the jungle and rainforest, both the weather and the landscape conspired with a resourceful and dedicated opponent to defeat the French. Acclimatized to the monsoon weather, the Viet Minh suffered considerably fewer nonbattle casualties than did the French. Ironically, it was the French who carefully chose the battlefield but it was the Viet Minh who shrewdly used its characteristics to win the engagement.

A strategic position undoubtedly. For it is true that any enemy coming from the east and making toward the Mekong will be tempted to pass through it in order to rest there, and take in supplies and that whoever controls Dien Bien Phu will control the whole region and part of South-East Asia, *provided he holds both the basin and the heights commanding it.*[15]

Conclusions

The dissimilarities between World War II fighting in Papua New Guinea and the siege at Dien Bien Phu are many. Separated by only a decade, the adversaries differed completely. The Papua New Guinea campaigns ranged widely across the varied terrain on an island at the periphery of the Orient. The battle at Dien Bien Phu was, from beginning to end, fought in an isolated section of a

FIGHTING IN THE TROPICS TODAY

U.S. operations in Southeast Asia, especially Vietnam, took advantage of the Allied military experiences there during World War II and knowledge gained from the French-Indochina conflict. Various prophylactics were developed to reduce the incidence of malaria. A combination of pills, taken on a prescribed regimen, all but eliminated malaria in U.S. units operating in the area. Nothing, however, could eliminate the problem of moisture and heat and its impact on the strength and will of fighting forces. Acclimatization for U.S. forces began with their arrival at the replacement centers. During their introductory training phase most arriving soldiers adjusted to the climate before they went into the field. No amount of training, however, could completely prepare the troops for the hardships and privations of prolonged duty in remote areas of the humid tropics. And the fact is, when operating in a rainforest or jungle, whether it be in New Guinea, Vietnam, or a place of the future, nature is largely in control.

small valley in the southeast part of the world's largest continent. In one situation the native people were extraneous, even obscure, to the operation as Japan and the Allies fought over foreign land far from home. In the other, two groups claimed a single dominance, as the ancient indigenous population struggled against the French in their attempt to reassert colonial rule that had been in place generations before World War II.

It is, however, the striking similarities that best reveal the military significance of these two operations. Aside from air transport for some American troops and the French paratroopers, just about all movement was on foot, which must have been most obvious to the Japanese who climbed over the Owen Stanley Range toward Port Moresby only to, if they survived, return by the same route weeks later. Climate, terrain, disease, parasites and pests, medical facilities, mobility, logistics, and communication were persistent problems for all participants in both campaigns. In contrast, cover and concealment offered by the dense vegetation was advantageous for unobserved buildup, clandestine movement, small-unit action, and surprise and ambush. Militarily, these conditions are commonly present in all of the humid tropics. Collectively they present one of the most difficult combat environments imaginable, as shown so vividly in Papua New Guinea and Dien Bien Phu.

CHAPTER 12

Heat, Rock, and Sand

The Western Desert and the Sinai

To the newcomer, the desert could be a place of real fear: no roads, no houses, no trees, no landmarks of any sort for hundreds of miles. He experienced for the first time real fear of being completely lost and meeting his end either in the arms of the enemy or in the malevolent wastelands of the deep empty deserts.

C. F. LUCAS PHILIP

Arid Regions

Deserts and steppes, together the most extensive of the five major climatic groups, cover about 30 percent of the world's land area and exist wherever the annual potential evaporation exceeds actual precipitation received. With such scarcity of moisture, most streams tend to flow intermittently and lakes are generally saline. The often cloudless sky favors rapid daytime heating and excess nighttime cooling, which results in the highest daily temperature ranges on the planet.[1] Although the dryness of a desert is everywhere similar, the terrain varies greatly from angular slopes of high barren mountains, to smooth contours within extensive dune fields, to low flats and salt marshes. Wind carries material derived from mountainsides and plains to deposit it eventually to form shifting sand hills and widespread layers of eolian silt. And, to survive, desert flora and fauna have developed special moisture-conserving traits that enable them to endure in this austere world.

The most extensive deserts of our planet are concentrated in the subtropics, from about 20 to 30 degrees North and South latitude. The largest of these, which stretches across North Africa and well into southwest Asia, is larger than all others combined. Even so, the interior of Australia, the Sonoran and Mohave Deserts of North America, the arid Kalahari of southern Africa, and the incredibly dry Atacama of coastal Chile and Peru all represent subtropical deserts of considerable extent.

Three large tracts of comparable dryness also exist in the middle latitudes: the biggest is in central Asia, a second associated with parts of the intermountain plateaus of the western United States, and the third in Argentina's Patagonia. Because these subhumid regions are farther from the equator, they experience large seasonal variations in temperature, giving them cold winters and hot summers.

Although different in location and latitude, all of these dry regions have one thing in common: the absence of a natural atmospheric condition that regularly moves air upward to induce condensation and precipitation. Instead, these areas are almost always under the influence of subsiding or stable air, both conditions that prohibit the large-scale upward circulation necessary for rain and snow.

Thus it is not simply lack of water vapor in the air, but the absence of atmospheric uplift that accounts for the world's steppes and deserts. For example, the Atacama of South America, Baja California, the Moroccan coast of North Africa, and the Namib section of the Kalahari in southern Africa are all often overrun by moist air originating over adjacent oceans.[2] In fact, water vapor is so plentiful that coastal fog may be common, but rainfall is rare. This peculiarity occurs when the base of air subsiding from a subtropical high-pressure cell is chilled below dewpoint by a cool offshore ocean current. The result is a cloud at ground level, fog, but the absence of a large-scale lifting process so necessary for rain. Although these deserts illustrate that air in dry areas may be moist, the four are limited to narrow coastal tracts and are relatively small. What then explains the world's vast areas of deserts and steppes?

Two sets of circumstances account for most of the world's dry areas. The first involves the continued presence of a large high-pressure air mass, or anticyclone. High pressure forces air to descend and diverge. The subsidence, which is the opposite of cyclonic lifting that favors condensation and precipitation, brings fair weather. As long as atmospheric pressure remains high it will not rain, and some areas are nearly always under the influence of an anticyclone. The most notable permanent centers of high pressure in the subtropics are known as Hadley cells (Map 1.1). Rather than remaining stationary, these shift somewhat north and south with the seasons. Even with these fluctuations, some Hadley cells are so large that certain areas never escape their influence, resulting in year-round dryness, a situation that accounts for most of the world's low-latitude deserts, the best example being the great Sahara.

The second precipitation-limiting situation involves reduced access to large sources of water vapor because of continent size combined with land barriers that produce a downwind "rain shadow." Most water vapor originates through evaporation from low-latitude oceans. Winds carry some of this moisture inland where it may condense and precipitate. (For example, the Gulf of Mexico is the major source of moisture that falls as rain in central United States.) When moisture-laden air encounters a highland, it cools as it ascends the barrier, possibly to the point where condensation and precipitation occur. More important, as the air then begins its leeward descent, its pressure and temperature increase. This process eliminates any chance for precipitation, and subjects adjacent regions to a "rain shadow" (dry area downwind from the land barrier). Such conditions, at least in part, account for the subhumid Great Plains and dry intermountain plateaus of the American West, the extensive arid areas in the interior of Asia, and the Patagonian Desert of Argentina.

Weather in any of these dry areas can be extremely stressful. The surface temperature of sand in low-latitude deserts can reach 90°C (200°F). Here someone protected from the direct rays of the sun might experience an air temperature between 55° and 60°C (130–140°F). Trees and shrubs are generally sparse, making shade an uncommon escape from relentless and oppressive heat.

Nighttime conditions differ drastically. Clear skies and sparse vegetation favor rapid ground reradiation once the sun sets. A nighttime chill, with temperatures commonly lower than 10°C (50°F), replaces the heat of the day. In addition to this large daily range, temperatures vary seasonally with the middle-latitude dry climates having overly hot summers and cold, windy winters.

Water is precious in a desert. For example, a soldier operating there should have 6–8 liters (6–8 quarts) of water daily. Permanent fresh-water streams are rare and precipitation infrequent and undependable, varying greatly from place to place and year to year. When rain does occur it may come in torrents that produce flash floods and create temporary lakes and bogs that may be impassable to foot or vehicle traffic. Although rainfall is infrequent, its runoff remains the most powerful erosional agent in the desert, carving deep wadis, arroyos, and gulches into the barren countryside while elsewhere depositing the sediment as alluvial fans and bajadas.[3] Wind is common, with higher gusts removing the unshielded, lighter, finer material from the desert floor, leaving behind a residual pavement of pebbles and rocks. Windblown sediment carried across the vast open spaces may gradually wear exposed rock surfaces. The fiercest desert winds, the Ghibli of Libya and the Khamisu of Egypt, create sandstorms that darken the skies, eliminate visibility, and drive grit into every corner. As these disturbances pass, temperatures may drop dramatically.

In areas where sand is abundant, wind may move the particles to form linear dunes tens of meters high that can stretch for kilometers. Elsewhere migrating, asymmetrical, crescent-shaped dunes, with a steep leeward slope that coincides with the angle of repose for windblown sand (30–35 degrees or 67–78 percent), may be numerous. Either way, all dunes impede human movement to some extent. Both wheeled and tracked vehicles have difficulty operating on the noncohesive sand. Large dune fields may have to be bypassed and even smaller features can limit and guide vehicular movement by their nature and distribution. Traversing dunes by foot is always arduous and their instability disfavors bunker or foxhole construction because of collapse.

Steep-walled, water-carved wadis often present major barriers to cross-country movement in the desert. When moist, the alluvium within wadis may not provide adequate support for vehicles and equipment. Salt marsh and bog areas are often impassable when wet. Even in dry periods they might sustain only one or two vehicle passes before the surface crust becomes unstable. Repeated crossing of the preferred rocky, gravel areas can also expose the less-competent underlying soil.

Bedrock is commonly exposed in a desert. Although this rock gives solid support, its surface configuration varies greatly, being an asset in one situation and an insurmountable obstacle in another. These areas also present special problems for the engineer involved either in construction or in the search for groundwater.

Equipment designed to operate in humid regions often does not function well in the heat and dust of the desert. Blowing sand and dust glaze windows and grind at rotor and turbine blades, engine bearings, and exposed mechanical components. Due to poor traction, vehicles moving across the desert often struggle along in low gear with overheated engines and transmissions, consuming disproportionate amounts of fuel as their lubricants become fouled with grit.

An arid environment can also interfere with communications. Excessive heat may create anomalies in radio and other electronic transmissions. On the ground unpredictable clouds or storms of dust may limit visibility. When the sky is clear, distant objects may appear to be closer than they actually are. And strange mirages occasionally shimmer off the land's surface. Misconception and miscommunication can be added enemies in the desert.

War in the Desert

The most extensive arid region of the world stretches about 10,000 km (6200 mi) from northwest Africa to Pakistan. As a measure of its dryness, only two large permanent rivers, the Nile and the Tigris-Euphrates, traverse the desert and both of these originate in distant, more humid regions. Even so, this tract has been the scene of many ancient and recent battles. The terrain in this vast arid region ranges from high mountains and deeply dissected rocky plateaus to low sand plains and flat salt depressions. The variety of landforms appears endless, thus the problems of maneuver are often diverse and complex. Nowhere has this been more apparent than in the World War II battles in North Africa.

Between September 1940 and November 1942, the British first fought the Italians and then the Germans back and forth across the Western Desert, all gaining and losing ground in massive sweeps marked by a succession of hardfought battles on a narrow coastal plain and low plateaus separated in most places by steep escarpments. Less than 10 years later periodic desert warfare was renewed, this time east of the Nile and the Suez Canal, where between 1949 and 1973 four wars were fought in the arid wilderness known as the Sinai, a region that is anything but a flat, sandy desert. And in another 20 years, major conflict erupted in the deserts of Iraq and Kuwait.

BATTLES OF THE WESTERN DESERT: GREAT BRITAIN VERSUS ITALY AND GERMANY

To Great Britain in 1940, struggling to maintain control of its worldwide empire and faced with a possible German invasion, the maintenance of power

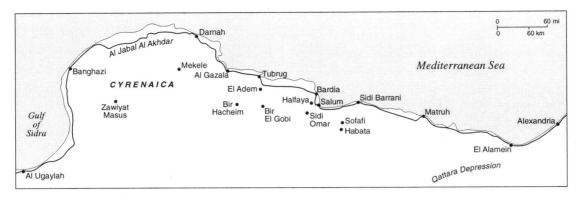

in the Middle East was of enormous strategic importance. England's presence in the eastern Mediterranean, including Egypt, blocked Axis expansion in the area, protected Allied overland supply routes to Russia by way of the Persian Gulf, gave access to the Balkans, and eventually provided a base of operations against Italian and German armies in North Africa. To counter the British presence, Italian forces in Libya, commanded by Marshall Rodolfo Graziani, moved eastward to the Egyptian border in preparation for an attack toward Alexandria, Cairo, and the Suez Canal. The invasion began on 13 September 1940, initiating 33 months of war in North Africa.

Most of the fighting took place along the coast and near-coastal zone of the Western Desert, that part of the Sahara that extends about 1600 km (1000 mi) from Tripoli in Libya to El Alamein just west of Alexandria on Egypt's Nile Delta. Eastward the area suitable for military operations narrows markedly because of the huge and impassable Qattara Depression, a feature the British eventually used to great tactical advantage (Map 12.1).

The most rugged topography in this desert extends about 150 km (90 mi) along the northeast coast of Cyrenaica. Known as Al Jabal Al Akhdar, these 50-km-wide (30-mi) mountains have summits approaching 900 m (3000 ft). In many places this range crowds the coastline so tightly that there is little room for even a narrow roadway. The only extensive green areas in the Western Desert are along this part of the coast. Here the vegetation is more abundant because the low mountain barrier induces rainfall and creates a local microclimate unlike adjacent desert areas.

To the southeast of the Akhdar lies the low Libyan Plateau, a relatively flat area that rises from the coastal lowlands to elevations of 150–230 m (500–750 ft). Near the Egyptian-Libyan border the plateau narrows as north- and south-bounding escarpments form a dagger of land pointing toward the heart of Egypt, Cairo. To the northeast a 180-m-high (600-ft) escarpment abruptly separates the plateau from the lower ground along the coast. More important, it presents a formidable obstacle to any vehicular movement except through passes at Salum and Halfaya.

The low ground along the coast between Salum and Alexandria is relatively smooth and is marked by small ridge lines and depressions that provide limited tactical advantage to those who maintain control of these terrain features. Adjacent to the south is the low Libyan Plateau which is, in turn, bordered on the south by a 180-m (600-ft) escarpment that marks the north edge of the Qattara Depression.

Trafficability in the region varies greatly. In Cyrenaica, south of the Akhdar, surface material ranges from silt, to sand and gravel, to pure gravel, to solid bedrock. Eastward the amount of silt becomes more abundant within the sand and gravel. In the lowlands of the Qattara Depression the soil, developed from fine-grained alluvium, forms an impassable morass for vehicles. North of the Qattara the desert is a mix of rolling dunes and thin sand over rock.

In 1940 there were few improved roads in the Western Desert. The main route near the coast extended from Al Ugaylah through Banghazi, Darnah, Al Gazala, Tubrug, Salum, Sidi Barrani, and El Alamein to Alexandria. Occasional tracks extend south from the highway only to disappear into the hills and shifting dunes on their way toward the few desert settlements. In Al Jabal Al Akhdar, the paucity and narrowness of north-south paths greatly limited movement across the range. Except for a few wadis and the widely scattered oases, drinkable surface water is nonexistent most of the time. During winter, however, southward-tracking middle-latitude wave cyclones occasionally bring rain, but the storms are generally short-lived and the water quickly runs off or is absorbed by the desert floor.

Inland, between the Akhdar and the Qattara, the largely open territory permits mechanized forces to move widely and rapidly. Although vegetative cover is almost totally absent, the vastness and variety of the area provided many opportunities to avoid observation. Proper camouflage plus numerous depressions and low hills offered some protection from ground observation during the day and enemy aerial reconnaissance could be minimized by maneuvering at night.

Land navigation off the roads in the desert could be difficult. Those who did not recognize the few cairns and other guiding landmarks of the region, or who relied carefully on a compass, were soon lost. Logistical difficulties were enormous, as the varying environment and extended troop movements created ever-changing conditions for men and resupply.

Suitable harbors along the Mediterranean shore are few and widely spaced. This situation tended to disfavor the aggressor because, with success, their supply lines grew longer, while the inverse was true for those withdrawing toward their nearest supply port. As fighting raged east and west across the desert, Tripoli and Alexandria served as the principal ports for the Germans and British, respectively; intermediate harbors at Banghazi and Tubrug changed hands frequently as the fighting progressed. All the while coastal roads linking the ports with the front lines were known and obvious targets for land, sea, and air forces of the belligerents.

To counter the desert's openness, extensive minefields were common and reinforcement of strongpoints essential. Once a position was occupied, barbed wire was strung and other defenses were quickly established in attempts to block an enemy advance or to channel them into killing zones. Major cities were fortified to protect supply depots and to provide fall-back positions for the defending forces. Along with logistics, communication and intelligence were of constant concern. Also, both sides continually used ground and air reconnaissance in order to prevent surprise.[4]

For more than two years the combatants battled back and forth, east to west and west to east, across the harsh Western Desert. As the campaigns evolved and intensified, casualties grew, as did demand for supplies. Each day's increased material requirements further burdened logistics and manpower systems. And it all started with an unwise and self-defeating offensive by the Italians.

When Marshall Graziani's Italian Tenth Army advanced eastward into Egypt in September 1940, their first objective was to seize the critical passes at Salum and Halfaya. After placing artillery fire on the lightly manned British positions atop the coastal escarpment, the Italians moved down these narrow paths and onto the coastal road to take control of the area. Faced with minimal opposition from withdrawing British reconnaissance elements, they reached Sidi Barrani in only three days. Graziani then established a position west of Sidi Barrani and stretched his forces from the coast southward to the escarpment in the vicinity of Sofafi. He established defensive positions there because the front was narrowed by the nature of the terrain. Although the Italians erected barriers and mined the area, their efforts were not well coordinated, resulting in a large unprotected gap between the forces nearest the coast and those holding the inland, or southern, flank.

On 9 December 1940, British forces under Lt. Gen. Richard O'Connor slashed through the opening. Two British divisions attacked five Italian divisions and, aided by the element of surprise, penetrated deep behind the Italian lines threatening their route of withdrawal. The road between Sidi Barrani and the frontier was quickly cut, and the outflanked Italian troops either fled or were captured. By early February 1941, the British had destroyed 10 Italian divisions and captured 140,000 prisoners.

Gen. Sir Archibald Wavell quickly recognized that to control Cyrenaica he needed only to seize the frontier passes and capture those places where the Italian forces had withdrawn—Bardia along the Bay of Salum, and Tubrug 130 km (80 mi) farther west along the coast. The fleeing Italians failed to establish a sizable defense at Salum, and British forces occupied the escarpment, invested the Italian forces at Bardia, and obtained their surrender (Figure 12.1).

At the same time Bardia was falling, the British began an attack on Tubrug, which was not only well fortified but also a vital supply base for the Italians. Supported by air and naval forces, Wavell took the city in late January 1941. Italian infantry then retreated west along the coast while their armored forces moved inland toward the road junction at Mekele, where the desert tracks joining Banghazi and Tubrug cross a north-south trail into Al Jabal Al Akhdar

FIG. 12.1. Part of a British Eighth Army supply convoy moving up the steep grade of Salum Pass with the Mediterranean Sea in the background. The "pass" is actually a road built into the escarpment that separates the coastal plain from a low inland plateau. Note the condition and width of this important transportation artery. (U.S. Army photo, photographer unknown, 29 December 1943.)

(see Map 12.1 above). Wavell's infantry pursued the fleeing Italians along the coast while his armor, following an inland route, raced cross-country through the open gravel-surfaced desert toward Banghazi.

Bypassing the scattered Italian forces, the British armor quickly reached the coastline west of Zawiyat Masus and cut off the withdrawal of enemy troops moving south from Banghazi. The strength of the British 7th Armored Division was by then evident, and on 7 February the Italians surrendered unconditionally to Wavell.

Following the surrender, Wavell's army continued their advance, setting up outposts facing Italian defensive positions across the Gulf of Sidra. By then, however, Wavell had extended his supply lines over 1000 km (620 mi) from his base at Alexandria. Plans to use Banghazi for resupply were thwarted by the German air force. The vulnerability of Wavell's forward position on the Cyrenaic bulge quickly became apparent as German ground forces prepared to enter the battle.

On 24 March 1941, Field Marshal Erwin Rommel and his Afrika Korps attacked, quickly pushing across the same open desert that only a few weeks before had offered opportunity to the British. Encounters were numerous but few could be termed battles. In only three weeks, the Germans had reached Al

Gazala and seized the critical passes atop the Salum escarpment. Their troops following the coast road captured Banghazi on 4 April and reached Tubrug on 30 April.

Rommel pushed the bulk of the British army out of Libya, but could not capture their garrison at Tubrug. Then commenced an ultimately unsuccessful seven-month siege of this key port. While the attack on Tubrug continued, Rommel strengthened his positions in the desert and improved the fortifications at Bardia, Salum, Halfaya, and Sidi Omar. To the German commander, passes at Halfaya and Salum were points of great tactical importance, for they were the two likely places between the coast and Habata where both wheeled and tracked vehicles could cross the escarpment.

Meanwhile the British, positioned across the frontier, took advantage of a pause in the fighting to resupply and upgrade their units. Then, in mid-May, the British tried but failed to recapture the two critical passes and relieve their garrison at Tubrug.

Months later, on 18 November 1941, climbing the escarpment west of Matruh, the British forces, now under Gen. Claude Auchinlek, made a wide inland swing into the open desert toward Tubrug, outflanking the German armor positioned at Salum and Halfaya passes. Supporting British troops also attacked

FIG. 12.2. Widely separated tanks crossing a level area in North Africa where the desert's surface is mainly gravel. Areas like this are widespread in the Western Desert and ideal for armor warfare. (U.S. Army photo, photographer unknown, 24 April 1942.)

FIG. 12.3. Tanks and troops of the British Eighth Army in a small North African wadi. These stream-eroded features could provide valuable cover in the barren desert, as seen in the background and in Figure 12.2. (U.S. Army photo, by Aarons, 26 January 1943.)

Salum along the coast. There were no massive attacks by either army because the open ground on the plateau was more suitable for independent, small-unit, armored operations. For two weeks English and Germans fought each other in small pockets throughout the open desert (Figures 12.2 and 12.3).

In late November Rommel made a desperate attempt to outflank his enemy but was foiled when the beleaguered British elements in Tubrug broke out and joined the advancing British army. Recognizing that his supply lines were in jeopardy, Rommel pulled back to positions extending from Al Adem to Bir El Gobi. Then, in the face of growing British pressure, he ordered a withdrawal through Banghazi to Al Ugaylah. On 6 January another pause began as both sides resupplied and refitted.

By 21 January 1942, Rommel, once again ready, launched an attack across the desert northeast toward Zawiyat Masus. Racing over sands previously traversed by both the British and his own armor, Rommel made excellent use of his growing knowledge of the region. The speed of his advance almost trapped the British forces at Banghazi, but they managed to slip away along the coast road and flee east across the desert in small groups. Within two days Rommel reached the center of the Cyrenaican Plateau. Auchinlek, recognizing the danger in falling back through the passes, quickly established a defensive position along a line running south from Al Gazala. Having nearly exhausted his supplies and not willing to fight in this relatively open area before establishing and securing his supply lines, Rommel delayed battle for four months.

While the German general prepared for his next offensive, the British continued to improve the defensive line from Al Gazala to Bir Hacheim. To reinforce the existing terrain, which was devoid of any prominent features, the British established numerous minefields, wire barriers, and fortified fighting positions in brigade "boxes." Recognizing the formidable opposition he faced, Rommel, in a flanking movement on 27 May, swung to the south of the southernmost minefield near Bir Hacheim to threaten the British left flank. For nearly two weeks a battle raged in and behind the Bir Hacheim line. In a classic engagement of tanks, perhaps the greatest in the war in North Africa, the British were soundly defeated (the British lost about 670 of their 740 tanks) and began to fall back toward Egypt, again leaving behind an isolated garrison at Tubrug. In spite of valiant attempts to hold out against overwhelming odds, on 21 June Tubrug fell to Rommel's forces and opened to him the port needed so urgently for an advance toward the Nile.

Recognizing that Rommel was in pursuit and that any hope of stopping him rested with the establishment of a strong defensive position somewhere west of Alexandria, Auchinlek moved all the way back to a position near El Alamein. Here the open desert was constricted to a width of about 50 km (35 mi), bounded on the north and south by the Mediterranean Sea and the Qattara Depression, respectively. The narrowness of the front made this an ideal place for Auchinlek (and Wavell beforehand) to develop a well-fortified position. East of El Alamein, however, there are no major natural barriers to movement toward Cairo or Alexandria. For the British this was the last good defensive position west of the Suez Canal.

Here, throughout June and July and into August, both sides made massive efforts to reinforce their positions and prepare for the offensives each expected of the other. With a much shorter supply line to Alexandria, the British were favored in upgrading their forces. In stark contrast, Rommel's logistics were especially difficult because of problems resulting from British naval and air action in the Mediterranean and along the coast, and a supply line that now extended about 500 km (300 mi) from Tubrug.

The most prominent feature near El Alamein is Alam Halfa, a 100-m-high (330-ft) ridge that trends east-west for nearly 24 km (15 mi). The newly appointed British commander, Gen. Bernard Montgomery, recognized the importance of this landmark and placed strong forces on and near that ridge. On 31 August 1942, Rommel launched the battle of Alam Halfa, attempting to outflank the strong El Alamein line by penetrating its southern flank. Using the ridge as a strongpoint, Montgomery slowed and finally stopped Rommel's advance. After two days of fighting, the Germans withdrew to their original line of battle.

According to Rommel, the British won at El Alamein because there was "no open desert fighting." The battle, fought on Montgomery's territory in the location he favored, not only improved British morale but also weakened Rommel's army, depleted his supplies, and threatened his security in Egypt.

For the next six weeks, both German and British forces rested and refitted. Rommel, still plagued by shortages and harassment from the British air force, could only tend his wounds. Meanwhile, the British, with shorter lines of communication and improving logistics, moved reinforcements and materiel to the front virtually unimpeded.

Montgomery massed his forces near the coast and along the shadow of one of the few terrain features in the region, the Miteriya Ridge. On 23 October 1942, his army attacked the German left flank for more than a week. Finally, on 2 November, British forces broke through and poured behind enemy lines. In response, Rommel ordered a westward withdrawal on 5 November; the British followed on his heels.

Rommel's retrograde maneuver was huge, continuing all the way to the Tunisian border, but it saved most of his army. Montgomery's one reasonable real chance to trap the Germans occurred on 6 November. The British were poised to cut off enemy movement along the coast road by dashing across the desert to Matruh. But in late autumn rain-producing, mid-latitude wave cyclones begin to track far enough south to affect the north coast of Africa. Although these storms are not common, that is exactly what happened on 6 November. Heavy rains quickly made a muddy plain of the land between the British forces and the coastal road, stopping all cross-country movement by armor. By 7 November, at about the time the ground dried out, Rommel had escaped.

War in the Western Desert raged for more than two years before migrating westward into Tunisia and before American involvement. Many large and small battles were fought and the desert was an important factor in each. Soil and surface material ranging from rock to mud guided and restricted off-road vehicular movement. In contrast, widespread gravelly areas favored massive armor battles where little restrained the tactician, but everything seemed a problem to the logistician. Both sides took advantage of the open terrain to execute classic flanking maneuvers. But, as at Sidi Barrani and El Alamein, each also used narrow desert corridors to restrict their enemy's maneuver.

Control of passes, depressions, ridge lines, and coastal roads in the North African desert was essential for success. At least five times the passes at Salum and Halfaya held the key to victory or failure. Transportation and logistics presented huge problems. For the most part, supplies had to move on the heavily traveled and exposed coastal road. Recognizing that the supply vehicles traveled on only a few routes, pilots of attacking aircraft knew exactly where to look in their searches for enemy columns. And wheeled support vehicles that needed to follow armored forces across the open desert often got bogged down and became easy targets for an encroaching enemy.

The distances in North Africa were enormous and the roads poor. It was

UNITED STATES MILITARY ACTION IN NORTH AFRICA

In late 1942 the variable nature of the desert became immediately apparent to U.S. soldiers landing in northwestern Africa during Operation Torch. Instead of flat, open, sandy land, they soon found themselves advancing through the rugged Western Dorsal Mountains whose summit exceeds 1000 m (3000 feet). Meanwhile, the Germans had established a defensive line northward from the huge Shott el Jerid (Salt Marsh) along the trend of the Eastern Dorsal Mountains to the Mediterranean Sea west of Bizerte. In this area, untested American troops first encountered German forces in World War II. The battle, begun in the Eastern Dorsals on 14 February, quickly turned into a confused, three-day American retrograde movement culminating in fierce fighting in and near Kasserine Pass. Here, with leadership problems, the U.S. Army failed to take advantage of a strong defensive position related to the mountainous terrain adjacent to the pass and suffered one of its worst defeats of the war.

Later, after some reorganization, American troops advanced northeastward with the British toward Tunis, only to find that their thrust at the Germans was measurably hampered by an inability to maintain efficient lines of communication across the desert of northwest Africa. When the U.S. forces reached El Guettar, they found that the narrow thread of troops that connected the tactical forces with their supply bases had been weakened so severely that they could not employ their full combat power.

U.S. military leaders learned much from Kasserine Pass and their advance towards Tunis. Better training, improved leadership, and effective reorganization molded the army into a powerful force proved more than effective a few months later in Sicily.

Over the next 50 years, the Armed Forces of the United States have been equipped with helicopters, high-mobility tracked and wheeled vehicles, and a wide variety of sophisticated technological equipment that markedly increases effectiveness and control in deserts and other difficult terrains. Yet in November 1980, when returning to the Sahara to train west of Cairo during Operation Bright Star, they discovered that the combination of heat, sand, and dust recreated many of the same problems for them that faced their grandfathers 38 years earlier.

The biggest difference between World War II and modern war in the desert involves intelligence and communications. With greatly improved aerial reconnaissance, precise remotely sensed data from satellites, and the startlingly accurate Global Position System (GPS), it is possible to develop, conduct, and coordinate military operations in a fashion never before possible as was so vividly demonstrated in the 1992 Operation Desert Storm in Iraq and Kuwait. Even so, when maneuvering in the desert the soldier still faces a hostile environment that can be as dangerous as the enemy.

1100 km (700 mi) from El Alamein to Al Ugaylah, and a round trip to Banghazi was about 2000 km (1240 mi) and required 14 days. Any long British advance westward stretched their lifeline to the limit. German and Italian moves to the east created the same problem and made the capture of Mediterranean ports of Tubrug and Banghazi vital for continued advances. When the

British eliminated enemy resupply through these ports, the length of the German supply line was immediately more than doubled.

The war in the Western Desert was the epitome of "hurry up and wait." The seven major campaigns waged in the 27 months between Graziani's attack on Sidi Barrani in 1940 and Rommel's withdrawal into Tunisia in 1942 required less than six months to carry out. The remaining time was spent in refitting or siege of garrisons at Tubrug, Bardia, and Salum. Most of the time the logistician worked while the tactician paced. Interestingly, although the war sprawled across the Western Desert, many of the battles were fought within a linear distance of less than 160 km (100 mi); these included the siege of a few cities, fighting in the open desert south of Tubrug, and combat in and around Sidi Barrani and the Salum escarpment. In contrast, five-sixths of the desert hosted only chance contacts and meeting engagements.

CONFLICT IN THE SINAI, 1917–79

The nature of the territory dictated over the centuries the course of warfare in the Sinai.... The conflict is predetermined by the demands of the desert.

CHAIM HERZOG

In the form of a 400-km-long (250-mi) wedge pointing southward from the Mediterranean Sea, the Sinai forms a keystone Middle Eastern link between Asia and Africa (Map 12.2). Within its 60,000 km² (23,000 sq mi) lies some of the most rugged and driest land in the world. Yet all conquerors and vanquished wishing to move overland from Palestine into the fertile Nile Valley, or from Egypt to the riches of Syria and the Tigris-Euphrates, had to first cross the Sinai. The few trails, all traced to antiquity, reveal a careful adjustment to terrain.

During 1917 the British in Egypt anticipated that the Central Powers might move against the Suez Canal, the vital life-line linking Mediterranean and Asian parts of their empire. The waterway itself was a formidable obstacle to aggression from Palestine west toward Cairo, and the mission was not to defend the city but to keep the canal open and functioning. The British commander in Egypt concluded that in order to protect the waterway, he had to hold the enemy in the Sinai. This mission could best be accomplished by blocking the three major routes that crossed the peninsula: one paralleling the Mediterranean coastline and two through the north-central part of the land mass. The British success in controlling these routes, and their subsequent advance into Palestine along the coastal route, foreshadowed maneuvers that would be repeated time and again in the mid-twentieth century: movements dictated as much by the desert and coastal terrain as by the initiative of the commanders involved.

Despite being separated by the Gulf of Suez, the Sinai's climate is simply an eastward continuation of the Sahara's Western Desert. Centered on the 30th parallel, it is beneath the descending air of a persistent Hadley cell. As a result,

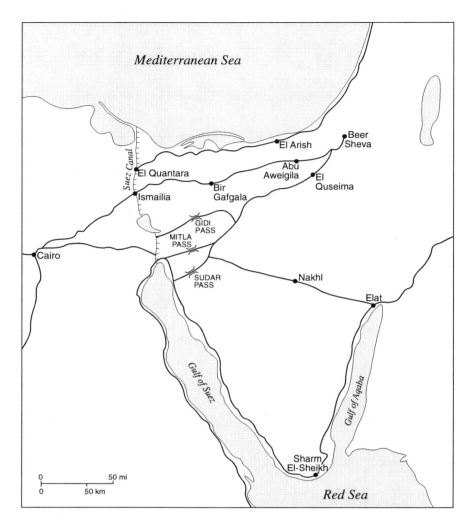

MAP 12.2. The Sinai and its major roads and passes.

Mediterranean Sea

Suez Canal

El Arish

Beer Sheva

El Quantara

Abu Aweigila

El Quseima

Ismailia

Bir Gafgala

GIDI PASS

MITLA PASS

Cairo

SUDAR PASS

Nakhl

Elat

Gulf of Suez

Gulf of Aqaba

Sharm El-Sheikh

Red Sea

0 50 mi
0 50 km

the total annual rainfall is less than 25 cm (10 in), with nearly all of it falling in the winter months, making the area exceptionally dry during the warmest time of the year when evaporation rates are highest. The summer temperature, which varies with altitude and latitude, averages from 20° to 30°C (70–90°F) but ranges greatly between day and night. Winter daytime temperatures are not nearly as hot but the daily range does remain high.

The Sinai Peninsula was uplifted by the same forces—all related to plate tectonics—that created the Syria–East Africa rift and formed the Jordan Valley and the Dead Sea. Of the peninsula's three topographic provinces the southern is the most mountainous, with some summits above 2400 m (8000 ft) and ranges that extend northward on both the east and west sides of the peninsula. Deep valleys and wadis bordered by precipitous cliffs greatly limit cross-country movement and make travel along the few primitive roads very difficult.

The interior center third of the peninsula is a high, deeply dissected lime-stone plateau, the Gebel el Tih, which slopes north toward the Mediterranean

Sea. This upland is bordered on the south by steep, 1500-m-high (5000-ft) cliffs that effectively separate this region from the southern mountains. Wadis and gorges that cut deeply into the surface of Gebel el Tih produce a vast assemblage of sharp, angular landforms. When the rare rainfall does occur, runoff drains north toward the Mediterranean in the El Arish River system that covers a third of the Sinai. This drainage system consists of an intricate system of innumerable tributaries that join together, like stems in a leaf, as they extend toward the Mediterranean Sea. Altogether this landscape presents many barriers and great friction for cross-country movement, especially in an east-west direction.

The northern province is unlike the other two in that it is lower, smaller, and less rugged. But most important, it forms an uninterrupted natural corridor between Egypt and Israel. The terrain here consists of a narrow coastal plain with numerous sand dunes among a few volcanic remnants and an occasional wadi, the largest being at the town of El Arish.

Nature, in this delineation of the Sinai, has dictated only a few east-west routes suitable for crossing the peninsula. One is a road along the coast extending westward through El Arish to El Quantara. Here sand dunes may encroach from both sides, occasionally cover the highway, and limit off-road mobility. A second route extends from Beer Sheva through Abu Aweigila and Bir Gafgala to Ismailia. It follows a sandy lowland, then crosses the north part of the central, dissected plateau. Because of the soils and terrain, wheeled-vehicle travel within this area is restricted to roads. A variant also starts at Beer Sheva, but once in the plateau it splits into three paths through the Gidi, Mitla, and Sudar Passes that extend to or near the Suez Canal. Here, too, off-road mobility is restricted by the rugged bordering uplands.

Foot, animal, and vehicle movement is also possible on a central road extending from Elat to Nakhl and then through a western pass to the Suez. However, this track has always been largely unimproved as it crosses desert highlands scarred by wadis and clefts. The final and by far longest route is a round-about road south into the coastal mountains while paralleling the Gulf of Aqaba to emerge near Sharm El-Sheikh; it then turns northwest along the Gulf of Suez to the canal.

Since 1948 Egypt and Israel have been at war four times. In each instance the combatants, in spite of their desire to develop new tactical approaches, were essentially tied to the few natural east-west routes through the Sinai. In 1948–49, during the Israeli War of Independence, the attacking Israeli forces seized the critical road junctions at El Arish, which controlled the coastal route, and Abu Aweigila and El Quseima, thus blocking routes leading to the Mitla and Sudar Passes. By controlling these key road junctions, the fledgling nation prevented the Egyptians from reinforcing their forward units in the Gaza and Negev regions, threatening them with annihilation.

During the Arab-Israeli War of 1956, the Israelis fought hard to again secure El Arish and, once done, proceeded rapidly along the coastal highway to El

Quantara. Other elements advanced from Abu Aweigila through Bir Gafgala to Suez and through the Mitla Pass to the Suez. The latter unit lost most of its armored vehicles (11 of 13) to the desert terrain rather than to the enemy. Another Israeli element moved through the Sudar Pass and southward along the Gulf of Suez coast to link up at Sharm El-Sheikh with a force that had wound its way south from Elat along the mountain roads adjacent to the Gulf of Aqaba. In the end, the Israeli army controlled just about all of the Sinai and secured the entrance to the Gulf of Aqaba.

Eleven years later, in the Six-Day War, the Israelis essentially repeated what they had done in 1956 by advancing along the coastal road, and then moving from Abu Aweigila, through Bir Gafgala to Ismailia and from El Quseima through Gidi, Mitla, and Sudal Passes to the Suez Canal and the coast. (Between 1956 and 1967 the Egyptians, recognizing the danger of having only one passage through these mountains, carved the Gidi Pass.) Unique to this war, a small amphibious force was also sent southward on the Gulf of Aqaba to Sharm El-Sheikh, where it was joined by Israeli paratroopers. Together they moved north along the coast to seize nearby oil fields.

The 1973 Arab-Israeli War started with an Egyptian attack against Israeli forces holding the east bank of the Suez Canal. After securing a foothold on the east bank, the Egyptians attempted to break out along the three routes: El Quantara toward El Arish, Ismailia to Bir Gafgala, and through the Mitla Pass. In the face of Egyptian pressure, the Israelis fell back, establishing defensive positions astride the coastal road, west of Bir Gafgala, and at the upland passes. With support from their air force and reinforcement by their mobilized ground forces, the Israeli elements in the Sinai held and soon launched a counterattack that took some of their troops across the canal to establish firm positions on its west bank.

It is important to note that the same roads were used in all four wars. In looking at these conflicts collectively, Arnon Sofer, professor of geography at the University of Haifa and a consultant to the Israeli army, concluded, "Basic topographic features that underlie the axes of advance in the distant and recent past are likely to be the principal features in conventional wars in the future."[5] The Camp David accords of 1979 recognized this situation and established a multinational force and observers to supervise the Sinai once the Israelis withdrew from the peninsula. The disposition of these buffer forces now separates Egyptian and Israeli positions. Even so, should conventional war again occur in the Sinai it is a near certainty that the same routes would be central to the battle.

Conclusions

What, then, is the connection between 33 months of sporadic World War II fighting among Europeans in North Africa, and four separate short, yet fierce conflicts between Arabs and Jews between 1948 and 1973? Common threads— including the importance of planning, logistics, leadership, and tactics—are tightly woven within a background of nearly constant aridity, challenging terrain, and a greatly limited road network. Furthermore, the vast Sahara and the relatively small Sinai are uniformly dry, making the availability of water a constant problem for all of the armies. And finally, at one place or another, whether in the Western Desert or the Sinai, the local terrain guided and funneled the opponents along predictable paths as the fighting repeated itself in the same area again and again. Meanwhile, the individual on the battlefield had to fight heat, cold, wind, dust, fatigue, and disease in a desolate area where vegetative cover is absent and thirst may be fatal. For the isolated soldier on the battlefield, aside from conflict itself, only the quick-killing cold of the Arctic is more threatening than being without water in the desert.

Conclusion

Weather and terrain have more impact on battle than any other physical factor, including weapons, equipment, or supplies.

U.S. ARMY *FIELD MANUAL 100-5* (1982)

Mastering environmental dimensions is vital to survival on the battlefield.

U.S. ARMY *FIELD MANUAL 100-5* (1993)

The commander who can best measure and take advantage of the weather and terrain conditions has a decided advantage over his opponent.

U.S. ARMY *FIELD MANUAL 100-5* (1993)

Relevance and Insight

Although the three quotations above are from modern U.S. military doctrine, regardless of country or century, their messages are true and applicable to armies since the beginning of warfare. But what are some actual measures of physical geography's role within the milieu of military conflict?[1] Collectively, the range of operations considered in this book shows that environment and war are so inextricable that the first commonly shapes the second, and in the end the reverse is often also true. Equally important, geographic factors can be abstruse, capricious, disadvantageous, fortuitous, formidable, opportune, problematic, and risk-inducing, all words that connote active involvement rather than remote passivity.

It is also clear that environmental effects often do not become apparent until after a military mission is determined and the operation is well underway. This was illustrated by the sudden storms that twice devastated Kublai Khan's army in its attempts to invade Japan and the evolving weather that favored Allied troop withdrawals in the 1940 "Miracle at Dunkirk." If atmospheric conditions in these cases had been reversed, favoring Kublai Khan and Adolf Hitler, to the point that the outcome of battle was markedly changed, there is little doubt that world history would be far different.

Commanders can gain great stature when they use regional aspects of the terrain wisely. They are also often vilified when their maneuvers fail. Early in the American Civil War, Stonewall Jackson's operations in the Shenandoah Valley and Robert E. Lee's two invasions of the North took full advantage of lowlands and mountains to conduct effective and extensive flanking maneuvers that threatened both the Union army and their capital. In contrast, McClellan's equally opportunistic Peninsular Campaign (which sidestepped a frontal confrontation, avoided a major river crossing, and placed a flanking army within striking distance of Richmond) failed as his forces lost momentum on the Coastal Plain. Regardless of outcome, geography was basic to all of these campaigns. It was up to each commander to recognize, as best they could, the options and limitations, develop a plan, and then execute it effectively. Although Jackson's success, McClellan's failure, and Lee's mixed results are obviously related to many factors, two of the most important in the shaping of their campaigns were the regional physiography and the local terrain on the battlefields.

When combined with prescribed tactics, the landscape can sometimes present a tenacious friction that constrains, or even curtails, operations. Examples include Flanders during World War I and Burnside's American Civil War Mud March of 1863. However, technology can quickly change that which previously seemed impossible, as during 1965 in the Ia Drang Valley of Vietnam where application of the helicopter-based, air-mobile operations established a new doctrine for warfare that lessened the obstacles and restraints on cross-country movement. However, Ia Drang also showed that although innovation may present new opportunities, it may also increase the level of peril, as illustrated by the incredible intensity of the fighting, the difficulties of withdrawal experienced by U.S. units, and the huge number of NVA casualties on and near the Chu Pong massif.

Geography may also repeatedly guide armies along the same path, as shown by three century-separated advances toward Moscow by different European powers, and successive Egyptian and Israeli maneuvers in the Sinai during four recent wars. Considering the terrain, objectives, and circumstances, it is unlikely that any of these invaders would have followed routes other than the ones chosen. But initiative and execution can also lead armies through places thought impenetrable or on an unexpected course, such as the 1940 and 1944 German advances in the Ardennes and the U.S. landings at Inchon in 1950.

Speculation and Reason

To appreciate the potential impact of environmental factors, consider the following questions related to two 1944 operations in Western Europe.

First, what might the results be if 1944 weather had forced repeated postponements of the Normandy landings, delaying them until a time when the Germans would be more alert and better prepared to defend the landing beaches? Or what are some reasonable implications if the invading troops were confined to their beachhead for months, similar to Anzio? Worst of all for the Allies, what assessments follow if, like Gallipoli in World War I, the invasion had failed altogether?

Assuming that any one of these three postulated events actually happened, how far westward would the USSR armies have advanced in Europe? Who would have occupied western Germany? What would be the role of Britain and the United States in concluding the war? Where would the continent's post–

World War II political boundaries be established? And how might a failure or long delay in the invasion influence the postwar extent of power and longevity of the former USSR? The answer to any one of the questions highlights the huge importance of the landings in Normandy on 6 June 1944.

Second, what would have happened if different Allied maneuvers in Holland during September 1944 had made Operation Market-Garden successful, thus placing British troops well in the Allied vanguard east of the Rhine? Would Eisenhower have ended his "broad front" European strategy in favor of a single deep thrust into Germany led by the British? Would a British advance toward the Ruhr, as planned, invite a large-scale Wehrmacht counterattack (one happened three months later in Belgium's Ardennes) that might have isolated Montgomery's army on the wrong side of not one but several large rivers with no close port secured for resupply? Last, assuming that Market-Garden and follow-up British attacks deep into Germany were successful, what role would the American armies play in the finalization of the war in Western Europe and how would that have affected U.S. postwar international presence and influence? Again, answering these queries is profoundly thought-provoking.

Although these questions are patently speculative, they are also both fair and worthy because each is based on possible, sometimes probable, alternatives. Furthermore, and more important, equally reasonable geography-related questions can be derived for all military operations. In fact, many battles were made distinct, not by their following doctrine and expectations, but because they were unpredictable and fought in unanticipated or unlikely surroundings. Who among us will presume to predict the exact nature and overall impact of environmental effects on any anticipated conflict? The point is, when considering geographic influences for the future, instead of expecting repetition, similarity, and predictability, experience shows that one should be alert for opposites, contrasts, and the unprecedented.

Consider the difference between the short view of the December 1944 German counterattack in the Ardennes and its paradoxical long-term implications for the United States. No speculation is needed here, the historical record is clear, and the fortunes of war proved dramatically ironic in the end.

For those familiar with the Battle of the Bulge, many believe that it was a major defeat where outnumbered U.S. troops were initially surprised, many captured, all badly battered, and some beaten back more than 100 km (62 mi) from the original front. Correct as this short-term understanding is, a more profound view is that American forces were at first supremely tested and then responded so valiantly and effectively that within about a month all of the lost territory was recovered, and as that happened the army appeared more powerful than ever before. At the same time the U.S. soldiers and generals (including Omar Bradley, George Patton, and Anthony McAuliffe, among others) clearly proved their military will and capability to all, certainly including the Germans. Probably even more important for the future, the quick and powerful U.S. recovery from the German counterattack enhanced trust and confidence by other nations, a view that served this country long and well in the geopolitical turmoil that followed the war. In the short term the terrain, vegetation, and weather aided the Germans, but in the end it was the United States that, at considerable cost (approximately 80,000 casualties), benefited most from the Battle of the Bulge.

Perspectives for the Present

As a measure of geography's importance in conducting a mission, it is possible that an accurate and detailed map in the hands of a skilled reader is more critical to success in battle than available weapons and relative strength. It is important, however, that the study of geographic influences on warfare does not lead to oversimplification and the assumption that the outcome of battle is predetermined or overly controlled by the environment. Such notions are often favored by hindsight because opposing forces are easily located and differentiated, fronts and flanks of armies well defined, and the strategic objectives clear to all involved. The problem is that before and during a battle none of the above may have been well understood by anyone on either side.

The fact is that rather than being increasingly orderly and predictable, the nature of warfare has grown more complex, militarily and otherwise. Using successive experiences of the United States as an example, it evolved from a relatively remote rebellion; experienced two border conflicts (War of 1812 and the Mexican War); suffered through an absolutely tragic two-theater civil war; and then became embroiled in a series of foreign conflicts including one with Spain, two enormous world wars, a bitter three-year struggle in Korea, no less than a decade of military involvement in Southeast Asia, and a short but huge role in Operation Desert Storm.[2] Meanwhile there have been many small-scale and increasingly perplexing military operations, recent examples including such places as Bosnia, Grenada, Haiti, Lebanon, Panama, and Somalia. Added to this are growing problems with domestic and foreign terrorism; policy dilemmas related to changing governments and power vacuums; and participation of U.S. armed forces in the name of humanity, public safety, drug-trafficking, and peace-keeping. In this progression, military operations have become more diverse and intricate, while sought-after order and simplicity are less and less attainable.

The effect of this tendency toward chaos makes it even more important that environmental parameters be accurately measured, analyzed in depth, and appropriately assessed, because the sum of these relationships provides an improved basis for understanding geographic variability and its potential impact.[3] Failure to do this, as shown in this book, invites poor decisions, added suffering, serious setbacks, and the increased possibility for failure or defeat.

The record shows that the outcomes of many battles are decided as much by the losers' errors as the winners' astuteness. In that process geographical factors often have, in one way or another, a multiplying effect on a military operation. The continuing problem is that no one can precisely predict how the environment will influence the progress or outcome of the next battle. All one can be sure of is that in some way they will be formidable. Then, as unknowns appear, evolve, and multiply, training, leadership, intelligence, and innovation become increasingly important.

Perspectives on the Future

The arrival of the "Information Age" and the advent of a "Revolution in Military Affairs" will not overcome the everlasting influences of terrain and weather. But they may significantly enhance a soldier's ability to cope with these facets of war. A global network of satellites, together with Earth-based systems, quickly provides a wide variety of reliable data to analysts on the ground, including detailed information on the magnitude and moisture content of storms. If Eisenhower's meteorologists had access to these systems, the D-day decision would have appeared to be less of a gamble. However, if Hitler had the same advantage, German forces along the Normandy coast would surely have been on alert and more quickly aware of the thrust and magnitude of the Allied invasion.

Satellites and other remote sensing systems provide an array of information about the battlefield environment that was previously not imagined. Even at the battalion or lower level, data on topography, soils, vegetation, surface moisture, and drainage regimes can now be combined with that for the movement and deployment of forces. The potential result is a more judicious identification of enemy and friendly lines of movement, enhanced positioning of troops, and a computerized view of what a fog-shrouded terrain would look like on a clear morning. This also means that imprecise impressions of terrain and weather are transformed into absolute reality.

The presence of sophisticated geographic information systems now permits the rapid transformation of maps and overlays into powerful analytical tools, such as ground and oblique views of the operational landscape from both the enemy's and friendly troops' perspective. These can be incredibly valuable in combat. With the added use of global positioning systems, soldiers of the twenty-first century will be capable of operating in a terrain- and weather-rich digital realm while knowing their precise location on the battlefield. Simply put, these technologies provide a basis for using geography to greater advantage in warfare.

Epilogue: Circumstance and Responsibility

Just as no two battles have ever been fought in precisely the same way, geography's inescapable influence on warfare is never identical but, instead, changes with site, situation, time, leadership, and technology. The power and importance of these influences are apparent at all levels, ranging from individual and small-unit action, through battles, operations, and campaigns, to war-time theaters, grand strategy, and worldwide geopolitics.

To the infantryman in combat a tiny hummock may be as protective as a mountain range; movement over a small stream more dangerous than crossing a large river; and a shallow hollow, small bush, or sudden rainstorm the difference between life and death. In these situations a soldier's geography may extend only as far as one can see, shout, shoot, throw, or reach. Meanwhile the mere randomness of position can become the paramount event of one's lifetime.

As the size of an operation grows, so does the impact in any command decision, certainly including those that relate to environmental effects. This was carried to the ultimate during World War II when the Supreme Allied Commander in Europe based one of the most momentous decisions in the history of warfare on a short-term weather forecast for the Normandy coast. At that time, in June 1944, Gen. Dwight Eisenhower *demonstrated* the powerful relationship between environmental conditions and human decisionmaking when he used judgment, experience, data, and advice to secure advantage from a rapidly passing wave cyclone. Fifteen years after ordering that D-day invasion to commence, then-President Eisenhower *stated* the following to the U.S. Military Academy Corps of Cadets: "Military operations are drastically affected by many considerations, one of the most important of which is the geography of the area."[4]

Whether foot soldier or commander, at all scales of action, and regardless of time, those who know more about the shape, nature, and variability of battleground conditions will always have at least one significant advantage over a less-knowledgeable enemy.

Notes

Introduction

1. Recently published military geography books by U.S. authors are few in number and all are of modest size. For examples, see Louis C. Peltier and G. Etzel Pearcy, *Military Geography* (New York: Van Nostrand, 1966); Patrick O'Sullivan and Jesse W. Miller, *The Geography of Warfare* (New York: St. Martin's Press, 1983); and O'Sullivan's *Terrain and Tactics* (New York: Greenwood Press, 1991).

Compilations of articles that relate aspects of geography to warfare include the U.S. Air Force Reserve Officers Training Corps Air University's *Military Aspects of World Political Geography* (Maxwell Air Force Base, 1959); U.S. Army Command and General Staff College, *Readings in Military Geography (RB 20-4)* (Fort Leavenworth, Kans., 1962); and the 1981, 1984, and 1990 editions of *Readings in Military Geography,* edited and compiled, and reprinted for cadet use by the Department of Geography and Computer Science at the U.S. Military Academy, West Point, New York, by John B. Garver and Gerald E. Galloway Jr.

The most comprehensive collection of geography-related references is U.S. Military Academy, Department of Geography and Computer Science, *A Bibliography of Military Geography* prepared by Eugene J. Palka and Dawn M. Lake (West Point, N.Y.: Gilbert W. Kirby Jr., n.d.).

Chapter 1. Storms, Fair Weather, and Chance

1. For examples, see Peveril Meigs, "Some Geographical Factors in the Peloponnesian Wars," *Geographical Review* 51 (1961): 370–80; David M. Ludlum, "The Weather of American Independence, 3: The Battle of Long Island," *Weatherwise* (June 1975): 118–21, 147; David Blumenstock, *The Ocean of Air* (New Brunswick, N.J.: Rutgers University Press, 1959), 319–46; and chapters 2, 3, 4, and 12 of this text.

2. For pertinent descriptions, see chapters 4 and 8 of this text.

3. Temperature changes are the result of the adiabatic process, which occurs through expansion or compression of air without loss or gain of outside energy.

4. The four precipitation-producing situations are: frontal, involving contact between two unlike masses of air; convergence and thence lifting as air moves into a low-pressure trough; convection, which is the upward movement of a warmer cell of air; and orographic lifting, which results from air moving across a topographic barrier.

5. A middle-latitude cyclone should not be confused with a tornado, which consists of a small vortex of very high velocity wind swirling about a central column with unusually low air pressure.

6. The type of front is determined by which air mass is the aggressor. Occlusion results when a cold front overtakes a warm front to force the warm air aloft.

7. "Typhoon" is the common name for such a storm over the Pacific Ocean; "hurricane" and "cyclone" are names for its generally smaller counterparts over the Atlantic Ocean/Caribbean Sea and Indian Ocean, respectively.

8. The upper-air westerlies with their higher-velocity jet streams flow eastward in a zone from near the poles to within 25 degrees of the equator in both hemispheres, while the tropical easterlies, flowing in an opposite direction, are confined to a narrower tract high above the earth's equatorial zone.

9. For example, differences in heating and cooling rates for land and water may induce convergence, resulting in clouds and possibly precipitation along or near coastal areas. Also, super-cooled, and thus heavier, air in contact with snow or ice may periodically rush downward in mountainous areas to produce a concentrated high-velocity wind in nearby valleys and lowlands.

10. S. R. Turnbull, *The Samurai: A Military History* (New York: Macmillan, 1977), 86–94.

11. James Murdoch, *A History of Japan*, vol. 1 (London: Kegan Paul, Trench, Trubner, & Co., Ltd., 1925), 491–532.

12. Ibid.

13. Viscount Kikujira Ishii, *Diplomatic Commentaries* (Baltimore: Johns Hopkins Press, 1936), 5–9.

14. Glenn T. Trewartha, *The Earth's Problem Climates* (Madison: University of Wisconsin Press, 1961), 180–99; E. Fukui, ed., *The Climate of Japan* (Tokyo: Kodansha, 1977), 65–84.

15. Vice-Admiral G. A. Ballard, *The Influence of the Sea on the Political History of Japan* (London: John Murray, 1921), 19–41.

16. J. Neumann, "Great Historical Events That Were Significantly Affected by the Weather, 1: The Mongol Invasions of Japan," *Bulletin of the American Meteorological Society* 56 (1975): 1167–71.

17. Ballard, *Influence of the Sea.*

18. Capt. F. Brinkley, editor, *Japan: Described and Illustrated by the Japanese*, vol. 2 (Boston: J. B. Millet, 1904), 38–102.

19. Ballard, *Influence of the Sea.*

20. Turnbull, *The Samurai.*

21. Murdoch, *History of Japan.*

22. Turnbull, *The Samurai.*

23. For a general review, see Trewartha, *Earth's Problem Climates*, 203–22.

24. Ibid.; W. G. Kendrew, *The Climates of the Continents* (London: Oxford University Press, 1961), 297–377; G. Manley, "Climate of the British Isles," in *Climates of Northern and Western Europe*, vol. 5 of *World Survey of Climatology*, ed. C. C. Wallen (New York: Elsevier, 1970), 81–133; R. Alery, "The Climate of France, Belgium, the Netherlands, and Luxembourg," in *Climates of Northern and Western Europe*, ed. Wallen, 135–93.

25. See, for examples, Maj. L. F. Ellis, "The War in France and Flanders, 1939–40," in *History of the Second World War*, United Kingdom Military Series, ed. J.R.M. Butler (London: HMSO, 1953); John Harris, *Dunkirk: The Storms of War* (London: David and Charles, 1980); Nicholas Harman, *Dunkirk: The Necessary Myth* (London: Hodder and Stoughton, 1980); Robert Jackson, *Dunkirk: The British Evacuation, 1940* (New York: St. Martin's Press, 1976); Walter Lord, *The Miracle of Dunkirk* (New York: Viking Press, 1982); Patrick Turnbull, *Dunkirk: Anatomy of Disaster* (New York: Holmes and Meier

Publishers, 1978); Peter Young, ed., *The World Almanac Book of World War II* (Englewood Cliffs, N.J.: Prentice-Hall, 1981).

26. Len Deighton, *Blitzkrieg: From the Rise of Hitler to the Fall of Dunkirk* (London: Jonathan Cape, 1979), 210.

27. Harris, *Dunkirk;* Jackson, *Dunkirk.*

28. For a description of the harbor and moles, see Harris, *Dunkirk,* and Lord, *Miracle of Dunkirk,* 93.

29. The evacuation, called "Operation Dynamo," was organized and carried out under the direction of the British Royal Navy.

30. Two excellent books covering wide-ranging aspects of the Normandy invasion are Gen. Dwight D. Eisenhower, *Crusade in Europe* (New York: Doubleday, 1948), and Gordon A. Harrison, *Cross-Channel Attack* (Washington, D.C.: Office of the Chief of Military History, U.S. Army, 1951).

31. For climatological aspects of the areas see Manley, "Climate of the British Isles"; and Alery, "Climate of France, Belgium, the Netherlands, and Luxembourg." Details on the meteorology related to the Normandy invasion see Roger H. Shaw and William Innes, eds., *Some Meteorological Aspects of the D-day Invasion of Europe 6 June 1944,* and John F. Fuller, *Thor's Legions: Weather Support to the U.S. Air Force and Army, 1937–1987* (Boston: American Meteorological Society, 1990), 85–99.

32. Much of this discussion is derived from James Martin Stagg's *Forecast for Overlord* (New York: Norton, 1971); Shaw and Innes, eds., *Some Meteorological Aspects of the D-day Invasion;* and Fuller's *Thor's Legions.*

33. Gen. Omar N. Bradley, *A Soldier's Story* (New York: Henry Holt, 1951), 266.

Chapter 2. Too Much and Too Wet

1. Frederic H. Lahee's comprehensive *Field Geology* (current edition), a good source for geologic terminology, definitions, and parameters, provides a useful guide for applying basic facts and concepts to real-world problems in the earth sciences. Most of the geological terms used in this book, such as *porosity* and *permeability,* are likely to appear in any comprehensive geological field manual.

2. On the basis of the Atteburg limits, fine-grained soils can exist in four states of consistency; solid, semisolid, plastic, and liquid. The differences are related to the nature of the sediment and the amount of water it contains. With mixing and the addition of water a solid soil may quickly evolve into a less competent phase.

3. Chemical weathering always involves water and takes place through oxidation, hydration, carbonation, hydrolysis, base exchange and chelation. Particles produced are often smaller than sand and some material may be in solution. Most mechanical weathering is the product of pressure-release fracturing, or sheeting, and the expansive effects of crystal growth in cracks and pores. Hydrofracturing from freezing also occurs but it almost always is limited to microscopic situations. Major temperature changes and plant growth have also been proposed as types of mechanical weathering but recent studies suggest that both are of little importance in the breakup of rock.

4. See chapter 12 for the effects of rain and runoff in arid regions.

5. See chapter 1 for a description of the middle latitude wave cyclone.

6. In *The Face of Battle: A Study of Agincourt, Waterloo, and the Somme* (New York: Vintage Books, 1977) John Keegan presents insightful and interesting comparisons between three battles (Agincourt, Waterloo, and the Somme) fought in the same area of northwest Europe. Although Keegan's main focus is on tactics and individual chal-

lenges of combat, he also describes the effects of rain and mud on each of these campaigns.

7. Although not as large as the James and Potomac Rivers, the Rappahannock River is arguably the most important stream in the Eastern Theater of the American Civil War. Located about midway between Washington and Richmond, Lee regularly and effectively used the river to the South's advantage, especially for defense. In contrast, several Northern offensive operations were complicated by the river, including fighting at Fredericksburg (1862) and Chancellorsville (1863).

8. See chapter 8 for details on the geologic characteristics of the Piedmont and Coastal Plain.

9. The U.S. Soil Conservation Service's *Soil Taxonomy,* U.S. Department of Agriculture Handbook No. 436 (1975) is a detailed and comprehensive description of soils and current taxonomy based on the widely recognized Seventh Approximation System with its 10 soil orders. D. Isgrig and A. Srobel Jr., *Soil Survey of Stafford and King George Counties, Virginia,* U.S. Department of Agriculture Soil Conservation Service (1974), provides detailed information, including large-scale, air-photo-based maps, on the soils in the area and along the line-of-march followed by Burnside's troops.

10. For details on hindcasting and the sources for January 1863 weather data see Harold A. Winters, "The Battle That Was Never Fought: Weather and the Union Mud March of January, 1863," *Southeastern Geographer* 31 (May 1991): 31–38.

11. See chapter 1 for a discussion of frontal storms.

12. In the Northern Hemisphere mid-latitude cyclones tend to move from west to east with winds that circulate in a counterclockwise fashion. Thus winds from the east indicate that a wave cyclone is somewhere to the west. A successive wind shift to the southeast, south, and southwest indicate that the storm passed to the north of one's location. Conversely, as what happened during the Mud March, winds shifting from east to northeast, to north, to northwest indicate that the storm center followed a track south of Fredericksburg.

13. Burnside was plagued with many problems and pontoon bridges were certainly one of them. A key element in his plans for attacking Fredericksburg in 1862 was the early arrival of several pontoon bridges needed to cross the Rappahannock River. The pontoons were delivered more than a week late, allowing more time for the Confederates to strengthen their defenses and reinforce their army.

14. For climatic details see R. Alery, "The Climate of France, Belgium, the Netherlands, and Luxembourg," in *Climates of Northern and Western Europe,* ed. Wallen, 135–93.

15. The chalk most likely formed during the Mid-Tertiary, making it less than 40 million years old. The overlying sediments are Upper Tertiary and Quaternary in age with the youngest being deposited during the Holocene, which is postglacial. For a more detailed description of Flanders' bedrock geology and the World War I combat there, see Douglas W. Johnson's *Battlefields of the World War: Western and Southern Fronts: A Study in Military Geography,* American Geographical Society Research Series 3 (New York: Oxford University Press, 1921).

Chapter 3. Clouds and Fog

1. Gen. Dwight D. Eisenhower, *Crusade in Europe* (Garden City, N.Y.: Doubleday, 1948), 240.

2. Arthur Conan Doyle, *The Hound of the Baskervilles* (New York: McClure, Phillips, 1902).

3. John F. Fuller, "Weather and War" (Scott Air Force Base, Ill.: Office of MAC History, Military Airlift Command, U.S. Air Force, 1974), 1.

4. Charles B. MacDonald, *A Time for Trumpets* (New York: Morrow, 1985).

5. Hugh M. Cole, *The Ardennes: Battle of the Bulge. The United States Army in World War II* (Washington, D.C.: Office of the Chief of Military History, 1965), 23.

6. Peter Elstob, *Hitler's Last Offensive* (New York: Macmillan, 1971), 34.

7. Cole, *The Ardennes,* 46.

8. MacDonald, *Time for Trumpets,* 26; John Pimlott, *Battle of the Bulge* (Englewood Cliffs, N.J.: Prentice-Hall, 1983), 12.

9. MacDonald, *Time for Trumpets,* 23, 35.

10. Cole, *The Ardennes,* 22.

11. Wilfred G. Kendrew, *The Climates of the Continents* (London: Oxford University Press, 1961), 308.

12. Ibid., 302.

13. MacDonald, *Time for Trumpets,* 28.

14. Marvin D. Kays, "Weather Effects during the Battle of the Bulge and the Normandy Invasion" (White Sands, N.M.: Atmospheric Science Laboratory, 1982).

15. Cole, *The Ardennes,* 75–86.

16. Elstob, *Hitler's Last Offensive,* 46.

17. Hasso von Manteuffel, "The Battle of the Ardennes," in *Decisive Battles of World War II,* ed. H. A. Jacobsen and J. Rohwer (New York: Putnam, 1965), 394.

18. Ibid., 395.

19. Kays, "Weather Effects," 10.

20. Cole, *The Ardennes,* 137.

21. Ibid., 213.

22. Ibid., 669.

23. Ibid., 84.

24. Ibid., 115–16.

25. Kays, "Weather Effects," 19.

26. Elstob, *Hitler's Last Offensive,* 40.

27. Manteuffel, "Battle of the Ardennes," 409.

28. Kays, "Weather Effects," 20.

29. Dave Richard Palmer, *Summons of the Trumpet* (New York: Random House, 1978), 215.

30. Lt. Gen. Willard Pearson, *The War in the Northern Provinces: 1966–1968* (Washington, D.C.: Department of the Army, 1975), 17.

31. Bernard C. Nalty, *Air Power and the Fight for Khe Sanh* (Washington, D.C.: Office of Air Force History), 8.

32. G. W. Ritter, "Climate and Visibility in Southeast Asia," Naval Missile Center Tech Memo TM-67-28 (Point Mugu, California: Naval Missile Center, 1967), 5.

33. John F. Fuller, "Air Weather Service Support to the United States Army: TET and the Decade After," AWS Historical Study No. 8 (Scott Air Force Base, Ill.: Office of MAC History, Military Airlift Command, 1979), 14.

34. Fuller, "Weather and War," 13.

35. Robert Pisor, *End of the Line: The Siege of Khe Sanh* (New York: Norton, 1982), 206.

36. Ibid.

37. Fuller, "Air Weather Service Support," 23.

38. William C. Westmoreland, *A Soldier Reports* (New York: Doubleday, 1976), 443.

39. Capt. Moyers S. Shore, *The Battle for Khe Sanh* (Washington, D.C.: History and Museums Division, Headquarters, U.S. Marine Corps, 1979), 29.

40. Nalty, *Air Power and the Fight for Khe Sanh,* 14.

41. Westmoreland, *A Soldier Reports,* 442.

42. Nalty, *Air Power and the Fight for Khe Sanh,* 23.

43. Shore, *Battle for Khe Sanh,* 53. For an analysis focused on TET and the fighting during 1968 see Ronald H. Spector's *After TET: The Bloodiest Year in Vietnam* (New York: Fress Press, 1993).

44. Pisor, *End of the Line,* 167.

45. Ray L. Bowers, *The United States Air Force in Southeast Asia: TAC Airlift* (Washington, D.C.: Office of Air Force History, 1983), 299.

46. Nalty, *Air Power and the Fight for Khe Sanh,* 58.

47. John Morrocco, *Thunder from Above* (Boston: Boston Publishing, 1984), 179.

48. Ibid., 178.

49. Fuller, "Air Weather Service Support," 23.

50. Shore, *Battle for Khe Sanh,* 72.

51. Westmoreland, *A Soldier Reports,* 421.

52. Fuller, "Air Weather Service Support," 24.

53. Lt. Gen. John J. Tolson, *Airmobility Studies: 1961–1971* (Washington, D.C.: Department of the Army, 1973), 179–80.

54. John F. Fuller, "Weather and War" (Scott Air Force Base, Ill.: Office of MAC History, Military Airlift Command, 1974), 13.

55. Tolson, *Airmobility Studies,* 192.

Chapter 4. Invading Another Climate as Seasons Change

Epigraph: Count Philippe-Paul de Segur, *Napoleon's Russian Campaign,* trans. J. David Townsend (Boston: Houghton Mifflin, 1958), 19.

1. The most common climate classifications in use today are modifications of a system derived by Vladimir Koppen, a nineteenth-century German climatologist and plant geographer. Koppen based his system on average temperature and moisture requirements of major vegetative regimes. For example, humid tropical climates require ample precipitation year-round and average temperatures every month above 18°C (64.4°F) in order to support lush growth of broadleaf evergreen forests and palms. Climates and vegetation in the former Soviet Union vary widely. Subarctic regimes of the far north have only brief periods in summer above freezing and support short-grass tundra, whereas farther south vast forests of needle-leaf evergreen and broadleaf deciduous trees thrive in the longer, wetter summers. Still farther south and away from major water bodies, forests give way to grasslands, and then desert, as dry conditions begin to dominate.

2. For a good text on Soviet geography see Paul E. Lydolph, *Geography of the U.S.S.R.* (Elkhart Lake, Wis.: Misty Valley Publishing, 1979, 1984).

3. Under the Koppen climate-classification system, most of the former European USSR is considered Humid Continental (temperatures of the coldest month average below freezing, and temperatures of the warmest month above 10°C [50°F]). West of a line running near the Vistula River, most of Central and Western Europe is classified as Marine West Coast (temperatures of the coldest month average above freezing and the warmest month above 10°C [50°F]). The transition is gradual, as areas to the east are farther removed from the moderating effects on temperature of the Atlantic Ocean.

The dividing line near the Vistula River does not mark a sudden change in climate, but it serves as a transitional indicator of conditions farther east.

4. Being solid, land has no ability to mix surface and subsurface material. Land is also a poor conductor and a good insulator. Solid earth has a low specific heat (the amount of heat energy [calories] needed to raise the temperature of 1 gram of a substance 1 degree centigrade) so that when the sun's rays strike, the land heats quickly but warms only to shallow depths. Furthermore, heat is not stored well and is instead released quickly into the atmosphere permitting land to cool rapidly in the absence of solar radiation. In contrast, water mixes well and has a high specific heat. It takes more energy to warm the water, but it can hold the heat longer, store it in greater quantities at greater depth, and circulate it from one place to another.

5. Under normal conditions, air temperature decreases as altitude increases. The reverse is sometimes true when land cools rapidly (especially during long winter nights in Siberia), causing the air temperature immediately above the surface to fall and create a low-level inversion or thermal high-pressure system. The colder surface air is more dense, or heavier, than the air above and therefore cannot rise. High pressure generally produces clear, stable weather and little precipitation. The Siberian high is the most intense high-pressure system on earth. At the height of winter, air in valleys is extremely cold and still, so that ice crystals formed from the breath of animals and humans may appear to hang in the air. Higher up, above the inversion, air may have greater movement and clouds may form. Because surface winds blow from areas of high pressure toward areas of low pressure, the effects of the Siberian high radiate outward over much of the continent in the form of cold, dry weather.

6. Data for Maps 4.3 through 4.6 comes from U.S. Air Force weather service, *Worldwide Airfield Climatic Data,* Defense Technical Information Center, Cameron Station, Va., April 1973; Central Intelligence Agency, *National Intelligence Survey, U.S.S.R.,* Washington, D.C., December 1952; Paul Lydolph, *Climates of the Soviet Union,* vol. 7 of *World Survey of Climatology,* ed. C. C. Wallen (New York: Elsevier, 1977); C. C. Wallen, ed., *Climates of Central and Southern Europe,* vol. 6 of *World Survey of Climatology* (New York: Elsevier, 1977); W. G. Kendrew, *The Climates of the Continents,* 4th ed. (New York: Oxford University Press, 1953).

7. Department of the Army, "Effects of Climate on Combat in European Russia," Department of the Army Pamphlet 20-291 (Washington, D.C.: U.S. Army, February 1952), 29.

8. Due to greater amounts of snow and less severe temperatures, snow-cover density (measured in grams per cubic meter [g/cm^3]) in the European Russia is higher than anywhere in the country, except parts of the Arctic and Pacific coasts where high winds correlate with high snow-cover density. Density increases from November through March and is 8–14 percent less in forested areas than in exposed fields. Dense snow cover is difficult to remove from roads and airfields, and any snow, when packed by traffic into ice, provides poor traction. Additionally, high snow-cover density increases the insulation value of the snow, so that the ground below freezes only to shallow depths or does not freeze at all. While this effect benefits agriculture, heavy vehicular or other traffic breaking through the snow immediately encounters saturated unfrozen ground (mud) to add to other difficulties. See Michael A. Billelo, *Regional and Seasonal Variation in Snow-Cover Density in the U.S.S.R.,* Cold Regions Research and Engineering Laboratories (CRREL) Report 84-22 (Hanover, N.H.: U.S. Army Corps of Engineers, CRREL, August 1984).

9. U.S. Army Corps of Engineers, CRREL, *Notes for the Winter Battlefield,* Field Manual 31-71 (Hanover, N.H.: CRREL, January 1984), 14.

10. Robert G. Jensen, Theodore Shabad, and Arthur W. Wright, *Soviet Natural Resources in the World Economy* (Chicago: University of Chicago Press, 1983), 21.

11. Carl von Clausewitz, *The Campaign of 1812 in Russia* (Westport, Conn.: Greenwood Press, 1977), 5.

12. Martin Van Creveld, *Supplying War: Logistics from Wallenstein to Patton* (Cambridge: Cambridge University Press, 1977), 62–64.

13. Ibid., 63.

14. Von Clausewitz, *Campaign of 1812 in Russia,* 60.

15. De Segur, *Napoleon's Russian Campaign,* 26.

16. Ibid., 87.

17. Ibid., 161 and 175.

18. De Segur, *Napoleon's Russian Campaign,* 149–54; von Clausewitz, *Campaign of 1812 in Russia,* 199–200.

19. Von Clausewitz, *Campaign of 1812 in Russia,* 34.

20. De Segur, *Napoleon's Russian Campaign,* 242.

21. James Lucas, *War on the Eastern Front, 1941–45: The German Soldier in Russia* (New York: Bonanza Books, 1982).

22. "The German Campaign in Russia, Planning and Operations (1940–1942)," in *World War II Military Studies: A Collection of 213 Special Reports on the Second World War Prepared by Former Officers of the Wehrmacht for the United States Army,* ed. Donald S. Detwiler, 24 vols. (New York: Garland, 1979), 15–16.

23. Field Marshal Erich von Manstein, *Lost Victories,* trans. and ed. Anthony G. Powell (Novato, Calif.: Presidio Press, 1982), 177.

24. Alan Clark, *Operation Barbarossa: The Russian-German Conflict, 1941–45* (New York: Morrow, 1985), 188.

25. For a thorough examination of the manner and extent of improvisations the Wehrmacht made on the Eastern Front, see Detwiler, ed., *World War II Military Studies,* vol. 18, and Department of the Army, "Military Improvisations during the Russian Campaign," DA Pamphlet 20-201 (Washington, D.C.: U.S. Army, August 1951).

26. Department of the Army, "Effects of Climate on Combat in European Russia," 48.

27. Ibid., 43–60.

28. Hans von Greiffenberg, "Battle of Moscow," Historical Division, Headquarters U.S. Army Europe, reprinted in *World War II German Military Studies,* ed. Detwiler, 16:16–19. General Greiffenberg detailed problems of weather and roads in this post–World War II report written for the U.S. Army.

29. Department of the Army, "Effects of Climate on Combat in European Russia," 29–38; von Greiffenberg, "Battle of Moscow," 49–60.

30. Department of the Army, "Effects of Climate on Combat in European Russia," 31.

31. For a comprehensive and detailed history of the Soviet drive to Berlin see John Erickson, *The Road to Berlin* (Boulder, Colo.: Westview Press, 1983).

32. Michael C. Deprie, U.S. Army, *Moscow: The Principle of the Objective* (Fort Leavenworth, Kan.: U.S. Army Command and General Staff College, 1977), 84.

33. Von Greiffenberg, "Battle of Moscow," 76–77.

34. Department of the Army, "Effects of Climate on Combat in European Russia," 5–7.

35. Ibid., 6.

36. U.S. Army Corps of Engineers, CRREL, *Notes for the Winter Battlefield*, 5.

37. For a concise summary of how the Dnieper River ice railway bridge was constructed, see Lucas, *War on the Eastern Front*, 93–104. The bridge served its purpose; it permitted resupply of German forces east of the river for several weeks before warm weather intervened.

38. Lucas, *War on the Eastern Front*, 12.

39. Department of the Army, "Effects of Climate on Combat in European Russia," 79.

Chapter 5. Forests and Jungles

Epigraph: Morris Schaff, *The Battle of the Wilderness* (Boston: Houghton Mifflin, 1910), 107–8.

1. The term *biomass,* which is often used to describe the richness of vegetation, is the dry weight of living organic matter (plant and animal) within a given surface area and is generally measured in grams per square meter.

2. Vegetation may be classified by its water needs. Xerophytes are adapted to a very limited water supply, either seasonal or perennial. Hydrophytes live in water or saturated soil. Tropophytes have adjusted to adversities (drought or cold) by seasonal changes in their structures (e.g., deciduous plants). Vegetation in areas with more than adequate water supply are classed as hygrophytes. Mesophytes require a moderate amount of moisture to sustain life.

3. Care should be taken in evaluating the original nature of battlefields in the Wilderness. Although efforts are being made to maintain parts of the Wilderness in the form that existed during the Civil War, more than 100 years of vegetative growth and a variety of human impacts have considerably altered the landscape.

4. For descriptions of the Mud March see Bruce Catton, *Glory Road: The Bloody Route From Fredericksburg to Gettysburg* (New York: Doubleday, 1952), chap. 3, and Edward J. Stackpole, *The Fredericksburg Campaign* (Harrisburg, Pa.: Military Services Publication Co., 1957).

5. See Matthew Forney Steele, "American Campaigns," War Department Document (Washington, D.C.: GPO, 1909).

6. Abner Doubleday, *Chancellorsville and Gettysburg* (New York: Scribner's, 1882), 32.

7. Andrew A. Humphreys, *The Virginia Campaigns of '64 and '65* (New York: Scribner's, 1883), 44.

8. Doubleday, *Chancellorsville and Gettysburg*, 12.

9. Humphreys, *Virginia Campaigns of '64 and '65*, 44.

10. For readings on 1864 fighting in the Wilderness representative of both the North and the South see ibid., and Douglas S. Freeman's *Lee's Lieutenants: A Study in Command*, 3 vols. (New York: Scribner's, 1942–44).

11. This account of November 1965 military operations in the Ia Drang Valley is drawn from both reports written in the period immediately after the engagement and those developed in the 30 years since the battle. One of the most detailed accounts is provided in *We Were Soldiers Once . . . and Young* by H. G. Moore, a battalion commander in the battle, and journalist J. L. Galloway (New York: Random House, 1992).

12. "Once We Will Win," in Vo Nguyen Giap, *The Military Art of Peoples War,* ed. Russell Stetler (New York: Monthly Review Press, 1970), 259.

13. Lt. Gen. Harry W. O. Kinnard, "Victory in the Ia Drang: The Triumph of a Concept," *Army* (September 1967).

14. Ibid., 84.

15. Lt. Gen. Harry W. O. Kinnard, as quoted in George C. Herring, "The 1st Cavalry and the Ia Drang Valley, 18 October–24 November 1965," in *America's First Battles: 1776–1965*, ed. Charles E. Heller and William A. Stofft (Lawrence: University of Kansas Press, 1986), and Dave R. Palmer, *Summons of the Trumpet* (San Raphael, Calif.: Presidio Press, 1978), 93.

16. Harry G. Summers, "The Bitter Triumph of Ia Drang," *American Heritage* 35 (Feb.–Mar. 1984): 58.

Chapter 6. Terrains and Corridors

1. The foundation of a continent is a stable region known as the craton. Deep erosion of the craton can result in continental uplift through isostasy. This process permits large land masses to persist above sea level for very long periods of geologic time.

2. It is generally believed that this movement is produced by forces internal to the earth. For a review of the earth's dynamics, including behavior of the crust, see *Scientific American* 249, no. 3 (September 1983): 114–42.

3. Broad crustal uplift and subsidence are referred to as epeirogenic activity, while intense compressive crustal folding and faulting are orogenic in nature. Both can be geomorphically important, as illustrated by the Appalachian terrain, which formed in the last 10 million years through epeirogenic uplift, resulting in erosion that exposed ancient orogenic structures that had formed hundreds of millions of years previously.

4. This process of convergence gradually closed an ancient sea named Iapetus. Subsequent separation recreated its present counterpart, the Atlantic Ocean.

5. Pangaea, meaning "all lands," was largely formed by 300 million years ago and began to separate into a number of continents and islands about 180 million years before the present.

6. Two major themes are generally involved in the explanation of the earth's outermost crust and the configuration of its surface. The first, geology, is concerned with the planet's history as revealed by rocks and their structures. The second, geomorphology, is concerned with the origin and development of surface features. When one studies the Appalachians, it becomes clear that the geology reveals a vast record of conditions and events that date back more than a billion years. In contrast, most geomorphic development of the area is recent in terms of geologic time. Thus, it is important not to confuse the age of the rocks and their structures with the evolution of the area's geomorphology. Such is true for practically all erosional landscapes worldwide. Therefore, both geologic and geomorphic references are cited here. For a brief but well-written summary of Appalachian geology, see Philip B. King, *The Evolution of North America* (Princeton, N.J.: Princeton University Press, 1977), 42–54. For regional geomorphology of the area, see Nevin M. Fenneman, *Physiography of the Eastern United States* (New York: McGraw-Hill, 1938), 121–342; and William D. Thornbury, *Regional Geomorphology of the United States* (New York: Wiley, 1965), 72–151. Systematic geomorphology texts that describe processes affecting the Appalachians include Arthur L. Bloom's *Geomorphology,* 2nd ed. (Englewood Cliffs, N.J.: Prentice-Hall, 1991); William D. Thornbury's *Principles of Geomorphology,* 2nd ed. (New York: Wiley, 1969); and *Process Geomorphology* by Dale F. Ritter, R. Craig Kochel, and Jerry R. Miller (Dubuque, Iowa: Wm. C. Brown, 1995).

7. Thornbury, *Regional Geomorphology of the United States,* 100–108; Fenneman, *Physiography of the Eastern United States,* 163–94.

8. Thornbury, *Regional Geomorphology of the United States,* 109–29; Fenneman, *Physiography of the Eastern United States,* 195–278.

9. Thornbury, *Regional Geomorphology of the United States,* 130–51; Fenneman, *Physiography of the Eastern United States,* 279–342.

10. Thornbury, *Regional Geomorphology of the United States,* 90–92; Fenneman, *Physiography of the Eastern United States,* 145–55.

11. Such evolutionary changes are known as both stream piracy and stream capture. This process is nicely explained and illustrated by A. K. Lobeck, *Geomorphology: An Introduction to Landscapes* (New York: McGraw-Hill, 1939), 198–201.

12. Such gaps, originally carved by streams whose drainage was captured, are technically called wind gaps, even though wind had nothing to do with their formation. If still occupied by streams, such features are referred to as water gaps.

13. The oldest Coastal Plain sediments were deposited as the North American continent became involved in the breakup of Pangaea, and successively younger formations have accumulated ever since. They consist mainly of detrital material from the continent and shallow water marine sediments. Sediments of the Atlantic Coastal Plain have never been subjected to intense crustal deformation. Where elevated above sea level, the movement has been epeirogenic, thus these formations are little disturbed, generally inclined gently from land to sea as they were originally deposited.

14. Geomorphically, shorelines have been classified in many ways. One long-established approach that has wide application differentiates shore zones that are experiencing either long-term uplift or subsidence. In nonorogenic areas the first results in a smooth shore developed on the emergence of the nearly flat ocean bottom. These strands are commonly backed by coastal lagoons because of natural beach construction processes. Submergent shore zones are highly irregular because the drowning of subaerially eroded land precisely follows irregularities in the terrain. For details see Lobeck, *Geomorphology,* 447–57.

15. For some fairly recent studies see Frank E. Vandiver, *Mighty Stonewall* (New York: McGraw-Hill, 1957), 165–285; Robert G. Tanner, *Stonewall in the Valley* (New York: Doubleday, 1976); and Bryon Farwell, *Stonewall: A Biography of General Thomas J. Jackson* (New York, Norton, 1992).

16. For details, see Douglas Southall Freeman, *Lee's Lieutenants* (New York: Scribner's, 1946), 166–202.

17. Because of their differing positions and respective practices of naming combat sites after terrain features or towns, this 17 September 1862 battle was called Antietam by the North and Sharpsburg by the South. Interesting books on this battle include James V. Murfin, *The Gleam of Bayonets* (New York: Thomas Yoseloff, 1965), and Stephen W. Sears, *Landscape Turned Red: The Battle of Antietam* (New Haven, Conn.: Ticknor and Field, 1983).

18. Changes in sea level (so common during the Pleistocene Epoch) or the altitude of the land (because of degradation or tectonic activity) may have a profound effect upon stream systems. For example, even slight uplift of the land immediately provides more potential energy for a stream system to erode downward. Elegant relationships among volume, gradient, and velocity determine a stream's erosive potential. Generally this factor is greatest in the larger rivers and decreases progressively to the smallest tributaries. Thus, down-cutting and valley development tend to be proportional to

stream size. See Ritter, Kochel, and Miller, *Process Geomorphology,* 169–307; and Marie Morisawa, *Streams* (New York: McGraw-Hill, 1968).

19. Edward J. Stackpole, *The Fredericksburg Campaign* (New York: Bonanza Books, 1957), 64–238.

20. John Bigelow, *The Campaign at Chancellorsville* (New Haven: Yale University Press, 1910).

21. For additional geologic details, see Andrew Brown, "Geology and the Gettysburg Campaign," *Geotimes* 6, no. 1 (1961): 8–12, 40–41.

22. George R. Stewart, *Pickett's Charge* (Boston: Houghton Mifflin, 1959), 172–73.

23. For descriptions of the Wilderness (chapter 5) and analysis of Grant's leadership, see J.F.C. Fuller, *The Generalship of Ulysses S. Grant* (Bloomington: Indiana University Press, 1958).

24. For a summary of the structural geology of Europe, see Jean Aubouin, "Les Grands Ensembles Structuraux de l'Europe," in *Geologie des Pays Europeans: France, Belgique, Luxembourg* (Paris: 26e Congress Geologique International, 1980), xxv–xxxiv.

25. For general information on the rocks and structure of the Paris Basin, see ibid., 433–74. A dated but useful detailed description of the geology and topography of the Verdun sector is presented by Douglas W. Johnson in *Battlefields of the World War: Western and Southern Fronts: A Study in Military Geography,* American Geographical Society Research Series 3 (New York: Oxford University Press, 1921), 316–414. Also, note that the Paris Basin is of epeirogenic origin in contrast to the orogenic formation of the major Appalachian structures.

26. Johnson, *Battlefields of the World War;* see discussion of the Somme, Marne, and Lorraine as well as Verdun.

27. Ibid., 349–66.

28. Henri Philippe Petain, *Verdun* (New York: Dial Press, 1930); Alistair Horne, *The Price of Glory* (New York: St. Martin's Press, 1962). These two references are interesting because the first is by the French general who was central to action and contains several foldout maps and two panoramic sketches that reveal much about the military geography of the area. The second is a prize-winning account of the battle but contains very few maps and diagrams.

29. Johnson, *Battlefields of the World War,* 349–66; *Verdun: Vision and Comprehension—The Battlefield and Its Surroundings,* 4th ed. (Drancy, France: Editions Mage, 1985), 58–78. After their defeat in the Franco-Prussian War (1870–71), the French improved their defenses on the German frontier. A group of fortresses on the cuesta near Verdun was central to this plan. Thought to be invulnerable prior to World War I, after the capture of similar installations elsewhere in 1914, the forts near Verdun were stripped of their large guns and lightly manned in favor of natural strongholds and trenches. Even so, battles at sites such as Fort Douaumont and Fort Vaux, among others, were incredibly hard-fought and both were lost and recaptured by the French in 1916.

30. Accounts of battle often convey vivid aspects of geography even though the author may not be an expert in that field. For subtle yet powerful impressions of terrain written by an American author who was a soldier in France during World War I, read Hervey Allen's *Toward The Flame* (repr., Pittsburgh, Pa.: University of Pittsburgh Press, 1968), which traces his movements from arrival in Europe to especially intense fighting at the village of Fismette in 1918. Also his two stories in *It Was Like This* (New York: Farrar and Rinehart, 1936) indirectly reveal differing environmental sensitivity, as the

first focuses on the combat experiences of a responsible and bright lieutenant who seeks to report to his superior while the second traces an evolving mindset of violence experienced by a young corporal "of rather limited mental ability" (9). When read from a geographic point of view, both books convey the harshness, danger, and chance of the battlefield in an especially effective fashion.

Chapter 7. Troubled Waters

1. Sun Tzu, *The Art of War*, trans. with an introduction by Samuel B. Griffith (New York: Oxford University Press, 1963), 116.

2. The chemical and physical processes are respectively referred to as corrosion and corrasion. Traction involves movement of material in the stream bed, referred to as bedload. All stream-deposited sediment, regardless of size, is alluvium.

3. River velocity tends to vary both from surface to bottom and side to side. In a straight stretch of river, the velocity is lowest in shallow water along both banks and immediately adjacent to the stream's wetted perimeter. Velocity tends to be greatest toward the lower center of the flow. In meanders the velocity is lowest on the inside of the curve—hence the deposition of point bar sediments—and highest on the outside where bank erosion is most likely. Average velocity will increase with an increase in both gradient and volume. Thus, even though a river's gradient remains the same, the velocity of the water will increase during times of high water, a condition of considerable concern in planning for the crossing of the Rhine River in 1945.

4. Some rivers that originate in mountainous areas extend through arid flatlands as they approach the sea. Such streams are commonly subjected to periodic flooding from added rainfall. Examples are the Nile, which begins in the hills of Sudan and Ethiopia and flows north through Egypt, and the Tigris-Euphrates system, which rises in Turkey and Syria and trends southward to the Persian Gulf. While both of these streams are permanent, flowing year-round, many others in arid regions are intermittent or ephemeral, meaning that they carry water seasonally or only during and after heavy rainfall.

5. U.S. Army Engineer Topographic Laboratories, *Synthesis Guide for River Crossings*.

6. Alistair Horne, *To Lose a Battle* (New York: Little, Brown, 1969).

7. The down-faulted wedge, in this case less resistant than adjacent rocks, is eroded more rapidly to create a wide, steep-boundaried lowland characterized today by an extensive floodplain and meandering river course.

8. "Market" refers to the airborne part of the operation, "Garden" was the codename for the concurrent ground maneuvers. This description is not intended to doubt or defend the decision to conduct the operation or evaluate the relationship between Montgomery and Eisenhower. Those judgments are left to the reader.

9. Browning was Deputy Commander, First Allied Airborne Army. Cornelius Ryan, *A Bridge Too Far* (New York: Simon and Schuster, 1974), 9.

10. Brian Horrocks, *Corps Commander* (New York: Scribner's, 1977).

11. Bernard Montgomery, *El Alamein to the River Sangro: Normandy to the Baltic* (New York: St. Martin's Press, 1974), 321.

12. Maurice Tugwell, *Arnhem: A Case Study* (London: Thorton Cox, 1975), 11.

13. U.S. Headquarters, 82nd Airborne Division, Field Order No. 11, Annex 1b, 13 September 1944.

14. Sir John Hackett, interview with Gerald E. Galloway, West Point, N.Y., 24 October 1984.

15. Horrocks, *Corps Commander,* 99.

16. Cyril Falls, "Geography and War Strategy," *Geographical Journal of the Royal Geographical Society* 112 (July–September 1948): 1–3; and British Airborne Corps, "Allied Airborne Operations in Holland, After-Action Report," n.d. (c. February 1945), 2.

17. U.S. Army Corps of Engineers, European Theater of Operations, "Historical Report No. 20," August 1945, 2.

18. Walter Bedell Smith, "Eisenhower's Six Great Decisions: Victory West of the Rhine," *Saturday Evening Post,* 29 June 1946, 40.

19. For an account of the decision by Brig. Gen. William Hoge to capture the bridge rather than follow existing orders, see Bruce Alsup, "The Chance Seizure of the Remagen Bridge Over the Rhine," in *Combined Arms in Battle since 1939,* ed. Roger J. Spiller (Fort Leavenworth, Kan.: U.S. Army Command and General Staff College Press, 1992).

20. There is a great deal of speculation that vibrations from exploding artillery shells and anti-swimmer mines, for example, caused the collapse. It is remarkable that it did not occur sooner.

21. General Montgomery's 21st Group was actually scheduled to lead the series of crossings on 23 March. When Montgomery awoke that day, he was told that Patton's Third Army had taken advantage of an opportunity and had crossed the Rhine at Oppenheim the night before, almost unopposed.

22. The velocity in this section of the Rhine was very swift and boats, even when not under fire, had difficulty maneuvering.

23. Albert von Kesselring, "Interrogation of Genlfldm Albert Kesselring," *ENTHIN 70 in World War II, German Military Studies,* vol. 3, Historical Division, U.S. Army Europe, 24 July 1985.

24. Smith, "Eisenhower's Six Great Decisions: Victory West of the Rhine," 47.

Chapter 8. Glaciers Shape the Land

1. Proposed explanations for the glacial and nonglacial episodes of the Pleistocene include the Milankovitch Theory, involving changes among three Earth-Sun relationships (precession of the equinoxes over a 21,000-year period, slight variations in obliquity with the plane of the ecliptic over a 40,400-year period, and the eccentricity of the earth's orbit with a periodicity of 91,800 years); changes in the amount of carbon dioxide and volcanic dust in the atmosphere that may increase absorption and reflection, respectively, of solar energy; differences in solar emission; and changes in oceanic circulation with special reference to water from the tropics gaining access to the Arctic Ocean. For elaboration, see Tage Nilsson, *The Pleistocene: Geology and Life in the Quaternary Ice Age* (Dordrecht, Holland: D. Reidel, 1983), 516–25.

2. Albedo is the percentage of available solar radiant energy reflected to space from Earth. Twenty percent albedo means that 20 percent of insolation from the sun is reflected. The lower the albedo, the higher the rate of energy absorption and surface heating. The albedo for forests and grasslands tends to be a relatively low 5 to 25 percent. But for snow and ice it may be very high, ranging from 45 to 85 percent which tends to perpetuate their existence.

3. For a detailed description of the Pleistocene Epoch, see Richard F. Flint, *Glacial and Quaternary Geology* (New York: Wiley and Sons, 1964), 1–266, 463–676.

4. Glacial ice is actually a metamorphic rock derived from snow, rain, and refrozen meltwater. New snow, with a specific gravity near 0.05 and porosity as high as 95 percent, gradually recrystallizes to form a granular material with a porosity of about 50

percent and 0.5 specific gravity. Continued compaction removes air, and when specific gravity exceeds 0.8, permeability becomes zero, and by definition the material is now ice. Eventually the specific gravity may reach about 0.9; at this point internal stress can produce deformation, creating a glacier.

5. The rate of glacial movement may vary from hardly measurable to possibly more than 100 m (300 ft) per day in a surging glacier. Most glaciers, however, have center line average velocities of about 0.3 to 0.7 m (1 to 2 ft) per day. The rate of movement is related to many factors, with the supply of new snow, the slope of the glacier's surface, frozen or unfrozen base, and ice temperature being of major importance.

6. *Drift* is a general name for all glacial deposits. The term originated in England as a prescientific biblical explanation, related to the flood of Noah, for the origin of unconsolidated material beneath the surface but overlying solid rock. *Till* is the name given to all drift deposited more or less directly from the ice; it consists of an unstratified, heterogeneous mixture of clasts ranging from clay-sized particles to large boulders. Glaciofluvial sediments are stratified deposits of sand and gravel, glacially derived, but deposited by streams flowing from the glacier. Glaciolacustrine sediments are generally thinly layered, clay and silt-sized clasts carried by streams and deposited in quiet water. Loess is windblown dust, largely of the silt-sized fraction. For detailed discussions, see Flint, *Glacial and Quaternary Geology,* 147–97, 243–66, and David E. Sugden and Brian S. John, *Glaciers and Landscape* (New York: Wiley and Sons, 1976), 213–336. Each of these sediment types may be associated with certain topographic forms, but names of landforms and sediment types should not be interchanged (see n. 7).

7. In a general way, glacial landforms and sediment types (see Figure 8.1 and n. 6) can be related meaningfully. Features composed partly, if not wholly, of till include: the end moraine (also called terminal and recessional moraines), which forms at the margin of an active glacier holding a constant position; ground moraine (or till plain), which is a low, rolling landscape uncovered with the retreat of an ice margin; and often the drumlin, which is a streamlined and elongated hill formed subglacially either by glacial molding of previously deposited drift or the plastering on of basal drift.

Glaciofluvial sediments form outwash plains, smooth but gently sloping surfaces extending outward from the glacial margin. If isolated masses of ice are buried by these sediments, the surface of the outwash plain will become pitted as the glacial remnants melt. Glaciofluvial deposits may also be deposited within and along preexisting valleys to form valley trains. And more localized features, including kames (conical hills) and eskers (winding, low ridges), are composed of glaciofluvial sediments deposited, respectively, in glacial cavities and by streams flowing upon, within, or beneath the glacier.

Glaciolacustrine sediments, which consist largely of clay and silt that settle out of still or slow-moving water, tend to level lake bottoms. If such a lake is drained during or after deglaciation, a nearly flat lacustrine plain emerges as a topographic feature and its border (former shore zone) may be marked by an ancestral beach ridge or wave-cut bluff. The technical name for dust deposited from wind is loess. For detailed discussions see Flint, *Glacial and Quaternary Geology,* 198–226, and Sugden and John, *Glaciers and Landscape,* 213–336.

8. See chapter 6 in this text for a brief review of possible plate movements.

9. For a review of the formation of the Alps and the Tethys Sea, see Rhodes W. Fairbridge, *The Encyclopedia of Geomorphology* (New York: Reinhold Book Corp., 1968), 752–58.

10. For a summary of the multiple glaciation of the Alps, see Flint, *Glacial and Quaternary Geology,* 606–7, 643–48, and Nilsson, *The Pleistocene,* 66–80.

11. For descriptions with some interesting illustrations, see E. Alexander Powell, *Italy at War and the Allies in the West* (New York: Scribner's, 1917), 59–90; Martin Hardie and Warner Allen, *Our Italian Front* (London: A. and C. Black, 1920), 58–104; Douglas W. Johnson, *Battlefields of the World War: Western and Southern Fronts: A Study in Military Geography,* American Geographical Society Research Series 3 (New York: Oxford University Press, 1921), 488–540; and Luigi Villare, *The War on the Italian Front* (London: Cobdere Sanderson, 1932).

12. Two Interesting but similar books on this topic are Lt. Col. W.G.F. Jackson, *Seven Roads to Moscow* (London: Eyre & Spottiswoude, 1957), and Leonard Cooper, *Many Roads to Moscow: Three Historic Invasions* (New York: Coward-McCann, 1968).

13. Flint, *Glacial and Quaternary Geology,* 592–601, 628–29, discusses the Scandinavian Ice Sheet and its predecessors in Europe, including the Dnieper and Don River glacial lobes.

14. For a general description of the Pripet Marshes and polesyes, see L. S. Berg, *Natural Regions of the Soviet Union* (New York: Macmillan, 1950), 52–53.

15. For a review of late glacial effects on Russia see A. A. Velichko, ed., *Late Quaternary Environments of the Soviet Union* (Minneapolis: University of Minnesota Press, 1984).

16. For general geographic descriptions of the northwest European section of the former Soviet Union, see Berg, *Natural Regions of the Soviet Union,* 22–59; Paul E. Lydolph, *Geography of the U.S.S.R.: Topical Analysis* (Elkhart Lake, Wis.: Misty Valley Publishing, 1979), 31–38; Leslie Symons, *The Soviet Union: A Systematic Geography* (London: Hodder and Stoughton, 1983), 21–27.

17. See Philippe-Paul de Segur, *Napoleon's Russian Campaign,* trans. J. David Townsend (New York: Time Inc., 1965); Alan Palmer, *Napoleon in Russia* (London: Andre Deutsch, 1967); Eugene Tarle, *Napoleon's Invasion of Russia, 1812* (New York: Oxford University Press, 1942); and Gen. Carl von Clausewitz, *The Campaign of 1812 in Russia* (Westport, Conn.: Greenwood Press, 1977). Also, see chapter 4 for descriptions of the weather and climate in this campaign.

18. For details on this battle see E. R. Holmes, *Borodino, 1812* (London: Charles Knight & Co., 1971), among many others.

19. Three among very many accounts of these operations are Paul Carell, *Hitler's War on Russia: The Story of the German Defeat in the East,* trans. Ewald Osers (London: George G. Harrap and Co., 1964); Albert Seaton, *The Russo-German War, 1941–45* (New York: Praeger, 1971); and Alexander Werth, *Russia at War, 1941–45* (New York: E. P. Dutton, 1964).

Chapter 9. Peninsulas and Sea Coasts

1. B. H. Liddell-Hart, *The Real War* (Boston: Little, Brown, 1930), 130.

2. Paul D. Komar, *Beach Processes and Sedimentation* (Englewood Cliffs, N.J.: Prentice-Hall, 1976), 123.

3. The period from about 1550 to 1850 is known as the Little Ice Age. Glaciers expanded, Norse colonies in Greenland were extinguished, and famine and sickness ravaged Europe where the climate was wetter and the temperature was about 5°C cooler than it is today.

4. Wave period is the time it takes for two successive wave crests to pass a fixed point. Typical wave periods of wind waves are 6 to 14 seconds.

5. Fred Sheehan, *Anzio: Epic of Bravery* (Norman: University of Oklahoma Press, 1964), vii.

6. Eric Linklater, *The Campaign in Italy* (London: His Majesty's Stationery Office, 1951), 62.

7. Teddy D. Bitner, "Kesselring: An Analysis of the German Commander at Anzio," master's thesis, Army Command and General Staff College, Fort Leavenworth, Kan., 1983, 27.

8. Linklater, *Campaign in Italy*, 30.

9. Martin Blumenson, *Salerno to Cassino: The Mediterranean Theater of Operations* (Washington, D.C.: Office of the Chief of Military History, U.S. Army, 1969), 194.

10. Sheehan, *Anzio*, 9.

11. Blumenson, *Mediterranean Theater of Operations*, 266.

12. Sheehan, *Anzio*, 6–7.

13. Samuel Elliot Morison, *History of the United States Naval Operations in World War II* (Boston: Little, Brown, 1960), 9:338.

14. Ibid., 339.

15. Ibid.

16. Blumenson, *Mediterranean Theater of Operations*, 354.

17. Sheehan, *Anzio*, 58.

18. A "wadi" is a normally dry and incised bed of an intermittent stream. This term is used primarily for such features in the Sahara and Middle East.

19. Bitner, "Kesselring," 52.

20. Blumenson, *Mediterranean Theater of Operations*, 362.

21. Mark W. Clark, *Calculated Risk* (New York: Harper, 1951), 290.

22. Winston S. Churchill, *Closing the Ring*, vol. 5 of *The Second World War* (Boston: Houghton Mifflin, 1951), 488.

23. Sheehan, *Anzio*, 27.

24. Ibid., 144.

25. Lt. Col. Chester G. Starr, ed., *From Salerno to the Alps: A History of the Fifth Army, 1943–1945* (Washington, D.C.: Infantry Journal Press, 1948).

26. Roy E. Appleman, *South to the Naktong, North to the Yalu: The United States Army in the Korean War* (Washington, D.C.: Office of the Chief of Military History, 1961), 19.

27. Joseph C. Goulden, *Korea: The Untold Story of the War* (New York: McGraw-Hill, 1982), 112.

28. T. R. Fehrenbach, *This Kind of War* (Washington, D.C.: Brassey's, 1994), 106.

29. William Manchester, *American Caesar* (Boston: Little, Brown, 1978), 574.

30. Malcolm W. Cagle and Frank A. Manson, *The Sea War in Korea* (Annapolis, Md.: U.S. Naval Institute, 1957), 78.

31. Fehrenbach, *This Kind of War*, 162.

32. Ibid., 165.

33. Ibid., 166.

Chapter 10. Island Battles

1. Peter A. Isely and Philip A. Crowl, *The U.S. Marines and Amphibious War* (Princeton, N.J.: Princeton University Press, 1951), 192.

2. Drilling on Eniwetok Atoll revealed a coral accumulation of as much as 1220 m (4000 ft).

3. Francis P. Shepherd, *Submarine Geology* (New York: Harper and Row, 1973), 348.

4. James P. Kennett, *Marine Geology* (Englewood Cliffs, N.J.: Prentice-Hall, 1982), 316.

5. Island-arc and hot-spot volcanoes (plumes) form many islands that have emerged from beneath the sea due to volcanism, but most other volcanic landforms of the ocean floor remain far beneath the surface. Submerged volcanoes that rise more than 1000 m (3000 ft) above the ocean floor are called seamounts. Physically they resemble their subaerial counterparts, characterized by fairly steep slopes peaking at the top. In the Pacific most of these volcanoes lie on the same lithospheric plate (Pacific plate) with chains generally parallel and similar in age. Young chains such as the Hawaiian Islands and Australs trend northwest-southeast, while older chains such as the Emperor, Marshall, and Gilbert Islands trend north-south. Submarine mountains resembling seamounts in size and shape, first discovered by echo sounding during World War II, but with flat tops, are called guyots. Evidence indicates that the flat tops were probably produced by wave erosion when the island was at or near sea level. Many seamounts and guyots are believed to originally form near a mid-oceanic ridge and gradually submerge as the surface of the diverging plates subsides with separation.

6. The Gilbert Islands belong to a type commonly called hot-spot, or plume, volcanoes. Plate tectonicists have theorized that there are areas of high heat flow in the upper mantle of Earth. This anomalously intense heat source beneath the rigid, but drifting, crustal plate generates magma (molten rock), which is periodically injected through the crust to form a volcano. Because Earth's lithospheric plates are moving, with time the related area of volcanic activity appears to move but the migration is actually related to the direction and rate of plate migration. This process produces a linearly trending chain of volcanoes as island after island emerges from the sea. Material from this type of volcano is largely high-density basalt.

This theory was first advanced by James Dana in 1885 and later developed by J. T. Wilson in 1963. Evidence supporting the hot-spot hypothesis is provided by the Hawaiian Emperor Chain. The Hawaiian Islands lie on the Hawaiian Ridge, extending approximately 2400 km (1500 mi) from the island of Hawaii to the coral atoll of Midway. The only presently active volcanoes in the chain are on the island of Hawaii, the southernmost and largest island in the group. The volcanic islands increase in age and degree of erosion toward the northwest, with Kanai, the farthest north of the Hawaiian Islands, being the oldest (5 million years old), indicating that the Pacific plate is now moving toward the northwest at a rate of approximately 9 cm (3.5 inches) a year. The island of Hawaii is currently atop the hot spot evidenced by the active volcanoes of Kilauea, Mauna Loa, and Hualalai.

7. Henry I. Shaw Jr., Bernard C. Nalty, and Edwin T. Turnbladh, *Central Pacific Drive,* History of the U.S. Marine Corps Operations in World War II 3 (Washington, D.C.: Historical Branch, G-3 Headquarters, U.S. Marine Corps, 1966), 30.

8. Shepherd, *Submarine Geology,* 363.

9. In general, waves may form due to three primary natural causes and may be classified as either free or forced waves. A free wave is generated by the instantaneous application of a disturbing force. Once formed the wave is free to travel without influence of the initially applied force. The most common natural cause of free waves is an earthquake. When the earth beneath the sea is displaced, a wave called a tsunami is generated as the sea regains its equilibrium height. This is the fastest ocean surface wave, traveling across the open sea at speeds in excess of 800 km (500 mi) per hour. Forced waves are generated when the disturbing force is applied continuously, and the wave takes on the characteristics of the forcing functions. The tides are forced waves.

The third natural cause of waves is the wind, which generates neither entirely forced nor entirely free waves. In the storm-generating area, wind waves are forced, but as they leave this area they become free.

10. There are three types of breakers: spilling, plunging, and surging. Waves break because they become over-steepened; that is, the ratio of wave height to wave length increases until the wave becomes unstable at an approximate ratio of 1 : 7. Over-steepening occurs when the shoaling wave increases in height and shortens in wavelength. Spilling breakers gradually peak until the crest becomes unstable as foam progresses down the wave front. They are short, steep waves that are characteristically found on low-gradient beaches. Plunging breakers are waves of intermediate steepness and are more common on steeper beaches. As these waves approach the shore, the crest travels faster than the base of the wave, curling over the top and "plunging" into the surf zone. Surging breakers peak as if to break; then the base surges up the beach ahead of the wave front. The long, flat waves are found on beaches of very steep slopes. While these three wave forms have distinctive characteristics, there is in actuality a continuum of breaker types at the shore line, with one type grading into the other. Nevertheless, beaches of a particular steepness can be expected to have a characteristic breaker type. There is also a clear relationship between beach slope and sediment size. Steep beaches are associated with larger materials while more gently sloping beaches usually have a finer sediment size.

11. Capt. James R. Stockman, USMC, *The Battle for Tarawa* (Washington, D.C.: Historical Section, U.S. Marine Corps, 1947), 1.

12. Earl J. Wilson, *Betio Beachhead* (New York: Putnam, 1945), 13.

13. Development for much of Marine Corps amphibious doctrine for fighting in the Pacific is credited to Maj. Earl Ellis. Ellis foresaw the potential of island warfare if Japan were to be the enemy. For this reason he disagreed with existing doctrine, which was largely defensive, planning instead for offensive operations against Japanese-held islands (Shaw, Nalty, and Turnbladh, *Central Pacific Drive,* 4–5).

14. J. D. Hittle, "Jomini and Amphibious Thought," *Marine Corps Gazette,* 1946. Antoine H. Jomini, "Precis de l'art de la guerre, IIe partie," Paris, 1838.

15. Martin Russ, *Line of Departure: Tarawa* (Garden City, N.Y.: Doubleday, 1975), 6.

16. William Manchester, *Goodbye Darkness* (Boston: Little, Brown, 1979), 223.

17. Ibid., 230.

18. Samuel E. Morison, *History of the United States Naval Operations in World War II* (Boston: Little, Brown, 1960), 14:4.

19. Ibid., 3–5.

20. John Pethick, *An Introduction to Coastal Geomorphology* (London: Edward Arnold, 1984), 23. A slope of 1 : 100 indicates a 1 m change in the vertical with each 100-meter change in the horizontal.

21. Raymond Henri, *Iwo Jima: Springboard to Final Victory* (New York: U.S. Camera Publishing, 1945), 3.

22. Y. Horie, "Japanese Defense of Iwo Jima," *Marine Corps Gazette* (February 1952): 22.

23. Frank O. Hough, *The Island War* (Philadelphia: Lippincott, 1947), 345.

24. Whitman S. Bartley, *Iwo Jima: Amphibious Epic* (Washington, D.C.: Historical Branch, U.S. Marine Corps, 1954), 221.

Chapter 11. Hot, Wet, and Sick

1. The term "tropics" generally refers to the continually warm, humid regions between the Tropics of Cancer and Capricorn (23½° North and South latitude, respec-

tively). However, several other climates exist within this realm; these include low-latitude deserts and steppes, and a mix of cooler regimes in high-altitude mountains and plateaus.

2. The monsoon of Asia (monsoon, from the Arabic word *munsim* [meaning *season*]) is most commonly thought of as the annual, May-to-October rainy period that affects coastal India, Burma, Thailand, and Southeast Asia. Actually there are two monsoons: one involving air moving outward from Asia, which occurs during the low-sun period (Northern Hemisphere winter); and the reverse, when the elevation of the noon sun is relatively high. A rainy season tends to result wherever and whenever these winds are onshore. Thus the wet season for most areas of monsoon climate in Asia extends from about April into September but there are some exceptions such as along the north-central coast of Vietnam (see chapter 3).

3. In low areas and along the coasts the high water table supports abundant vegetation year round, regardless of the dry season.

4. For an expert and detailed account of World War II fighting in Papua New Guinea see Samuel Milner's *Victory in Papua: The United States Army in World War II* (Washington, D.C.: Office of the Chief of Military History, U.S. Army, 1989).

5. See U.S. War Department, Supreme Commander for Allied Powers, *Reports of General MacArthur: The Campaigns of MacArthur in the Pacific* (c. 1949), 70, for a detailed description of General Horii's predicament.

6. Robert L. Eichelberger, "Report of the Commanding General, Buna Forces on the Buna Campaign, December 1, 1942–January 23, 1943."

7. Robert L. Eichelberger and Milton MacKaye, "Our Bloody Jungle Road to Tokyo," *Saturday Evening Post,* August 13, 1949.

8. Eichelberger changed commanders of several units. The combination of casualties and reliefs created a new and more effective command for the Buna Task Force.

9. Eichelberger and MacKaye, "Our Bloody Jungle Road to Tokyo," 108.

10. Robert L. Eichelberger and Milton MacKaye, "Nightmare in New Guinea," *Saturday Evening Post,* August 20, 1949, 107.

11. Vo Nguyen Giap, *People's War, People's Army* (Hanoi: Foreign Languages Publishing House, 1961). For a recent U.S. analysis of the situation see James R. McLean, "Assessing the Adversaries at Dien Bien Phu," in *Combined Arms in Battle since 1939,* ed. Roger J. Spiller (Fort Leavenworth, Kan.: U.S. Army Command and General Staff College Press, 1992).

12. Supreme Command Far East, France, "Lessons from the Indochina War," vol. 3.

13. Ibid.

14. Bernard Fall, *Hell in a Very Small Place: The Siege of Dien Bien Phu* (New York: Lippincott, 1966), 193.

15. Emphasis added; Roy Jules, *La Bataille de Dien Bien Phu* (Paris: Rene Julliard, 1963).

Chapter 12. Heat, Rock, and Sand

1. An increase in sun elevation angle increases potential solar radiation. Normally clear skies and sparse vegetation in dry areas permit most of this radiant energy to reach the earth's surface only to be re-radiated in a form that can be absorbed as heat by the atmosphere. Air is also heated through direct absorption of incoming solar radiation, release of latent heat from condensation of evaporated moisture, and conduction. However, at night when incoming solar radiation ceases, the earth continues to re-radiate energy with a corresponding decrease in ground temperature. Thus, in deserts

the clear skies and sparse vegetation that favor rapid heating during the day have an opposite effect at night. The result is that the largest range in daily temperature tends to be associated with the world's non-marine dry climates.

2. Only the high-altitude polar ice caps are drier than the deserts along the west coast of Africa and South America, even though coastal deserts are adjacent to moisture and are frequently bounded by fog banks. The ocean currents bordering these deserts are especially cool because of their polar source and upwelling (the upward movement of cold water from the ocean's depths to the surface) induced by prevailing winds along the coast. The cool offshore conditions create a breeze from sea to land that moderates the temperatures along the coast, but eliminates convective action needed to produce rainfall. It is reported that one area in the Atacama desert of South America went without a trace of rain for 14 years.

3. Alluvial fans are numerous and widespread in rugged dry areas. Being so common, at many places they coalesce to form bajadas. These features consist of alluvium deposited outward from mountain canyons. Where sediment laden streams exit confining canyons water velocity may decrease. This permits streams to both deposit sediments and spread laterally, the long term result being a landform shaped like a fan with the higher apex at the mouth of the canyon. As a result the coarser material is deposited closest to the mountains, the finer most distant. Alluvial fans and bajadas are often important sources for ground water. Trafficability across these fans ranges from marginal to fair.

4. Rommel was always interested in the terrain of a potential battlefield. In North Africa he frequently used a light reconnaissance plane (STORK) or a special captured British armored vehicle to determine the nature of the landscape ahead.

5. See Arnon Sofer, "The Wars of Israel in Sinai: Topography Conquered," *Military Review* (April 1982): 60–72. The authors are indebted to Professor Sofer for introducing them to the challenges of the Sinai and his continuing discussions of military geography and water resources in the Middle East.

Conclusion

1. Among others, two recent useful analyses of conflict are Gwynne Dyer's *War* (New York: Crown, 1985), and John Keegan's *A History of Warfare* (New York: Knopf, 1993). Also see the second edition of Michael Howard's *The Causes of Wars* (Cambridge: Harvard University Press, 1984). Whether war is an extension of politics, a need for survival, a cultural trait or response, a product of philosophical beliefs, an outcome of the environment or social structure, or the dark product of the soul, once started geography is always a factor in its conduct.

2. For an analysis of U.S. conflicts see Russell Weigley's *The American Way of War* (Bloomington: Indiana University Press, 1973).

3. The concept of war evolving toward chaos is based in part on a lecture by Richard J. Eaton presented at the U.S. Military Academy, West Point, N.Y., in April 1993.

4. Dwight David Eisenhower, the White House, 22 April 1959. Excerpted from a statement by the USMA Corps of Cadets.

General Bibliography

Agar-Hamilton, J.A.I., and L.C.F. Turner. *Crisis in the Desert: July 1942*. Capetown: London University Press, 1952.

Alery, R. "The Climate of France, Belgium, the Netherlands, and Luxembourg." In *Climates of Northern and Western Europe*, ed. Wallen, 135–93.

Allen, Hervey. *It Was Like This*. New York: Farrar and Rinehart, 1936.

———. *Toward the Flame: A War Diary*. Reprint. Pittsburgh, Pa.: University of Pittsburgh Press, 1968.

Alsup, Bruce. "The Chance Seizure of the Remagen Bridge Over the Rhine." In *Combined Arms in Battle since 1939*, ed. Roger J. Spiller, 139–46. Fort Leavenworth, Kan.: U.S. Army Command and General Staff College Press, 1992.

Appleman, Roy E. *South to the Naktong, North to the Yalu: The United States Army in the Korean War*. Washington, D.C.: Office of the Chief of Military History, 1961.

Army Map Service. *Continental Europe—Strategic Maps and Tables*. Pt. 1, *Central West Europe*, and pt. 2, *South Central Europe*. U.S. Army, 1943 (maps by A. K. Lobeck).

Arnold, H. H. "Narrative of Market Operation." Letter to U.S. Army Command and General Staff College, 4 November 1944.

Aubouin, Jean. "Les Grands Ensembles Structuraux de l'Europe." In *Geologie des Pays Europeans: France. Belgique, Luxembourg*. Paris: 26e Congress Geologique International, 1980.

Aurthur, Robert A., Kenneth Cohlmia, and Robert T. Vance. *The Third Marine Division*. Washington, D.C.: Infantry Journal Press, 1948.

Badri, Hassan el, Tahael Magdoub, and Mohammed Dia el Zohdy. *The Ramadan War, 1973*. New York: Hippocrene Books, 1978.

Baldwin, Hanson W. "The Battle of the Bulge—A Case History of Intelligence." In Baldwin, *Battles Lost and Won*, 315–67. New York: Harper and Row, 1966.

Ballard, Vice-Admiral G. A. *The Influence of the Sea on the Political History of Japan*. London: John Murray, 1921.

Baron, Mordechai, ed. *Israeli Defense Forces: The Six Day War*. Philadelphia: Chilton, 1968.

Bartley, Whitman S. *Iwo Jima: Amphibious Epic*. Washington, D.C.: Historical Branch, U.S. Marine Corps, 1954.

Bascom, Willard. *Waves and Beaches*. New York: Doubleday, 1964.

Bauer, Cornelius. *The Battle of Arnhem: The Betrayal Myth Refuted*. London: Hodder and Stoughton, 1966.

Belloc, Hilaire. "The Bridge-Heads of the Rhine." *Land and Water* (London) 72 (December 1918): 3–6.

Berg, L. S. *Natural Regions of the Soviet Union*. New York: Macmillan, 1950.

Bigelow, John. *The Campaign at Chancellorsville*. New Haven: Yale University Press, 1910.

Billelo, Michael A. *Regional and Seasonal Variations in Snow-Cover Density in the U.S.S.R.* CRREL Report 84-22. Hanover, N.H.: U.S. Army Corps of Engineers, Cold Regions Research and Engineering Laboratories, August 1984.

Bitner, Teddy D. "Kesselring: An Analysis of the German Commander at Anzio." Master's thesis, Army Command and General Staff College, Fort Leavenworth, Kan., 1983.

Blond, Georges. *Verdun*. New York: Macmillan, 1964.

Bloom, Arthur L. *Geomorphology*, 2nd ed. Englewood Cliffs, N.J.: Prentice-Hall, 1991.

Blumenson, Martin. *Salerno to Cassino: The Mediterranean Theater of Operations*. Washington, D.C.: Office of the Chief of Military History, 1969.

Blumenstock, David I. *The Ocean of Air*. New Brunswick, N.J.: Rutgers University Press, 1959.

Bowers, Ray L. *The United States Air Force in Southeast Asia: TAC Airlift*. Washington, D.C.: Office of Air Force History, 1983.

Bradley, Gen. Omar N. *A Soldier's Story*. New York: Henry Holt, 1951.

Brinkley, Capt. F., ed. *Japan: Described and Illustrated by the Japanese*, vol. 2. Boston: J. B. Millet, 1904.

Brown, Andrew. "Geology and the Gettysburg Campaign." *Geotimes* 6, no. 1 (1961): 8–12, 40–41.

Burchfiel, B. Clark. "The Continental Crust." *Scientific American* 249, no. 3 (September 1983): 130–42.

Burris, 1st Lt. L. D., ed. *The Ninth Marines: A Brief History of the Ninth Marine Regiment*. Washington, D.C.: Infantry Journal Press, 1946.

Busch, Noel F. *Winter Quarters: George Washington and the Continental Army at Valley Forge*. New York: Liveright, 1974.

Cagle, Malcolm W., and Frank A. Manson. *The Sea War in Korea*. Annapolis, Md.: U.S. Naval Institute Press, 1957.

Carell, Paul. *Hitler's War on Russia: The Story of the German Defeat in the East*. Translated by Ewald Osers. London: George G. Harrap and Co., 1964.

Carver, Michel. *El Alamein*. London: Batsford, 1962.

Cash, John A. "Fight at Ia Drang, 14–16 November 1965." In *Seven Firefights in Vietnam*, ed. John Albright, John A. Cash, and Allen W. Sandstrom, 3–40. Washington, D.C.: Office Chief of Military History, U.S. Army, 1970.

Catton, Bruce. *Glory Road: The Bloody Route From Fredericksburg to Gettysburg*. New York: Doubleday, 1952; New York: Pocket Books, 1964.

Central Intelligence Agency. *National Intelligence Survey. U.S.S.R.—I. European U.S.S.R.; Section 23, Weather and Climate*. Cameron Station, Va.: Defense Technical Information Center, Defense Logistics Agency, December 1952.

Cerjan, Paul, and Theodore G. Stroup. "Employment of the Engineer System in Arid Mountainous and Desert Areas—A Concept Paper." Study Project. U.S. Army War College. August 1981.

Chandler, David G. *The Campaigns of Napoleon*. New York: Macmillan, 1966.

Churchill, Winston S. *Closing the Ring*. Vol. 5 of *The Second World War*. Boston: Houghton Mifflin, 1951.

Clark, Alan. *Operation Barbarossa: The Russian-German Conflict, 1941–45*. New York: Morrow, 1985.

Clark, Mark W. *Calculated Risk*. New York: Harper, 1951.

Cole, Hugh M. *The Ardennes: Battle of the Bulge. The United States Army in World War II*. Washington, D.C.: Office of the Chief of Military History, U.S. Army, 1965.

Cooper, Leonard. *Many Roads to Moscow: Three Historic Invasions*. New York: Coward-McCann, 1968.

Davis W. M. "Glacial Erosion in France, Switzerland, and Norway." In *Proceedings of the Boston Society of Natural History* 29 (1900): 273–321.

de Segur, Count Philippe-Paul. *Napoleon's Russian Campaign*. Translated by J. David Townsend. Boston: Houghton Mifflin, 1958.

Deighton, Len. *Blitzkrieg: From the Rise of Hitler to the Fall of Dunkirk*. London: Jonathan Cape, 1979.

Department of the Army. "Effects of Climate on Combat in European Russia." Department of the Army Pamphlet 20-291. Washington, D.C.: GPO, February 1952.

——. "Military Improvisations During the Russian Campaign." Department of the Army Pamphlet 20-201. Washington, D.C.: August 1951.

Department of the Army, Headquarters, U.S. Army, Japan. *18th Army Operations*. Japanese Monograph No. 37. Washington, D.C.: Office of the Chief of Military History, 1946.

Detwiler, Donald S., ed. *World War II Military Studies: A Collection of 213 Special Reports on the Second World War Prepared by Former Officers of the Wehrmacht for the United States Army*. 24 vols. New York: Garland, 1979.

Duncan, David Douglas. "Retreat Hell." In Duncan, *This Is War*. New York: Harper and Brothers, 1951.

Dupuy, R. Ernest. *World War II: A Compact History*. New York: Hawthorne Books, 1969.

Dupuy, T. N., G. P. Hayes, Paul Martell, V. E. Lyons, and John A. C. Andrews. "A Study of Breakthrough Operations." Report for the Defense Nuclear Agency, DNA 4124F. Dunn Loring, Va., October 1976.

Dyer, Gwynne. *War*. New York: Crown, 1985.

Dziuban, Stanley W. "Rhine River Flood Prediction Service." *Military Engineer* 37, no. 239 (September 1945): 348–53.

Eichelberger, Robert L. "Report of the Commanding General, Buna Forces on the Buna Campaign, December 1, 1942–January 23, 1943."

Eichelberger, Robert L., and Milton MacKaye. "Our Bloody Jungle Road to Tokyo." *Saturday Evening Post,* August 13, 1949.

Eisenhower, Gen. Dwight D. *Crusade in Europe*. Garden City, N.Y.: Doubleday, 1948.

Eisenhower, John S. D. *The Bitter Woods*. New York: Putnam's Sons, 1969.

Ellis, Maj. L. F. "The War in France and Flanders, 1939–40." In *History of the Second World War,* ed. J.R.M. Butler. United Kingdom Military Series. London: HMSO, 1953.

Elstob, Peter. *Hitler's Last Offensive*. New York: Macmillan, 1971.

Erickson, John. *The Road to Berlin*. Boulder, Colo.: Westview Press, 1983.

——. *The Road to Stalingrad*. New York: Harper and Row, 1975.

Fairbridge, Rhodes W. *The Encyclopedia of Geomorphology*. New York: Reinhold Book Corp., 1968.

Fall, Bernard. *Hell in a Very Small Place: The Siege of Dien Bien Phu*. New York: Lippincott, 1966.

——. *Street Without Joy*. New York: Schocken Books, 1961.

Falls, Cyril. "Geography and War Strategy." *Geographical Journal of the Royal Geographical Society* 112 (July–September 1948): 1–3.

Farwell, Bryon. *Stonewall: A Biography of General Thomas J. Jackson*. New York: Norton, 1992.

Fehrenbach, T. R. *This Kind of War*. Washington, D.C.: Brassey's, 1994.

Fenneman, Nevin M. *Physiography of Eastern United States*. New York: McGraw-Hill, 1938.

Flint, Richard F. *Glacial and Quaternary Geology*. New York: Wiley and Sons, 1964.

Foote, Shelby. *The Civil War: A Narrative*. 3 vols. New York, 1954–78.

Francheteau, Jean. "The Ocean Crust." *Scientific American* 249, no. 3 (September 1983): 114–29.

Franzana, Gregory M., and William J. Ely. *Leif Sverdrup: Engineer Soldier at His Best*. Gerald, Mo.: Patrice Press, 1980.

Freeman, Douglas S. *Lee's Lieutenants: A Study in Command*. 3 vols. New York: Scribner's, 1942–44.

Fugate, Bryan I. *Operation Barbarossa: Strategy and Tactics on the Eastern Front, 1941*. Novato, Calif.: Presidio Press, 1984.

Fukui, E., ed. *The Climate of Japan*. Tokyo: Kodansha, 1977.

Fuller, J.F.C. *The Generalship of Ulysses S. Grant*. Bloomington: Indiana University Press, 1958.

Fuller, John F. "Air Weather Service Support to the United States Army: TET and the Decade After." AWS Historical Study No. 8. Scott Air Force Base, Ill.: Office of MAC History, Military Airlift Command, U.S. Air Force, 1979.

——. *Thor's Legions: Weather Support to the U.S. Air Force and Army, 1937–1987*. Boston: American Meteorological Society, 1990.

——. "Weather and War." Scott Air Force Base, Ill.: Office of MAC History, Military Airlift Command, 1974.

Galloway, Gerald E., Jr., comp. *Readings in Military Geography*. West Point, N.Y.: Department of Geography, 1990.

Ganoe, William A. *History of the U.S. Army*. New York: Appleton-Century, 1942.

Garver, John B., comp. and ed. *Readings in Military Geography*. West Point, N.Y.: Department of Geography and Computer Science, 1981.

Garver, John B., and Gerald E. Galloway Jr., comps. and eds. *Readings in Military Geography*. West Point, N.Y.: Department of Geography and Computer Science, 1984.

Giap, Vo Nguyen. *The Military Art of Peoples War: Selected Writings of General Vo Nguyen Giap*. Edited by Russell Stetler. New York: Monthly Review Press, 1970.

——. *People's War, People's Army*. Hanoi: Foreign Languages Publishing House, 1961.

Goulden, Joseph C. *Korea: The Untold Story of the War*. New York: McGraw-Hill, 1982.

Graf, William L., ed. *Geomorphic Systems of North America*. Boulder, Colo.: Geological Society of America, 1987.

Gregg, Charles T. *Tarawa*. New York: Stein and Day, 1984.

Greiner, Helmuth. "Africa 1941." Manuscript MS #C-065f. Historical Division, U.S. Army European Command, Military History Institute, Carlisle Barracks, Pa., 1946.

Gugeler, Capt. Russell A. *Combat Actions in Korea*. Washington, D.C.: Combat Forces Press, 1954.

Hackett, Sir John. Interview with Gerald E. Galloway Jr. West Point, N.Y., October 24, 1984.

Hannah, Dick. *Tarawa: The Toughest Battle in Marine Corps History*. New York: Duell, Sloan and Pearce, 1944.

Hardie, Martin, and Warner Allen. *Our Italian Front*. London: A. and C. Black, 1920.

Harman, Nicholas. *Dunkirk: The Necessary Myth*. London: Hodder and Stoughton, 1980.

Harris, John. *Dunkirk: The Storms of War*. London: David and Charles, 1980.

Harrison, Gordon A. *Cross-Channel Attack*. Washington, D.C.: Office of the Chief of Military History, Department of the Army, 1951.

Hebrew University. *Atlas of Israel*. New York: Macmillan, 1985.

Hechler, Ken. *The Bridge at Remagen*. New York: Ballantine Books, 1971.

Heinl, Robert Debs. *Dictionary of Military and Naval Quotations*. Annapolis, Md.: U.S. Naval Institute Press, 1966.

——. *Victory at High Tide*. Philadelphia: Lippincott, 1968.

Henri, Raymond. *Iwo Jima: Springboard to Final Victory*. New York: U.S. Camera Publishing, 1945.

Herren, John D. Telephone interview with Gerald E. Galloway Jr. October 1988.

Herring, George C. "The 1st Cavalry and the Ia Drang Valley, 18 October–24 November 1965." In *America's First Battles: 1776–1965,* ed. Charles E. Heller and William A. Stofft. Lawrence: University of Kansas Press, 1986.

Herzog, Chaim. *The Arab-Israeli Wars*. New York: Random House, 1982.

Hibbert, Christopher. *The Battle of Arnhem*. New York: Macmillan, 1962.

Hittle, J. D. "Jomini and Amphibious Thought." *Marine Corps Gazette* (May 1946): 35–38.

Holmes, E. R. *Borodino, 1812*. London: Charles Knight and Co., 1971.

Horie, Y. "Japanese Defense of Iwo Jima." *Marine Corps Gazette* (February 1952): 18–27.

Horne, Alistair. *The Price of Glory*. New York: St. Martin's Press, 1962.

——. *To Lose a Battle*. New York: Little, Brown, 1969.

Horrocks, Brian. *Corps Commander*. New York: Scribner's, 1977.

Hough, Frank O. *The Island War*. Philadelphia: Lippincott, 1947.

Howard, Michael. *The Causes of Wars*. 2nd ed. Cambridge: Harvard University Press, 1984.

Hoyt, Edwin P. *Storm Over the Gilberts*. New York: Van Nostrand Reinhold, 1978.

Humphreys, Andrew A. *The Virginia Campaigns of '64 and '65*. New York: Scribner's, 1883.

Isely, Peter A., and Philip A. Crowl. *The U.S. Marines and Amphibious War*. Princeton, N.J.: Princeton University Press, 1951.

Isgrig, D., and A. Srobel Jr. *Soil Survey of Stafford and King George Counties, Virginia*. Washington, D.C.: U.S. Department of Agriculture Soil Conservation Service in cooperation with Virginia Polytechnic Institute and State University, 1974.

Ishii, Viscount Kikujira. *Diplomatic Commentaries*. Baltimore: Johns Hopkins Press, 1936.

Jackson, Robert. *Dunkirk: The British Evacuation, 1940*. New York: St. Martin's Press, 1976.

Jackson, W.G.F. *The Battle for North Africa 1940–43*. New York: Mason/Charter, 1975.

——. *Seven Roads to Moscow*. London: Eyre and Spottiswoude, 1957.

Jensen, Robert G., Theodore Shabad, and Arthur W. Wright. *Soviet Natural Resources in the World Economy*. Chicago: University of Chicago Press, 1983.

Johannessen, Karl R. "Hindcasting the Weather for the Normandy Invasion." In *Some Meteorological Aspects of the D-day Invasion of Europe 6 June 1944,* ed. Shaw and Innes, 39–81.

Johnson, Douglas W. *Battlefields of the World War: Western and Southern Fronts: A*

Study in Military Geography. American Geographical Society Research Series 3. New York: Oxford University Press, 1921.

Johnson, Robert W. *Battles and Leaders of the Civil War*. New York: Century Co., 1888.

Johnston, Richard W. *Follow Me! Second Marine Division*. New York: Random House, 1948.

Jomini, Antoine H. *Precis de l'art de la guerre, IIe partie*. Paris, 1838.

Kays, Marvin D. "Weather Effects During the Battle of the Bulge and the Normandy Invasion." White Sands, N.M.: Atmospheric Science Laboratory, 1982.

Keegan, John. *The Face of Battle: A Study of Agincourt, Waterloo, and the Somme*. New York: Vintage Books, 1977.

——. *A History of Warfare*. New York: Knopf, 1993.

Kelly, George A. *Lost Soldiers: The French Army and Empire in Crisis, 1947–1962*. Cambridge, Mass.: MIT Press, 1965.

Kendrew, W. G. *The Climates of the Continents*. London: Oxford University Press, 1961.

Kennett, James P. *Marine Geology*. Englewood Cliffs, N.J.: Prentice-Hall, 1982.

Kesselring, Albert von. "Interrogation of Genlfldm Albert Kesselring." In *ENTHIN 70 in World War II, German Military Studies*. Vol. 3. Historical Division, U.S. Army Europe, July 24, 1985.

King, Philip B. *The Evolution of North America*. Princeton, N.J.: Princeton University Press, 1977.

Kinnard, Lt. Gen. Harry W. O. "Victory in the Ia Drang: The Triumph of a Concept." *Army* (September 1967): 71–91.

Kinsman, Blair. *Wind Waves*. Englewood Cliffs, N.J.: Prentice-Hall, 1965.

Komar, Paul D. *Beach Processes and Sedimentation*. Englewood Cliffs, N.J.: Prentice-Hall, 1976.

Kopanev, I. D., and V. I. Lipovskaya. *Distribution of Snow Depth in the U.S.S.R.* Draft Translation 687. Hanover, N.H.: U.S. Army Corps of Engineers, Cold Regions Research and Engineering Laboratory, May 1978.

Lahee, Frederic H. *Field Geology*. New York: McGraw-Hill, 1952.

Langer, Paul H., and J. J. Zaslott. *North Vietnam and the Pathet Lao*. Cambridge: Harvard University Press, 1970.

Langley, Michael. *Inchon Landing*. New York: Time-Life Books, 1979.

Laux, Herbert B. "Observer's Report from Headquarters Army Ground Forces to Southwest Pacific Theater." Washington, D.C.: Headquarters Army Ground Forces, Army War College, 1943.

Liddell-Hart, B. H. *The Real War*. Boston: Little, Brown, 1930.

——. *The Rommel Papers*. New York: Harcourt Brace, 1957.

Linklater, Eric. *The Campaign in Italy*. London: His Majesty's Stationery Office, 1951.

Lobeck, A. K. *Geomorphology: An Introduction to Landscapes*. New York: McGraw-Hill, 1939.

Lord, Walter. *The Miracle of Dunkirk*. New York: Viking Press, 1982.

Lucas, James. *War on the Eastern Front, 1941–1945: The German Soldier in Russia*. New York: Bonanza Books, 1982.

Ludlum, David M. "The Weather of American Independence 3: The Battle of Long Island." *Weatherwise* (June 1975): 118–21, 147.

Lydolph, Paul. *Climates of the Soviet Union*. Vol. 7 of *World Survey of Climatology*, ed. C. C. Wallen. New York: Elsevier, 1977.

——. *Geography of the U.S.S.R.: Topical Analysis*. Elkhart Lake, Wis.: Misty Valley Publishing, 1979, 1984.

MacDonald, Charles B. "The Decision to Launch Operation Market-Garden." In *Command Decisions,* ed. U.S. Army Office of Military History, 329–41. New York: Harcourt Brace, 1951.

——. *The Last Offensive: The United States Army in World War II.* Washington, D.C.: Office of the Chief of Military History, U.S. Army, 1973.

——. *The Siegfried Line Campaign: The United States Army in World War II.* Washington, D.C.: Office of the Chief of Military History, U.S. Army, 1963.

——. *A Time for Trumpets.* New York: Morrow, 1985.

McKay, Karen. "Operation Ramon: Historic Move to the Negev." *Army* (April 1982): 46–57.

McKee, Edwin D., ed. *A Study of Global Sand Seas.* USGS Professional Paper 1052. Washington: U.S. Government Printing Office, 1979.

McLean, James R. "Assessing the Adversaries at Dien Bien Phu." In *Combined Arms in Battle since 1939,* ed. Roger J. Spiller, 121–29. Fort Leavenworth, Kan.: U.S. Army Command and General Staff College Press, 1992.

Maclean, Michael. *The Ten Thousand Day War: Vietnam 1945–1975.* New York: St. Martin's, 1981.

Manchester, William. *American Caesar.* Boston: Little, Brown, 1978.

——. *Goodbye Darkness.* Boston: Little, Brown, 1979.

Manley, G. "Climate of the British Isles." In *Climates of Northern and Western Europe,* ed. Wallen, 81–133.

Manteuffel, Hasso von. "The Ardennes." In *The Fatal Decisions,* ed. Seymour Freidin. New York: William Sloan Associates, 1956.

——. "The Battle of the Ardennes." In *Decisive Battles of World War II,* ed. H. A. Jacobsen and J. Rohwer. New York: Putnam's Sons, 1965.

Marshall, S.L.A. *Sinai Victory.* New York: Morrow, 1958.

Mathews, Allen R. *The Assault.* New York: Simon and Schuster, 1947.

Meigs, Peveril. "Some Geographical Factors in the Peloponnesian Wars." *Geographical Review* 51 (1961): 370–80.

Merriam, Robert E. *Dark December.* Chicago: Ziff-Davis, 1947.

Milner, Samuel. *Victory in Papua: The United States Army in World War II.* Washington, D.C.: Office of the Chief of Military History, 1989.

Montgomery, Bernard. *El Alamein to the River Sangro: Normandy to the Baltic.* New York: St. Martin's Press, 1974.

Montross, Lynn, and Nicholas A. Canzona, *The Inchon-Seoul Operation.* Vol. 2 of *U.S. Marine Operations in Korea, 1950–1953.* Washington, D.C.: Historical Branch G-3, Headquarters, U.S. Marine Corps, 1954.

Moore, Frank, ed. *The Rebellion Record.* 11 vols. New York: Putnam, 1863.

Moore, H. G., and J. L. Galloway. *We Were Soldiers Once . . . and Young.* New York: Random House, 1992.

Morisawa, Marie. *Streams.* New York: McGraw-Hill, 1968.

Morison, Samuel E. *History of the United States Naval Operations in World War II.* Vol. 7: *Aleutians, Gilberts, and Marshalls: June 1942–April 1944.* Boston: Little, Brown, 1960.

——. Vol. 9: *Sicily, Salerno, Anzio, January 1943–May 1945.* Boston: Little, Brown, 1960.

——. Vol. 14: *Victory in the Pacific, 1945.* Boston: Little, Brown, 1960.

Morrocco, John. *Thunder from Above.* Boston: Boston Publishing, 1984.

Murdoch, James. *A History of Japan.* Vol. 1: *From the Origins to the Arrival of the Portuguese in 1542* A.D. London: Kegan Paul, Trench, Trubner, & Co., Ltd., 1925.

Murfin, James V. *The Gleam of Bayonets*. New York: Thomas Yoseloff, 1965.

Nalty, Bernard C. *Air Power and the Fight for Khe Sanh*. Washington, D.C.: Office of Air Force History, 1973.

Neumann, J. "Great Historical Events That Were Significantly Affected by Weather, 1: The Mongol Invasions of Japan." *Bulletin of the American Meteorological Society* 56 (1975): 1167–71.

Newcomb, Richard F. *Iwo Jima*. New York: Holt, Rinehart, and Winston, 1965.

Nilsson, Tage. *The Pleistocene: Geology and Life in the Quaternary Ice Age*. Dordrecht, Holland: D. Reidel, 1983.

Nobecourt, Jacques. *Hitler's Last Gamble*. New York: Schocken Books, 1967.

O'Sullivan, Patrick. *Terrain and Tactics*. New York: Greenwood Press, 1991.

O'Sullivan, Patrick, and Jesse W. Miller. *The Geography of Warfare*. New York: St. Martin's Press, 1983.

Orni, Efraim, and Elisha Efrat. *Geography of Israel*. Philadelphia: Jewish Publication Society of America, 1976.

Orpen, Ned. "War in the Desert." *South African Forces, World War II*. Capetown: Purnell, 1971.

Palmer, Alan. *Napoleon in Russia*. London: Andre Deutsch, 1967.

Palmer, Dave Richard. *Summons of the Trumpet*. New York: Random House, 1978.

Pearson, Willard. *The War in the Northern Provinces: 1966–1968*. Washington, D.C.: Department of the Army, 1975.

Peltier, Louis C., and G. Etzel Pearcy. *Military Geography*. New York: Van Nostrand, 1966.

Petain, Henri Philippe. *Verdun*. New York: Dial Press, 1930.

Pethick, John. *An Introduction to Coastal Geomorphology*. London: Edward Arnold, 1984.

Petrie, W. M. *Researches in Sinai*. London: John Murray, 1906.

Phillips, C. E. Lucas. *Alamein*. Boston: Little, Brown, 1962.

Pimlott, John. *Battle of the Bulge*. Englewood Cliffs, N.J.: Prentice-Hall, 1983.

Pisor, Robert L. *The End of the Line: The Siege of Khe Sanh*. New York: Norton, 1982.

Powell, E. Alexander. *Italy at War and the Allies in the West*. New York: Scribner's, 1917.

Pratt, Clayton A. "Military Use by Warsaw Treaty Organization Forces of 20th Century Operational Routes in the Benelux and Northern Germany." Final Report AD-AO74000. Laramie, Wyoming: University of Wyoming–Laramie, April 1977.

Rand Corporation. "A Translation from the French: Lessons of the War in Indochina," vol. 2. Memorandum RM-5271-PR. May 1967.

Reeder, Russell Potter. "Fighting on Guadalcanal." Report. Washington, D.C.: Government Printing Office, 1943.

Reinhart, G. G., and E. F. Sharkey. "Air Interdiction in Southeast Asia." Rand Memorandum RM 5283-PR, Rand Corporation, November 1967.

Rice, Thomas. "Burnside's Mud March: Wading to Glory." *Civil War Times Illustrated* 206 (1981): 16–27.

Ritter, Dale F., R. Craig Kochel, and Jerry R. Miller. *Process Geomorphology*. Dubuque, Iowa: Wm. C. Brown, 1995.

Ritter, G. W. "Climate and Visibility in Southeast Asia." Naval Missile Center Tech Memo TM-67-28. Point Mugu, Calif.: Naval Missile Center, 1967.

Roy, Jules. *La Bataille de Dien Bien Phu*. Paris: Rene Juillard, 1963.

Russ, Martin. *Line of Departure: Tarawa*. Garden City, N.Y.: Doubleday, 1975.

Ryan, Cornelius. *A Bridge Too Far*. New York: Simon and Schuster, 1974.

Schaff, Morris. *The Battle of the Wilderness.* Boston: Houghton Mifflin, 1910.

Schmidt, Heinz W. *With Rommel in the Desert.* London: George G. Harrap and Company, 1951.

Sears, Stephen W. *Landscape Turned Red: The Battle of Antietam.* New Haven, Conn.: Ticknor and Field, 1983.

Seaton, Albert. *The Russo-German War, 1941–45.* New York: Praeger, 1971.

Shamshurov, V. K. "Special Features of the Engineer Support of Troop Combat Operations in Mountains. Special Features of Engineer Support of Troop Combat Operations in Deserts." Soviet Monograph. U.S. Army Foreign Science and Technology Center, October 1981.

Shaw, Henry I., Jr., Bernard C. Nalty, and Edwin T. Turnbladh. *Central Pacific Drive.* History of the U.S. Marine Corps Operations in World War II 3. Washington: Historical Branch, G-3 Headquarters, U.S. Marine Corps, 1966.

Shaw, Roger H., and William Innes, eds. *Some Meteorological Aspects of the D-day Invasion of Europe, 6 June 1944.* American Meteorological Society Symposium Proceedings, 19 May 1984, Fort Ord, California.

Sheehan, Fred. *Anzio: Epic of Bravery.* Norman: University of Oklahoma Press, 1964.

Shepherd, Francis P. *Submarine Geology.* New York: Harper and Row, 1973.

Shore, Moyers S. *The Battle for Khe Sanh.* Washington, D.C.: History and Museums Division, Headquarters, U.S. Marine Corps, 1979.

Smith, Holland M., and Percy Smith. *Coral and Brass.* New York: Harper and Row, 1973.

Smith, Walter Bedell. "Eisenhower's Six Great Decisions: Encirclement of the Rhine." *Saturday Evening Post,* 6 July 1946.

———. "Eisenhower's Six Great Decisions: Victory West of the Rhine." *Saturday Evening Post,* 29 June 1946.

Sofer, Arnon. "Topography Conquered: The Wars of Israel in Sinai." *Military Review* (April 1982).

Spector, Ronald H. *After TET: The Bloodiest Year in Vietnam.* New York: Free Press, 1993.

———. *Eagle Against the Sun.* New York: Free Press, 1985.

Sprietsma, Maj. C. F. "Analysis of the Battle of Iwo Jima." Maxwell AFB, Ala.: Air Command and Staff College. Report No. ACSC-84-2470. 1984.

Stackpole, Edward J. *Drama on the Rappahannock.* New York: Bonanza Books, 1957.

———. *The Fredericksburg Campaign.* Harrisburg, Pa.: Military Services Publication Co., 1957.

———. *The Fredericksburg Campaign.* New York: Bonanza Books, 1957.

Stagg, James Martin. *Forecast for Overlord.* New York: Norton, 1971.

Stanley, Arthur P. *Sinai and Palestine.* New York: Redfield, 1957.

Starr, Lt. Col. Chester G., ed. *From Salerno to the Alps: A History of the Fifth Army, 1943–1945.* Washington, D.C.: Infantry Journal Press, 1948.

Steele, Dennis. "Spanning the Sava." *Army* (February 1996).

Steichen, Edward, ed. *U.S. Navy War Photographs: Pearl Harbor to Tokyo Bay.* New York: Crown, 1956.

Stetler, Russell, ed. *The Military Art of Peoples' War: Selected Writings of General Vo Nguyen Giap.* New York: Monthly Review Press, 1970.

Stewart, George R. *Pickett's Charge.* Boston: Houghton Mifflin, 1959.

Stockman, Capt. James R., USMC. *The Battle for Tarawa.* Washington, D.C.: Historical Section, U.S. Marine Corps, 1947.

Stolfi, Russel. "Barbarossa Revisited: A Critical Reappraisal of the Opening Stages of the Russo-German Campaign (June–December 1941)." *Journal of Modern History* 54 (1982): 27–46.

Strahler, Alan, and Arthur Strahler. *Introducing Physical Geography*. New York: Wiley and Sons, 1994.

Strawson, John. *The Battle for North Africa*. New York: Scribner's, 1969.

——. *The Battle for the Ardennes*. New York: Scribner's, 1972.

Strother, K. C., and R. E. Koon, "The Military Geography of Laos: A Barrier to Communist Aggression." Memorandum RM 2986-PR, Rand Corporation, March 1962.

Sugden, David E., and Brian S. John. *Glaciers and Landscape*. New York: Wiley and Sons, 1976.

Summers, Harry G., Jr. "The Bitter Triumph of Ia Drang." *American Heritage* 35 (Feb.– Mar. 1984): 50–58.

Sun Tzu. *The Art of War*. Translated with an introduction by Samuel B. Griffith. New York: Oxford University Press, 1971.

Symons, Leslie. *The Soviet Union: A Systematic Geography*. London: Hodder and Stoughton, 1983.

Tanner, Robert G. *Stonewall in the Valley*. New York: Doubleday, 1976.

Tarle, Eugene. *Napoleon's Invasion of Russia, 1812*. New York: Oxford University Press, 1942.

Thornbury, William D. *Principles of Geomorphology*. 2nd ed. New York: Wiley, 1969.

——. *Regional Geomorphology of the United States*. New York: Wiley and Sons, 1965.

Toland, John. *Battle: The Story of the Bulge*. New York: Random House, 1959.

Tolson, John J. *Airmobility Studies: 1961–1971*. Washington, D.C.: Department of the Army, 1973.

——. *Airmobility: 1962–1971*. Washington, D.C.: GPO, 1973.

Trewartha, Glenn T. *The Earth's Problem Climates*. Madison: University of Wisconsin Press, 1961.

Tugwell, Maurice. *Arnhem: A Case Study*. London: Thorton Cox, 1975.

Turnbull, Patrick. *Dunkirk: Anatomy of Disaster*. New York: Holmes and Meier Publishers, 1978.

Turnbull, S. R. *The Samurai: A Military History*. New York: Macmillan, 1977.

U.K. 21st Army Group. "Operation Market-Garden: 17–26 September 1944." After-Action Report, c. January 1945.

U.K. British Airborne Corps. "Allied Operations in Holland, After-Action Report," c. February 1945.

U.K. British Airborne Division. "Report on Operation Market," 10 January 1945.

U.K. First Allied Airborne Army, Historian. "Narrative of Operations in Holland." Memorandum for Chief of Staff, 9 October 1944.

U.S. Air Force Air Weather Service. *Worldwide Airfield Climatic Data, Eastern Europe*. Doc. No. AD776611. Cameron Station, Va.: Defense Technical Information Service, Defense Logistics Agency, April 1973.

——. *Worldwide Airfield Climatic Data, U.S.S.R.* Doc. No. AD776612. Cameron Station, Va.: Defense Technical Information Center, Defense Logistics Agency, April 1973.

U.S. Air Force Reserve Officers Training Corps Air University. *Military Aspects of World Political Geography*. Maxwell AFB, 1959.

U.S. Army. *Field Manual 100-5*. Washington, D.C.: Department of the Army, 1982.

U.S. Army. *Field Manual 100-5*. Washington, D.C.: Department of the Army, 1993.

U.S. Army Command and General Staff College. *Readings in Military Geography (RB 20-4)*. Fort Leavenworth, 1962.

U.S. Army Corps of Engineers, CRREL. *Notes for the Winter Battlefield*. Field Manual 31-71, Northern Operations. Hanover, N.H.: CRREL, January 1984.

U.S. Army Engineer Topographic Laboratories. *Synthesis Guide for River Crossings*. 1983.

U.S. Army Map Service. Map: "Libya." Scale 1:50,000. Series P761, c. 1964.

——. Map: "Libya and United Arab Republic." Scale 1:250,000. Series 1501, c. 1966.

——. Map. "United Arab Republic." Scale 1:50,000. Series P7773, c. 1961.

U.S. Engineer School. "The Rhine Crossing: Twelfth Army Group Engineer Operations." Brig. Gen. P. H. Timothy, July 1946.

U.S. First Army. "Report of Rhine River Crossings." May 1945.

U.S. Headquarters 82nd Airborne Division. "Field Order No. 11." 13 September 1944.

U.S. Military Academy, Department of Geography and Computer Science. *A Bibliography of Military Geography*. Compiled by Eugene J. Palka and Dawn M. Lake. Senior Editors Gerald E. Galloway Jr., William J. Reynolds, and Harold A. Winters. West Point, N.Y.: Gilbert W. Kirby Jr. (n.d.).

U.S. Ninth Army. "Engineer Operations in the Rhine Crossing." 30 June 1945.

U.S. Soil Conservation Service. *Soil Taxonomy: A Basic System of Soil Classification for Making and Interpreting Soil Surveys*. Agricultural Handbook 436. Washington, D.C.: U.S. Soil Conservation Service, 1975.

U.S. Third Army, Engineer. "Crossing of the Rhine River by Third Army." August 1945.

U.S. War Department, Military Intelligence Division. *Papuan Campaign: The Buna-Sanananda Operation, 16 November 1942–23 January 1943*. Washington, D.C.: U.S. Government Printing Office, 1943.

U.S. War Department, Office of Chief Engineer, Headquarters 82nd Airborne Division, "FO No. 11," 13 September 1944.

U.S. War Department, Supreme Commander for Allied Powers. *Reports of General MacArthur: The Campaigns of MacArthur in the Pacific*. Superintendent of Documents, Washington, D.C., c. 1949.

U.S. War Department, European Theater of Operations, Office of Chief Engineer. "Historical Report 20 Forced Crossing of the Rhine, 1945."

Valkenburg, Samuel Brian. "Geographical Aspects of the Defense of the Netherlands." *Economic Geography* (April 1940).

Van Creveld, Martin L. *Supplying War: Logistics from Wallenstein to Patton*. Cambridge: Cambridge University Press, 1977.

Vandiver, Frank E. *Mighty Stonewall*. New York: McGraw-Hill, 1957.

Velichko, A. A., ed. *Late Quaternary Environments of the Soviet Union*. Minneapolis: University of Minnesota Press, 1984.

Verdun: Vision and Comprehension—The Battlefield and Its Surroundings. 4th ed. Drancy, France: Editions Mage, 1985.

Villare, Luigi. *The War on the Italian Front*. London: Cobdere Sanderson, 1932.

von Clausewitz, Carl. *The Campaign of 1812 in Russia*. Westport, Conn.: Greenwood Press, 1977.

von Greiffenberg, Gen. Hans. "Battle of Moscow." Historical Division, Headquarters U.S. Army Europe. Reprinted in *World War II German Military Studies,* vol. 16, ed. Donald S. Detwiler. New York: Garland Publishing, 1979.

von Manstein, Field Marshal Erich. *Lost Victories*. 4th printing. Trans. and ed. Anthony G. Powell. Novato, Calif.: Presidio Press, 1982.

Wallen, C. C., ed. *Climates of Central and Southern Europe*. Vol. 6 of *World Survey of Climatology*. New York: Elsevier, 1977.

——, ed. *Climates of Northern and Western Europe*. Vol. 5 of *World Survey of Climatology*. New York: Elsevier, 1970.

Ward, Robert D. "Weather Controls Over the Fighting in Mesopotamia, in Palestine, and Near the Suez Canal." *Scientific Monthly* 6 (April 1918): 291–304.

Warlimont, Walter. "The Decision in the Mediterranean 1942." In *Decisive Battles of World War II: The German View*, ed. Hans-Adolt Jacobsen and Jurgen Rohwer. London: Andrew Deutsch, 1965.

Warner, Denis. *The Last Confucian*. New York: Macmillan, 1963.

Weigley, Russell F. *The American Way of War*. Bloomington: Indiana University Press, 1973.

Werth, Alexander. *Russia at War, 1941–45*. New York: E. P. Dutton, 1964.

Westmoreland, William C. *A Soldier Reports*. New York: Doubleday, 1976.

Westphal, Siegfried. "Notes on the Campaign in North Africa, 1941–1943." *Royal United Services Institute Journal* 61 (February 1960): 105.

Wheeler, Richard. *Iwo*. New York: Lippincott and Crowell, 1980.

——. *A Special Valor: The U.S. Marines and the Pacific War*. New York: Harper and Row, 1983.

Wilson, Earl J. *Betio Beachhead*. New York: Putnam, 1945.

Winters, Harold A. "The Battle That Was Never Fought: Weather and the Union Mud March of January, 1863," *Southeastern Geographer* 31 (May 1991): 31–38.

Wolff, Leon. *In Flanders Fields: The 1917 Campaign*. New York, 1958.

Young, Peter, ed. *The World Almanac Book of World War II*. Englewood Cliffs, N.J.: Prentice-Hall, 1981.

Index

New Guinea, 2, 220, 233, 235–42, 236f, 240f, 246–47

Niagara, Operation (Vietnam War), 70

Nijmegen (Netherlands), 147–53

Nile River (Egypt), 144, 251, 258, 261, 285n4

Nimitz, Adm. Chester W., 220, 225, 227

99th Division (U.S.), 52, 53, 54

9th Armored Division (U.S.), 56, 60

Ninth Army (U.S.), 48, 160

9th Division (U.S.), 53, 156

Normandy (France), 1, 9, 96, 192, 207; Allied landings at, 15–17, 23–32, 267, 268, 270, 271; maps, 2f, 17f; photos, 29f, 30f, 31f

North Africa, 2, 5, 205, 230, 249, 251–61, 255f, 256f

North America: deserts in, 248, 249; glaciation in, 162, 163f, 164, 165, 166

North Atlantic Drift, 16, 50

North Korean People's Army, 191, 207–14

North Vietnamese Army (NVA), 63, 68, 69, 70–73, 98, 99, 267; at Ia Drang, 106–11

Oak Ridge (Battle of Gettysburg), 127

occlusion, 7, 8f

Ogasawara Gunto (Bonin Islands; Nanpu Shoto Group), 226

Okinawa, 71, 96, 192, 208, 211, 215, 220, 231

Omaha Beach (Normandy), 29f

101st Airborne Division (U.S.), 56, 60, 147–54

106th Infantry Division (U.S.), 53, 55, 56

109th Infantry Division (Japanese), 228

110th Infantry Regiment (U.S.), 56

128th Infantry (U.S.), 239

Owen Stanley Range (New Guinea), 236f, 237, 239, 247

Pacific Ocean, 14f, 216f

Pacific Theater (World War II), 191–92, 230; climates in, 96, 232–33, 234. See also particular campaigns and locations

Pakistan, 164, 169, 251

Palestine, 261

Palmer, Gen. Dave, 110

Pangaea, 117, 118, 127, 134, 282n5, 283n13

Paris Basin, 134, 135

Patagonia (Argentina), 248, 249

Patton, Gen. George S., 48, 52, 60, 156, 268, 286n21

Pegasus, Operation (Vietnam War), 72, 73

Peiper, Joachim, 54

Peloponnesian Wars (431–404 B.C.), 5

Peninsular Campaign (American Civil War), 112, 113, 122–24, 192, 266, 267

perigee, 195

permeability, 33–34

pests, 235, 239–40, 247

Peter Beach (Anzio, Italy), 202f, 203

Peter the Great (Russia), 180, 181

Philip, C. F. Lucas, 248

Philippines, 12, 232, 234

"Phony War" (World War II), 18, 145

Pickett's Charge (Battle of Gettysburg), 129, 132

Piedmont (Va.), 115–19, 122, 124, 127

Pleiku (Vietnam), 107–8

Plei Me (Vietnam), 107–8, 109

Pleistocene Epoch, 162, 164–69, 170–71, 190, 286n1

Plunder, Operation (World War II), 156

Poland, 74, 80, 145, 180

polar fronts, 6, 7, 11, 34

Poltava (Russia), 181, 189

Pontine Marshes (Italy), 204

Po River (Italy), 170, 173, 174–75, 176

porosity, 33–34

Port Moresby (New Guinea), 236f, 238, 247

Potomac River (U.S.), 118–21, 123–25, 127, 132–33

precipitation, 5, 7; at Anzio, 198, 201, 206; in the Ardennes, 51; and climate, 75, 278n1; and clouds, 45, 46; in deserts, 248, 249, 250, 252, 259, 262, 263; and erosion, 115; in Europe, 24f, 25f, 26f, 50; in Flanders, 40, 42; and flooding, 157; and glaciation, 165; in humid tropics, 233, 234, 235; and monsoons, 64–65; and mud, 33, 34, 36; in New Guinea, 236, 239; and rivers, 143–44; and temperature, 79, 80; and vegetation, 97, 98, 99, 101; at Verdun, 138; in Vietnam, 73, 243–44, 245; and weather modification, 69. See also snow

Pripet Marshes (Russia), 177, 179, 181, 182, 185, 186

Punic Wars (218 B.C.), 164

Pusan Perimeter (Korea), 191, 201, 205, 208, 210, 211, 213, 214

Pyrenees Mountains, 169

Qattara Depression (North Africa), 252, 253, 258

"Race to the Sea" (World War I), 42

rain. See precipitation

rainforest, 233–34, 236, 243, 246

Rapido River (Italy), 142, 199, 201

About the Authors

Harold A. Winters (Ph.D., Northwestern University, 1960) joined the Department of Geography at Michigan State University in 1965 and retired as professor emeritus in 1995. With prior teaching experience at Northern Illinois University and Portland State University, while at MSU he was also a visiting professor at seven other institutions of higher learning including three separate appointments with West Point. His research focuses on physical geography, including its influence on military operations, and awards include Honors of the Association of American Geographers and two U.S. Army Outstanding Civilian Service Medals.

Gerald E. Galloway Jr. (Ph.D., The University of North Carolina at Chapel Hill, 1979) was commissioned in 1957 when he graduated at the U.S. Military Academy, retiring from the U.S. Army thirty-nine years later as a brigadier general. An engineer, he completed two tours in both Vietnam and Germany plus service in the Office of the Chief of Staff of the Army. After being professor and head of the Department of Geography and Environmental Engineering at West Point he served as the Academy's Academic Dean for five years. Galloway's research concentrates on military geography and water resources development. Currently he is Dean of Faculty and Academic Programs at the Industrial College of the Armed Forces in Washington, D.C.

William J. Reynolds (Ph.D., Rutgers University, 1982) is a retired colonel and Vietnam veteran. Graduating in 1964 from West Point, his last nine years in the army were also at the Academy as he advanced to professor of geography and acting head of the Department of Geography and Environmental Engineering. Meanwhile, Reynolds completed two tours of duty in Southeast Asia plus assignments in Europe and the U.S. His research specialty is coastal geomorphology. Dr. Reynolds is now Northwest Regional Manager for Science Applications International Corporation.

David W. Rhyne (M.S., The Pennsylvania State University, 1979; M.T., Virginia Commonwealth University, 1995) is a 1971 West Point graduate. After twenty-two years as an officer he retired in 1993 at the rank of lieutenant colonel. With a variety of assignments, including two tours in Korea and attendance at the Army's Command and General Staff College, Rhyne taught for three years in the Academy's Department of Geography and Computer Science. He is now a member of the faculty at Stonewall Jackson Middle School in Hanover County, Virginia, and has research interests in relationships between historical and physical geography.

Library of Congress Cataloging-in-Publication Data

Winters, Harold A.
    Battling the elements : weather and terrain in the conduct of war / Harold A.
Winters with Gerald E. Galloway Jr., William J. Reynolds, and David W. Rhyne.
        p.   cm.
    Includes bibliographical references and index.
    ISBN 0-8018-5850-X (alk. paper).
    1. Military geography—Case studies.   2. Military geography.   I. Title.
UA990.W45   1998
355.4′7—dc21   98-5983
                                                                                    CIP